T0231236

MANIPULATIVE TENANTS

BACTERIA ASSOCIATED with ARTHROPODS

MANIPULATIVE TENANTS

BACTERIA ASSOCIATED with ARTHROPODS

Edited by
EINAT ZCHORI-FEIN
KOSTAS BOURTZIS

EUROPEAN COOPERATION
IN SCIENCE AND TECHNOLOGY

FA0701

Arthropod—Symbiosis

EUROPEAN
SCIENCE
FOUNDATION

CRC Press
Taylor & Francis Group
Boca Raton London New York

CRC Press is an imprint of the
Taylor & Francis Group, an **informa** business

Cover design: Mirit Assaf

Neither the COST Office nor any person acting on its behalf is responsible for the use which might be made of the information contained in this publication. The COST Office is not responsible for the external websites referred to in this publication.

CRC Press
Taylor & Francis Group
6000 Broken Sound Parkway NW, Suite 300
Boca Raton, FL 33487-2742

First issued in paperback 2018

© 2012 by Taylor and Francis Group, LLC
CRC Press is an imprint of Taylor & Francis Group, an Informa business

No claim to original U.S. Government works

ISBN-13: 978-1-4398-2749-9 (hbk)
ISBN-13: 978-1-138-37433-1 (pbk)

This book contains information obtained from authentic and highly regarded sources. Reasonable efforts have been made to publish reliable data and information, but the author and publisher cannot assume responsibility for the validity of all materials or the consequences of their use. The authors and publishers have attempted to trace the copyright holders of all material reproduced in this publication and apologize to copyright holders if permission to publish in this form has not been obtained. If any copyright material has not been acknowledged please write and let us know so we may rectify in any future reprint.

Except as permitted under U.S. Copyright Law, no part of this book may be reprinted, reproduced, transmitted, or utilized in any form by any electronic, mechanical, or other means, now known or hereafter invented, including photocopying, microfilming, and recording, or in any information storage or retrieval system, without written permission from the publishers.

For permission to photocopy or use material electronically from this work, please access www.copyright.com (http://www.copyright.com/) or contact the Copyright Clearance Center, Inc. (CCC), 222 Rosewood Drive, Danvers, MA 01923, 978-750-8400. CCC is a not-for-profit organization that provides licenses and registration for a variety of users. For organizations that have been granted a photocopy license by the CCC, a separate system of payment has been arranged.

Trademark Notice: Product or corporate names may be trademarks or registered trademarks, and are used only for identification and explanation without intent to infringe.

Library of Congress Cataloging-in-Publication Data

Manipulative tenants : bacteria associated with arthropods / editors, Einat Zchori-Fein, Kostas Bourtzis.
  p. ; cm. -- (Frontiers in microbiology ; 1)
  Includes bibliographical references and index.
  Summary: "In efforts to summarize the most up-to-date information available on bacterial symbionts of arthropods, this text provides an overview of primary symbionts in addition to 10 of the most abundant secondary symbionts known to date. The editors have brought together entomogists and microbiologists to create a full picture of the complex systems, thus encouraging the integration of theory and practice in efforts to find innovative routes to pest and disease management. The book provides comprehensive knowledge and a unique perspective on a fast-growing field. Figures within the text include diagrams, tables, graphs, pictures, and chemical structures"--Provided by publisher.
  ISBN 978-1-4398-2749-9 (hardcover : alk. paper)
  1. Arthropoda--Microbiology. 2. Endosymbiosis. I. Zchori-Fein, Elinat. II. Bourtzis, Kostas. III. Series: Frontiers in microbiology ;1.
  [DNLM: 1. Arthropods--microbiology. 2. Symbiosis. QX 460]

  QR325.M36 2012
  576.8'5--dc22
                                                                    2011003360

Visit the Taylor & Francis Web site at
http://www.taylorandfrancis.com

and the CRC Press Web site at
http://www.crcpress.com

This book is dedicated to the memory of
Prof. David Rosen (1936–1997)

Contents

Preface

In the English edition of his landmark book *Endosymbiosis of Animals with Plant Microorganisms* (1965), Prof. Paul Buchner, probably the most prominent founder of systematic symbiosis research, included a short description of the history of the research in the field and wrote: "I too soon fell victim to the spell of this subject, and from 1911 on devoted myself to it."

Almost half a century has passed, and the impact arthropod-bacteria symbiosis has on virtually all aspects of the biology of both host and symbiont is being recognized by a growing number of entomologists. Because the richness of each particular system always provides ample questions and new discoveries, the discussion of this subject tends to be system based, with primary emphasis on the insect host. However, recent screening studies have revealed that the diversity of bacteria associated with arthropods may not be as wide as initially expected, and some genera are constantly being found in hosts that belong to distantly related taxa.

In the preface to his book Buchner wrote that although much progress has been made in understanding the diversity and morphological basis of many associations, "more intensive teamwork between biochemistry and microbiology" is required in order to advance the field. In recognition of the importance of interdisciplinary collaborations, over 120 scientists from more than 20 European countries, as well as from the United States, Australia, Korea, and Tunisia, have joined forces under the umbrella of EU COST Action FA0701, *Arthropod Symbioses: From Fundamental Studies to Pest and Disease Management*. The main goal of the action was to unravel basic aspects of arthropod symbiosis and promote the transformation of data into novel, effective, and environment-friendly tools for the control of pests and diseases.

The current book is aimed at introducing the fascinating topic of bacteria-arthropod associations to researchers who are not familiar with it, enlarging the scope of knowledge of those who are, and providing a textbook for students, mainly in microbiology, but also in other branches of biology. The book thus provides an overview of primary (obligatory) symbionts as well as the most abundant secondary (facultative) symbionts currently known. An effort has been made to summarize the most up-to-date information available on each symbiont and present a synopsis of the field from the bacterial angle.

We thank the many colleagues who have made the publication of this book possible, and the peers who provided thorough reviews on all the chapters. It is our hope that the readers of this book will also fall "victims to the spell of this subject," which, in Buchner's words, is "often bordering the fantastic."

Einat and Kostas

The Editors

Kostas Bourtzis, PhD, is a professor of molecular biology, genetics, and biochemistry in the Department of Environmental and Natural Resources Management, University of Ioannina, Greece. His research is focused on the interactions between insect pests/disease vectors and symbiotic bacteria, with special emphasis on *Wolbachia*-mediated cytoplasmic incompatibility, genetic manipulation of *Wolbachia*, the molecular mechanism of cytoplasmic incompatibility, and *Wolbachia* genomics. His group has recently shown that *Wolbachia*-induced cytoplasmic incompatibility can be used as a means to suppress insect pest populations.

Einat Zchori-Fein, PhD, is a researcher in the Department of Entomology, Newe Ya'ar Research Center, Agricultural Research Organization, Israel. Her research is focused on the interactions between insect pests and symbiotic bacteria, with special emphasis on the diversity and phenotypes of secondary symbionts of the sweet potato white-fly *Bemisia tabaci*, horizontal transmission of secondary symbionts, and the multitrophic interactions among plants, plant pathogens, arthropod vectors, and natural enemies.

Per their research interests and personal beliefs, Prof. Bourtzis and Dr. Zchori-Fein are the chair and vice chair of the EU COST Action FA0701, *Arthropod Symbioses: From Fundamental Studies to Pest and Disease Management*, dedicated to promoting the use of endosymbiotic bacteria as a tool for the development of environmentally friendly approaches for the control of arthropods of medical and agricultural importance.

Contributors

Serap Aksoy
Yale University
USA

Alberto Alma
University of Turin
Italy

Geofrrey M. Attardo
Yale University
USA

Claudio Bandi
University of Milan
Italy

Didier Bouchon
University of Poitiers
France

Kostas Bourtzis
University of Ioannina
Greece

Henk R. Braig
University of Wales Bangor
UK

Hans Breeuwer
University of Amsterdam
The Netherlands

Aemur Cherif
University of El Manar
Tunisia

Elad Chiel
Agricultural Research Organization
Israel

Emanuela Clementi
University of Pavia
Italy

Richard Cordaux
University of Poitiers
France

Elena Crotti
University of Milan
Italy

Daniele Daffonchio
University of Milan
Italy

Alistair C. Darby
University of Liverpool
UK

Olivier Duron
University of Montpellier
France

Sara Epis
University of Milan
Italy

Guido Favia
University of Camerino
Italy

Elena Gonella
University of Turin
Italy

Yuval Gottlieb
Hebrew University of Jerusalem
Israel

Pierre Grève
University of Poitiers
France

Tom V.M. Groot
De Groene Vlieg
The Netherlands

Roy Gross
University of Wurzburg
Germany

Abedalaziz Heddi
University of Lyon
France

Gregory D.D. Hurst
University of Liverpool
UK

Václav Hypša
University of South Bohemia
Czech Republic

Nathan Lo
University of Sydney
Australia

Mauro Mandrioli
University of Modena and Reggio
Emilia
Italy

Isabel Martinez-Sanudo
University of Padova
Italy

Luca Mazzon
University of Padova
Italy

Wolfgan J. Miller
Medical University of Vienna
Austria

Matteo Montagna
University of Milan
Italy

Eva Nováková
University of South Bohemia
Czech Republic

Massimo Pajoro
IZSLER
Sezione di Pavia
Italy

Steve J. Perlman
University of Victoria
Canada

Dario Pistone
University of Milan
Italy

Irene Ricci
University of Camerino
Italy

Markus Riegler
University of Western Sydney
Australia

Vera I.D. Ros
Wageningen University
The Netherlands

Luciano Sacchi
University of Pavia
Italy

Davide Sassera
University of Milan
Italy

Claudia Savio
University of Padova
Italy

Daniela Schneider
Medical University of Vienna
Austria

Mauro Simonato
University of Padova
Italy

Andrea Squartini
University of Padova
Italy

Fabrice Vavre
University of Lyon
France

Jingwen Wang
Yale University
USA

Brian L. Weiss
Yale University
USA

Timothy E. Wilkes
University of Liverpool
UK

Tom L. Wilkinson
University College Dublin
Ireland

Einat Zchori-Fein
Agricultural Research Organization
Israel

Introduction

Primary and Secondary Symbionts, So Similar, Yet So Different

Fabrice Vavre and Henk R. Braig

Studies on the interactions microbes have with other organisms, and especially eukaryotes, have generally focused on pathogens. The main reason for this is of course the impact microorganisms have on human, animal, and plant health and their dramatic consequences to human individuals, societies, and economies. Diverse aspects of pathogenicity have been studied with particular attention, including the molecular mechanisms underlying these interactions, the evolution of genomes and gene content of the microbes, and analyses of the coevolutionary dynamics of these associations, notably in relation to the evolution of virulence. Despite important progress made in all these fields, it covers only a very small fraction of the biodiversity of host-microbe associations.

In contrast to pathogens, other microbes allow their hosts to exploit previously hostile environments by providing new metabolic capabilities. As such, microbes may directly participate in the extension of their host's niche and be at the origin of the radiation of entire taxa. Through this mechanism, these mutualistic interactions have undoubtedly allowed some of the major transitions in the history of life, as exemplified by the emergence of the eukaryotic or plant lineages. Beside these two astonishing examples of mutualistic interactions, numerous others could be provided for the considerable impact beneficial microbial partners that live in intimate association with their hosts had on the life history of multicellular organisms. The increasing attention that is currently paid to the bacterial flora in higher eukaryotes, and notably of plant rhizomes and the human gastrointestinal tract, shows that the scientific community is becoming aware of microbes not only being pathogenic, but also providing benefit to their hosts. Between pathogenic associations and their mutualistic counterparts, all intermediates exist and fall all along a parasitism-mutualism continuum. Understanding the evolution of these associations, including host and microbial mechanisms that regulate interactions and transitions between different phenotypes, requires consideration of this entire spectrum of possibilities. There is no doubt that microbiology would benefit from a unified field of research, integrating the whole diversity of the microbial world and its interactions with other organisms (McFall-Ngai, 2008; Schwemmler, 1989). The use of the term *symbiosis* as it was originally defined by Albert Frank and Anton de Bary, i.e., the living together of organisms belonging to different species, is a first step in this effort of unification

(Committee on Taxonomy, 1937; de Bary, 1879; Frank, 1877; Sapp, 1994). We would like to take the opportunity here to dispel some confusion that has recently arisen regarding the origin of the term *symbiosis*. The *Oxford English Dictionary* traces the introduction of the word *symbiosis* into biology back to Alfred W. Bennett, who translated and edited a German textbook of botany into English in 1877 (OED, 2009; Thome, 1877). This view has meanwhile been adopted by several popular websites, including Wikipedia. Neither the original German edition nor the first English edition mention symbiosis (Thome, 1869). Only the sixth English edition from 1885 (p. 267) makes use of the term *symbiosis* (Thome, 1885).

One group of interest for studying host-symbiont relationships is the phylum Arthropoda, which has the following characteristics:

1. Arthropods are very prone to microbial infections, and especially with vertically transmitted (from parents to offspring) symbionts, with which they establish stable associations, some of them being more than 275 million years old.
2. Host-symbiont associations are very diverse, in terms of both the phylogenetic position of the symbionts and the phenotypes symbionts induce in their hosts.
3. Within some groups, transitions between extreme phenotypes have been reported.
4. Genome peculiarities identified in some of these symbionts have allowed great progress in the understanding of microbial genome evolution, and recent results suggest that there is still a lot to be discovered.

All together, these characteristics make symbiosis in Arthropoda a propitious ground for developing an integrated field of research dedicated to the understanding of mechanisms and evolution of microbes associated with hosts. These associations have mainly been studied by researchers with a background in zoology, ecology, or evolutionary biology, and as a consequence, questions that have been addressed are more oriented to the understanding of their influence on the host phenotype and evolution than toward molecular microbiology or cell biology. Yet, these associations may also provide model systems for microbiology. In addition, recent developments in the field show that manipulation of symbiotic systems in invertebrates may allow the development of innovative methods for controlling pathogens, making these systems not only of heuristic value but also of great applied interest.

This book provides a comprehensive overview of the diversity of eubacterial symbionts identified to date as frequent partners of terrestrial arthropods. With a stronger emphasis on the host, some topics of insect symbiosis are also covered in a contemporary entomology series (Bourtzis and Miller, 2003, 2006, 2009). Within each chapter of this book, a detailed review on the biology of these diverse symbionts is given. The first objective of this introduction is to provide the reader with some general features of these associations, some definitions, and an overview on what he or she might find in the subsequent chapters. The second objective is to identify general questions that emerge from the study of these particular systems.

PRIMARY AND SECONDARY SYMBIONTS:
GENERAL FEATURES AND DEFINITIONS

Arthropod symbiosis has now been studied for more than one century, with pioneering works by, among others, Friedrich Blochmann, Karel Šulc, and Umberto Pierantoni, who provided the first descriptions of these symbionts with hypotheses on their functional role (Blochmann, 1887, 1892; Perru, 2007a, 2007b; Pierantoni, 1909, 1951; Šulc, 1906, 1931). However, the most influential researcher of the discipline is undoubtedly Paul Buchner. His seminal 1965 treatise has established the nature of modern arthropod symbiosis, and his descriptions and their interpretation are still remarkably up-to-date (Buchner, 1965). The next revolution came with the advancement of molecular technologies, which allowed easier systematic surveys for symbionts. During the last 20 years, considerable progress has been achieved in uncovering the diversity, incidence, and prevalence of microbial partners associated with insects. Out of these studies, a clear picture emerges: most, if not all, arthropods are associated in a permanent manner with microbial symbionts. Together with the description of the various effects symbionts have on their host's biology, the entire field of arthropod studies, including physiology, reproduction, ecology, evolution, and speciation, has been revolutionized.

Symbiosis in arthropods can occur in a multitude of forms. Some symbionts are not strictly speaking harbored within the host. For example, symbionts can be found within the gut lumen, or, like *Erwinia*, in ducts in the head of olive flies, or even within the antennal gland as for antibiotic-producing streptomycetes in South American digger wasps (Capuzzo et al., 2005; Kaltenpoth et al., 2010). Particularly evident in some groups such as termites, gut symbionts may provide nutritional functions to their hosts, where the symbiotic flagellates have ecto- and endosymbiotic Bacteroidetes bacteria themselves (Strassert et al., 2010). This is mirrored in wood-eating mammals such as beavers but not in wood-eating fish. Gut symbionts also play an important role in controlling infection by pathogens (Ryu et al., 2010). Despite the evident and growing interest for gut symbionts (e.g., *Asaia, Acetobacter, Serratia, Sodalis*), most studies have been performed on symbionts that are found within their hosts and together form permanent associations. These symbionts may be extracellular, but they are most often found within the cells of their hosts. The term *endosymbiont* is preferentially used for an intracellular symbiont, but some symbionts such as *Sodalis* can be found intracellular in the midgut epithelium, as well as free in the gut lumen. Some symbionts are found both extra- and intracellularly: primary endosymbionts such as *Riesia* in lice, *Wigglesworthia* in tsetse flies, and candidate primary endosymbiont *Erwinia* in olive flies, as well as secondary endosymbionts like *Sodalis* in tsetse flies or *Hamiltonella* and *Serratia* in aphids (Attardo et al., 2008; Estes et al., 2009; Moran et al., 2005; Perotti et al., 2006). When intracellular, they can live free in the cytoplasm or be embedded in host vacuoles of the host. Some are restricted to specific tissues, while others show a more systemic distribution. Many of these symbionts are transmitted vertically (from parents, generally mother, to offspring), but many of them also show horizontal transmission (between individuals that are not necessarily related, or even belonging to different species) under some circumstances (Bright and Bulgheresi, 2010; Jones et al., 2010; Vautrin and Vavre, 2009). Variations on all these themes seem infinite, and one should keep in mind that closely related symbionts, often belonging to the same genus, may actually differ considerably.

All these characteristics are important, as they often set conditions for the evolution of the association along the parasitism-mutualism continuum. Traditionally, arthropod symbiosis researchers refer to the classical dichotomy coined by Buchner between "primary" and "secondary" symbionts. As specified below, this distinction captures many characteristics of the association. Recent results, however, tend to obscure the limits between P- and S-symbionts, but they open fantastic opportunities to understand the evolution of symbiotic associations.

PRIMARY (P-) SYMBIONTS

Primary symbionts are indispensable partners of their hosts and are thus fixed in populations. It is estimated that at least 10 to 15% of insect species depend on obligate microbial partners for their own development and reproduction, but with considerable variations among taxa. P-symbionts generally provide their eukaryotic partner with essential nutrients that are absent or poorly represented in their food, and thus form nutritional symbioses with their hosts (Clark et al., 2010; Gosalbes et al., 2010). As such, they are most commonly found in plant sucking insects developing on unbalanced diets such as sugar-rich phloem or mineral-rich xylem, where they provide essential amino acids, or obligate hematophagous arthropods, where they provide B vitamins. Sometimes the nutritional deficiency is not so obvious. Weevils feeding on grain like rice or maize still contain endosymbionts that provide both amino acids and vitamins to their hosts. Omnivores like many ant species have primary endosymbionts too, such as *Blochmannia*. Here, the endosymbiont supplements the nitrogen, sulfur, and lipid metabolism of its host. A complete urease cluster allows *Blochmannia* to recycle ammonia with the help of glutamine synthetase into the amino acid glutamine (Gosalbes et al., 2010). *Blattabacterium*, the primary endosymbiont of cockroaches, performs a complete urea cycle, a phenoype that is only shared with a cellulolytic soil-inhabiting *Cytophaga* bacterium. Its contribution is not anabolic but catabolic by breaking down urea (Lopez-Sanchez et al., 2009). P-symbionts have allowed their hosts to exploit new niches and have played a major role in the radiation of some insect groups and the fantastic ability of insects to colonize any kind of environment.

Typically P-symbionts are intracellular and their distribution in the host body is restricted to specialized cells called bacteriocytes or mycetocytes, which sometimes group and form a specialized organ (or organs) called bacteriome or mycetome. One of the starting events of symbiosis research was the realization by Šulc that a previously described organ of homopteran insects in fact represented the living quarters of symbiotic yeast; he called the organ mycetome (Šulc, 1910). This shows also that yeasts can be primary as well as secondary endosymbionts of arthropods. Bacterial endosymbionts are currently given more prominence, but this might not reflect their importance. Indeed, the primary endosymbiont of longhorned beetles, which are threatening to destroy a wide variety of tree species in North America and Europe, is a mycetomic yeast (Gruenwald et al., 2010). It is noteworthy that bacteriomes are found in diverse arthropod taxa, suggesting that this host provision is a classical adaptation to nutritional symbiosis. However, *Rickettsia*, as nonnutritional symbionts, are found in bacteriomes as well (Braig et al., 2009; Perotti et al., 2006). P-symbionts are also found in the reproductive organs, the infection of which ensures vertical transmission to the next generation and is generally achieved through migration of the symbionts from the bacteriome. The strict vertical transmission of these P-symbionts has resulted in fantastic patterns of cospeciation between P-symbionts and their hosts. Some symbioses, such as the one between *Buchnera aphidicola* and aphids, have lasted for over 200 million years.

The above picture may, however, be biased. Wherever there is a bacteriome or mycetome, the identification and specification of the primary endosymbiont or symbionts is straightforward. However, blood-sucking flies and many other arthropods do not have bacteriomes, but their nutrition still depends on the bacterial and yeast symbionts in their gut, making them, by definition, primary (endo)symbionts. In many insect species, gut symbionts just increase the fitness but are not indispensable. The situation in stinkbugs is particularly interesting. Some stinkbug species harbor, as usual, their primary endosymbionts in colorful bacteriomes (Küchler et al., 2010). In other stinkbug species, symbiotic *Ishikawaella* are luminal gut bacteria that are transmitted in an encapsulated form to the next generation, resulting in the same host specificity and coevolution as transovarially transmitted endosymbionts (Hosokawa et al., 2006; Kikuchi et al., 2009). Such pseudovertical transmission of luminal bacteria is also found in other stinkbugs, but this time, the females smear or impregnate the symbionts on the eggs and safeguard their transmission this way (Kaiwa et al., 2010). Finally, in other stinkbug species, the essential bacteria are newly obtained from the environment at every generation (Kikuchi et al., 2007). Acquiring indispensable symbionts newly from the environment is more common in aquatic habitats; for example, this is the case for all obligate or primary light-producing symbionts of fish and squid. However, in medically important mosquitoes and flies, the complement of microbial gut symbionts seems to depend more on the environment. It changes depending on what is available. Many arthropods carry more than one symbiont. Tsetse flies have *Wigglesworthia* as a primary bacteriome-dwelling endosymbiont, *Sodalis* as a gut-based intra- and extracellular (endo) symbiont, as well as variable numbers and combinations of *Serratia*, *Enterobacter*, *Enterococcus*, and *Acinetobacter* species as gut symbionts (Geiger et al., 2009, 2010). In blood-sucking mosquitoes, the microbial composition seems to completely rely on the environment (Briones et al., 2008; Gusmao et al., 2010; Lindh et al., 2008; Rani et al., 2009; Zouache et al., 2009). This leaves a problem both for the definition of the indispensable symbionts of mosquitoes and for the application of gut symbionts in vector control approaches. The requirements of the host can be met by varying consortia of symbionts. Whether mosquitoes are evolutionarily young hematophagous species that just embarked upon a long path toward becoming bacteriomic (mycetomic) organisms, or whether they illustrate an alternative strategy of environmental exploitation, is a question that deserves attention. In any case, these few examples highlight that many forms of primary symbioses exist outside the classical bacteriocyte-associated symbioses.

Secondary (S)-Symbionts

Contrary to P-symbionts, S-symbionts are facultative for their hosts, and infection is often polymorphic within or between populations. Polymorphic means that some individuals or population of a species house one species of symbiont, while others accommodate different symbiotic species or none. S-symbionts have been categorized as either mutualistic or reproductive manipulators, based on the phenotypic effects identified. However, despite the practical use of these two terms in distinguishing different types of associations, one should keep in mind that mutualism and reproductive parasitism are not mutually exclusive, and that both can occur within a given association or be expressed by very closely related symbionts.

More and more cases of mutualistic S-symbionts are described. Exciting recent examples include *Fritschea*, chlamydial endosymbionts of whiteflies and scale insects (Everett et al., 2005); *Coxiella* and *Diplorickettsia*, both sister taxa of the *Rickettsiella*, in ticks (Clay et al., 2008; Heise et al., 2010; Mediannikov et al., 2010); Actinobacteria on wasps (Kaltenpoth, 2009); and *Stammerula* in tephretid fruit flies (Mazzon et al., 2008, 2010). The advantage provided to the host is generally associated with resistance to harsh environmental conditions such as heat, plant defense, or natural enemies (Clark et al., 2010). The latter is emerging as a frequent advantage conferred by S-symbionts, and the impressive range of natural enemies counteracted by symbionts includes viruses, bacteria, fungi, nematodes, and parasitoid wasps (Brownlie and Johnson, 2009; Panteleev et al., 2007). Such fitness advantages allow S-symbionts to spread within and among populations, but imperfect maternal transmission, costs associated with infection, and the conditional advantage they provide explain why their presence is often polymorphic within populations.

Another strategy used by S-symbionts to spread and maintain themselves within populations is to induce reproductive manipulations (Engelstaedter and Hurst, 2009). Initially thought to be restricted to the genus *Wolbachia*, it is now clear that numerous bacteria and microsporidia are able to manipulate their host's reproduction to their own advantage. Since most symbionts are strictly transmitted through females, any strategy favoring production of infected females relative to uninfected females can spread, even if this comes at the expense of males. Contrary to benefit provision, reproductive manipulations are thus selfish strategies that can lead to strong conflicts between the host and the symbiont (Stouthamer et al., 2010). The most direct way for a symbiont to increase its transmission is to transform nontransmitting males into transmitting females. This is achieved either through feminization of genetic males or through induction of thelytoky, a form of parthenogenesis or asexual reproduction where females can produce female offspring without being fertilized. While in the former case sexual reproduction is still needed, a feminized male (neofemale) mating with another male, this is not true in the latter, and in many cases entire populations or species have become asexual. Increased daughter production by a female can also be achieved by killing sons when competition occurs among siblings. This male-killing phenotype is induced by a variety of maternally transmitted symbionts. Finally, the last strategy used by reproductive manipulators is to reduce the production of daughters by uninfected females through induction of cytoplasmic incompatibility (CI). In its simplest form, CI occurs in crosses between uninfected females and infected males that lead to nonviable offspring, while all other crosses are fertile. This is the most common strategy in terms of affected host species.

One important difference between P- and S-symbionts is that in addition to vertical transmission, S-symbionts can be horizontally transmitted, with considerable variation in the frequency and modes of these events. This also means that secondary symbionts have to remain infectious, which in turn requires a certain level of virulence. Primary endosymbionts, with exceptions, have lost their virulence and the capacity to infect new hosts. When frequent, horizontal transmission can impact the epidemiology of the symbiont within its host species (Duron et al., 2010). In any case, it allows recurrent capture of new host species, blurring cospeciation processes. Absence of cospeciation also implicates frequent losses of infection within populations and species. As a consequence, associations between S-symbionts and a particular host are thought to be much more recent than the ones with P-symbionts. Thus, even though some S-symbionts, like *Wolbachia*, for example, have probably been associated with arthropods for at least 100 million years, particular lineages have probably interacted with numerous different hosts.

MUTUALISM AND DEPENDENCE

The clear dichotomy between old, obligate, stable primary symbioses and recent, dynamic secondary symbioses exposed above has to be tempered in the light of recent results, which highlight the limited knowledge we have on the microevolutionary processes involved in the evolution of obligate associations. Generally, primary symbionts are associated with new metabolic abilities conferred on their hosts that allowed them to exploit new niches. In other cases, the evolutionary trajectory of the association is not as clear. In particular, it is probable that dependence may evolve without being associated with exploitation of an alternative niche. Most of the essential vitamins required by human lice are found inside human erythrocytes. At one time, human lice lost the ability to lyse red blood cells in their gut, possibly a mutation in a key enzyme, requiring sudden supplementation of the vitamins by a primary endosymbiont, now *Riesia*, a member of the *Arsenophonus* genus (Kirkness et al., 2010; Perotti et al., 2009). It is tempting to speculate that this loss of function was permitted by a previous infection by the secondary symbiont, limiting the deleterious effect of the mutation, but leading to the dependence of the host on the symbiont. An example of endosymbiotic rescue of a metabolic failure outside arthropods is the ciliate *Euplotes*, where the primary endosymbiont *Polynucleobacter* reenables glycogenolysis and reestablishes the cell cycle of its host (Vannini et al., 2007a, 2007b). Similarly, in a *Drosophila melanogaster* strain, an otherwise fatal mutation is rescued by a secondary endosymbiont, *Wolbachia* (Starr and Cline, 2002). The hymenopteran wasp *Asobara tabida* depends on a *Wolbachia* strain for oogenesis, that thus acts as a primary endosymbiont (Kremer et al., 2009; Vavre et al., 2009b). Rescue by *Wolbachia* might or might not extend to preventing apoptosis in its host (Braig et al., 2009; Harris et al., 2010; Pannebakker et al., 2007; Vavre et al., 2009a). In this system it is proposed that host dependence is a consequence of the evolution of tolerance to the presence of the symbiont (Kremer et al., 2010). Similarly, *Rickettsia*, otherwise secondary endosymbionts, rescue in book- and barklice egg development of their hosts, making these *Rickettsia* primary endosymbionts (Braig et al., 2009; Perotti et al., 2006). Another interesting case is parthenogenesis induction by various symbionts in insects that led to the evolution of obligate asexuality, and thus dependence upon the symbiont (Braig et al., 2002; Koivisto and Braig, 2003). In this case, modeling suggests that the evolution of dependence is induced by the intergenomic conflict created by parthenogenesis induction (Bouchon et al., 2009; Majerus and Majerus, 2010; Stouthamer et al., 2010).

Once a host is dependent on a P-symbiont, the association is not fixed in this state. First, rescue is not limited to host metabolic complementation and P-endosymbionts are at least as prone to losing key metabolic capabilities as are their hosts. Secondary endosymbionts can fill in the gap and become stable coprimary endosymbionts. *Sulcia* and *Baumannia* are coprimary endosymbionts in leafhoppers and sharpshooters, *Serratia* is a coprimary endosymbiont in some aphid strains. A primary cosymbiont is proposed for *Carsonella* in psyllids but not yet identified. *Sodalis* in tsetse flies is involved in reciprocal metabolic complementation with the primary endosymbiont *Wigglesworthia*, and on its way to becoming a coprimary endosymbiont (Belda et al., 2010; Snyder et al., 2010). The detection of *Sodalis* in gonadal tissues of some specimens in a stinkbug might suggest a similar evolutionary trajectory (Kaiwa et al., 2010). While the presence of coprimary symbionts may be stable, they could also lead to endosymbiont replacement. This might be common in blood-sucking lice, where *Riesia* is a new arrival as a primary endosymbiont. In many weevils, *Nardonella* has been replaced by current endosymbionts; one of them, *Curculioniphilus*, might now be replaced by *Sodalis* (Toju et al., 2010).

PHYLOGENETIC DIVERSITY OF ARTHROPOD SYMBIONTS AND LIFESTYLE TRANSITIONS

Arthropod symbionts are phylogenetically very diverse (Moya et al., 2008). Some groups of bacteria seem much more prone to establish permanent associations with arthropods than others. Proteobacteria, and more precisely the alpha and gamma divisions, are by far the main associates of terrestrial arthropods revealed to date. A P-endosymbiont belonging to the beta division, *Dactylopiibacterium*, has just been described for cochineal scales (Ramirez-Puebla et al., 2010). Bacteroidetes is another branch of the *Eubacteria* where symbionts are frequently found. The reasons for this bias in the phylogenetic distribution of arthropod symbionts are not clear, but could involve ecological factors, such as the frequent occurrence of gamma-Proteobacteria in close contact with eukaryotic organisms (soil, oceans, and lakes), and genetic factors, such as higher frequency of horizontal gene transfer (HGT) in these organisms, allowing rapid adaptation to particular hosts (Toft and Andersson, 2010). These trends should be taken with caution, since the entire diversity of symbiotic microbes might not have been covered yet, and in many analyses, there is a separation into four subdiciplines which barely overlap: symbioses in terrestrial invertebrates, marine invertebrates, plants, and lower eukaryotes. While surveys of terrestrial arthropods for specific symbionts have been performed, we still lack surveys without a priori restrictions imposed by the employed methodology (Shi et al., 2010). First publications on new-generation sequencing techniques applied to arthropods are starting to appear (e.g., Bai et al., 2010; Zhang et al., 2010) and will give important information. However, many new sequencing approaches are still limited by the remaining specificity of the degenerate primers employed for the amplification.

Symbionts do not form monophyletic groups within each of the main lineages frequently found in arthropods, indicating that evolution toward symbiosis has occurred multiple times, even within these groups. Increasingly, primary endosymbiotic lineages can be linked to secondary endosymbiotic lineages, and secondary endosymbiotic lineages can be linked to free-living pathogens, suggesting rapid transitions and highlighting their evolutionary origin. Some symbionts, like *Wolbachia*, are particularly interesting in this respect since they exhibit remarkable evidence for rapid transitions among extreme phenotypes with their hosts. Indeed, *Wolbachia* is best known for inducing reproductive manipulations in arthropods, but it is also a P-symbiont in filarial nematodes, as in some arthropods like bed bugs. *Arsenophonus*, *Rickettsia*, and *Rickettsiella* show an even more extreme variation in the effects they induce in their hosts. All of them are known to be reproductive manipulators in some hosts, P-symbionts in others, and even pathogens vectored by arthropods in still others. *Erwinia* and *Arsenophonus* symbionts are closely related to plant pathogens (Bressan et al., 2009; Salar et al., 2010).

COMPARATIVE GENOMICS

The first genomes of arthropod symbionts that have been sequenced are those of various P-symbionts, in which remarkable and similar characteristics have been revealed. Most notably, the genomes of P-symbionts are extremely reduced, around 600 and 700 kb, but some are even smaller, with genome sizes below 200 kb, like *Hodgkinia cicadicola* in cicadas and *Carsonella ruddii* in psyllids. Within these small genomes, it is possible to read, as in an open book, what functions the symbionts provide to their hosts: in plant-sucking insects, conserved pathways are those involved in the synthesis of essential amino acids and cofactors, while in blood-feeding insects they are the ones involved in B vitamin synthesis. The loss of common genes experienced by P-symbionts depends on their intracellular locality and their mode of transmission to the next generation. *Buchnera* resides inside a host vacuole and has therefore lost genes for cell wall synthesis. *Blochmannia*, *Wigglesworthia*, and *Riesia*, which are free in the cytoplasm or even extracellular inside their bacteriome, have retained genes for cell wall synthesis. The *Wigglesworthia* genome suggests genes for mobility despite the lack of physical evidence for mobility, whereas in *Riesia* there is physical evidence for mobility. The only unifying feature is the loss of genes involved in DNA repair (Sharples, 2009). The streamlined genome of P-endosymbionts is also characterized by absence of repetitive sequences like IS or phages. However, P-endosymbionts still do carry plasmids. P-symbionts are also completely isolated from the environment, with no evidence for horizontal gene transfer to their chromosomes. These characteristics lead to the stasis of P-symbiont genomes, as exemplified by *Buchnera*, where various strains exhibit similar genome organization.

Interestingly, some P-symbionts, such as the *Sitophilus* primary endosymbiont (SPE) found in weevils of the genus *Sitophilus*, are of more recent origin. SPE is thought to have recently replaced the older P-symbiont *Nardonella*, usually associated with Cucurlionidae. The genome of SPE is still remarkably big for a P-symbiont (about 3 Mb), but is characterized by a small fraction of coding sequences and an important amount of transposable elements, like insertion sequences (ISs). Insertion sequences do not seem to be so important in the inactivation of genes on the way to primary endosymbiosis (Belda et al., 2010), but probably facilitate important deletion events. Recombination and genome erosion can become so massive that it becomes very difficult to identify the closest free-living relative or the likely ancestral clade to which the P-endosymbiont belongs, as is the case for old *Carsonella* (Williams et al., 2010). For younger P-endosymbionts like *Blochmannia*, this is much easier (Wernegreen et al., 2009). Genome erosion is not limited to arthropod symbionts. Vertically transmitted nitrogen-fixing endosymbiotic cyanobacterium undergoes the same fate (Ran et al., 2010).

Genomics of P-symbionts led to some hypotheses on the forces shaping symbiont genomes. Among them, the most popular was that the intracellular state of endosymbiont was responsible for these trends. First, intracellular lifestyle imposes isolation from the environment, leading to reduced exposure to HGT. Second, reduced selective pressures on some biological functions provided by the host or not under selection anymore in the host cell should facilitate genome erosion. Third, the effective size of intracellular bacteria is small, leading to reduced efficiency of selection and accumulation of slightly deleterious mutations, favoring the expression of mutational bias, genome erosion, and decay, even though strong selection tends to limit the degenerative effect of accumulating detrimental mutations (Allen et al., 2009). All three hypotheses are no longer sustainable. Intracellularity cannot be the sole factor explaining the organization of the P-symbiont genomes. First, the genomes of the P-symbionts *Ishikawella* and *Rosenkrantia* from stinkbugs show characteristics similar to those of intracellular P-symbionts despite being extracellular bacteria located within the gut. Second, the genomes sequenced from several S-symbionts exhibit very different characteristics from those of P-symbionts despite their intracellular location. Some of these genomes, like those of *Wolbachia*, show traces of reductive evolution and AT bias, but to a lesser extent than P-symbiont genomes, despite being restricted to an intracellular environment for a long time. Other genomes (like *Arsenophonus nasoniae* or *Hamiltonella defensa*) are still large, but this could also indicate recent origin of their symbiotic lifestyle and ability to colonize the extracellular environment. A striking feature of all these genomes compared to P-symbionts is their rapid dynamics, resulting from frequent recombination, a high number of HGT, and the presence of mobile elements like IS or phages (Sonthayanon et al., 2010). The transpositional activity of IS in secondary endosymbionts like *Wolbachia*, *Regiella*, and *Hamiltonella* is particularly impressive (Cordaux et al., 2008; Degnan et al., 2010). Despite being restricted to hosts, these genomes are clearly not isolated from other microorganisms like P-symbionts. In this context, additional hypotheses have to be invoked to understand what drives the evolution of symbiotic genomes and to link this evolution with the relationship symbionts have with their hosts. Making the difference between causes and consequences is, however, not an easy task. For example, rapid dynamics of S-symbionts' genomes could be a strategy for adaptation to new hosts after horizontal transmission. It could also be merely a consequence of the presence and activity of numerous repetitive elements favored by recurrent genetic exchanges between bacteria. Interestingly, while most P-symbionts have lost their recombination and DNA repair machinery, S-symbionts still possess them. It is tempting to speculate that this major difference is at the basis of the biological properties and evolution of these genomes. Indeed, frequent host shifts exhibited by S-symbionts through horizontal transmission could require rapid adaptation of the symbionts, which HGT and recombination between and within genomes could facilitate. This could explain why these genomes have conserved such machineries by maintaining sufficient selective pressures on them. In any case, these results demonstrate that we still have many things to learn from these genomes, since they include the most streamlined and degenerated genomes, but also the most dynamic ones.

IMMUNITY AND SYMBIOSIS

Immunity has been studied almost exclusively in pathogenic associations, and it was generally assumed that it played no role in the interaction between P- or S-symbionts and their hosts. This assumption is no longer valid, and the relationship between immunity and symbiosis is an emerging field that is directed into two main paths (Gross et al., 2009): (1) Which immunity-related mechanisms are involved in the control of permanent symbiosis? (2) What is the role of symbionts in the immunity of their hosts against their natural enemies, and what mechanisms do symbionts use for protecting their hosts against infections?

THE ROLE OF IMMUNITY IN THE REGULATION OF SYMBIONT POPULATIONS

There is growing evidence that immunity plays a central role in the control of symbiotic populations. As such, symbiotic associations may provide unique systems to decipher finely tuned mechanisms of immunity. Indeed, response to pathogens elicits an acute immune response with strong transcriptional response of some pathways (e.g., the IMD (immune deficiency) pathway against Gram-positive bacteria and fungi and the Toll pathway against Gram-negative bacteria and viruses), culminating in important synthesis of antimicrobial peptides. Because permanent associations do not elicit such a dramatic response, these systems may allow studying other components of immunity and corresponding bacterial determinants of the interaction. On the host side, it has been shown that cells and organs specialized in harboring symbionts have a particular immune profile that is most probably involved in maintaining symbiotic infections while keeping them in check (Anselme et al., 2008). On the bacterial side, some genes, like outer membrane protein A (OmpA), can almost be used as a predictor of the pathogenicity of a microorganism. Infection with *Escherichia coli* is lethal for tsetse flies, whereas an *E. coli* strain with a mutation in the *OmpA* gene is not. The symbiont *Sodalis* carrying the *E. coli OmpA* gene becomes a pathogen for tsetse flies (Weiss et al., 2008). Some *Rickettsia* lose their virulence through mutation in their *OmpA* genes (Braig et al., 2009; Ellison et al., 2008). The spectrum of interaction between a single host's innate immunity and a symbiont can cover evasion, suppression, and tolerance (Anbutsu and Fukatsu, 2010), and this, of course, might change with the environment, gender, and age of the host (Sicard et al., 2010). In the human louse system, where the host provides four different bacteriomes to its primary endosymbiont, *Riesia* still has to literally outrun circulating host immune cells when it leaves its major bacteriome to reach the ovaries (Perotti et al., 2007).

INTERACTIONS BETWEEN SYMBIONTS AND NATURAL ENEMIES

As said above, cases of protection conferred by symbionts against natural enemies are increasingly described. One of the very nice examples of adaptation via symbiosis concerns the spread of *Spiroplasma* in *Drosophila neotestacea*, where it confers resistance against a recently introduced nematode (Jaenike et al., 2010a, 2010b). There are numerous potential ways by which symbionts may interact with natural enemies. The most straightforward is that symbionts are directly interacting with natural enemies. This is probably the case for *Hamiltonella defensa*, where its phage APSE harbors toxin genes that could directly target attacking wasps (Oliver et al., 2009). In contrast, in tsetse flies, the symbiont *Wigglesworthia* elicits the transcription of an immune regulatory gene, PGRP-LB, which also has antiprotozoan activities. Thus, the presence of the symbionts indirectly affects protozoan growth and survival by affecting immune genes (Wang et al., 2009). Interestingly, PGR-LB has also proved to play a role in other symbiotic interactions. However, if it comes to parasites that are less costly, such as viruses transmitted by the host that are often asymptomatic in the insect, we see both inhibiting and facilitating effects of symbionts (Brownlie and Johnson, 2009; Gottlieb et al., 2010). From an evolutionary point of view, symbiont protection may reduce the intensity of the selective pressures exerted by parasites, which might in turn impact the evolution of the host immune system.

SYMBIOTIC CONSORTIA: TOWARD SOCIAL SYMBIOSIS?

Symbiont protection highlights the fact that symbiotic interactions have to be considered in a community context. Hosts are not dealing with a single partner, but with many of them simultaneously. In addition to the interactions permanent symbionts have with natural enemies, they can also interact with each other. In this book, symbionts are mainly treated independently of each other, but one should keep in mind that multiple infections with these symbionts are quite frequent. In particular, arthropods harboring P-symbionts frequently carry a cortege of S-symbionts. Aphids and whiteflies, where one P-symbiont and up to six S-symbionts are known, are good examples of the diversity of symbionts that can be found within a host. Microbiology has been revolutionized by the introduction of the social dimension of bacterial populations and communities. Symbiotic systems have played an important role in this emergence since phenomena like quorum sensing were demonstrated in mutualistic associations. This should not come as a surprise. Social strategies rely on communication via emission and diffusion of small molecules, and confinement within a host allows limiting the cost of producing these molecules, thus favoring the evolution of these mechanisms. Interspecific interactions between symbionts can also be envisioned for two main reasons. On the one hand, hosts provide a limited amount of space and resources, which could favor competitive interactions within hosts, while on the other hand, vertical transmission of bacterial symbiotic communities might favor the evolution of cooperation or dependence among members (Vautrin and Vavre, 2009). Although the importance of these phenomena in the evolution of multiple symbioses is still difficult to establish, some cases, like the specific erosion of the tryptophan pathway in the smallest *Buchnera* genome that is compensated by the S-symbiont *Serratia* in the aphid *Cinara cedri*, offer a good example that genome erosion of symbionts can be compensated by the presence of other symbionts. This phenomenon could also be the premise of endosymbiont replacement.

There is some compelling evidence that horizontal gene transfers may occur not only between symbionts, but also between symbionts and their hosts. Strikingly, the rate of these transfers appears to differ between different classes of symbionts. S-symbionts appear highly susceptible to HGT. For example, *Wolbachia* bacteria regularly acquire new genes from their surroundings. Recombination among *Wolbachia* strains seems to be quite frequent, but this is clearly not the only source of genetic information used by these symbionts. Transfers from *Wolbachia* to their hosts' genome have repeatedly been found, showing recurrent and possibly frequent HGT from bacteria to eukaryotes. This differs substantially from what is seen for P-symbionts. The human louse genome does not show evidence for bacterial genes. Analysis of the recently sequenced genome of the pea aphid *Acyrthosiphon pisum* revealed no gene transfer from *Buchnera* to the host genome despite their long coexistence. Nevertheless, different genes of bacterial origin have been detected in the pea aphid genome, and they are thought to be involved in the control of symbiosis. All these genes were phylogenetically related to genes found in alpha-Proteobacteria and, more precisely, to *Wolbachia*, which has been detected in few species of aphids (Gomez-Valero et al., 2004).

CONCLUSION AND PERSPECTIVES

P- and S-symbionts provide a fantastic basis for comparative studies to better understand the evolutionary history of host-symbiont interactions. Phylogenetic characterizations and genomes of these symbionts are now accumulating and provide the first step for comparative biology of symbiotic interactions. One emerging picture is that it might be quite easy for a pathogen to become an S-symbiont and for an S-symbiont to evolve toward or be trapped as a P-symbiont, and P-symbionts eventually will become extinct in the host through being replaced by a new P-symbiont. The transition between pathogenic and more benign associations is highly related to the evolution of the transmission mode. Indeed, even though numerous exceptions exist, stabilization of cotransmission of host and symbionts through vertical transmission plays a major role in the reduction of pathogenicity. The microevolutionary processes involved in lifestyle and symbiont virulence evolution have, however, to be clarified. Some symbionts, like *Arsenophonus*, are particularly appealing to study this question. Even though these evolutionary paths are frequent, they are probably not the only possible trajectories. In particular, pathogens and S-symbionts are highly susceptible to HGT, which can orient the association in any particular direction, depending on the nature of the transferred genes and their associated fitness gains.

In particular, numerous arthropods have close interactions with other organisms (notably plants and vertebrates), which may favor contact between symbionts and these other organisms, and favor the evolution of vector-borne diseases. While there are currently no examples of symbionts developing into new pathogens vectored by their former symbiont hosts, the situation in the Rickettsiales might suggest such transitions are possible. In particular, only a small fraction of *Rickettsia* have evolved pathogenicity in vertebrate hosts, and the ancestral state for *Rickettsia* is probably a benign association with arthropods. Outside bacteria, it has been proposed that vector-borne parasites like trypanosomes, causing diseases like sleeping sickness or Chagas disease in humans, evolved from gut symbionts of arthropods, but this has never been corroborated. Our lack of knowledge of eukaryotic symbionts of arthropods other than flagellates in termites is unfortunate. A huge diversity of eukaryotes has been reported from arthropods in a former, descriptive age. These associations have never been analyzed in a symbiotic context, but these eukaryotes might hold answers to various questions, notably the evolution of vectored pathogens from more benign associations.

The field of arthropod symbiosis has undergone major revolutions, notably when molecular techniques started to be applied to these systems. The next revolution is probably under way owing to next-generation sequencing techniques. They will first allow an easier and more exhaustive determination of the symbiont diversity within hosts. This can be done either by sequencing a PCR-amplified target locus (e.g., 16SrDNA) or through metagenomic approaches. Second, obtaining symbiont genomes has been inhibited by difficulties of securing sufficiently pure symbiont DNA in adequate amounts. New techniques like whole genome amplification associated with a large number of sequences that can be obtained or single molecule sequencing now open wider possibilities for genome sequencing. Third, transcriptomic analyses of host-symbiont systems are now possible even in nonmodel organisms. This increases dramatically the range of interesting systems that can be explored through these approaches. It also allows imagining the building of databases devoted to the sharing of this information and the extension of classification methods like gene ontology for genes associated to symbiotic interactions, which will facilitate the exploration of the functional bases of symbiosis. An exciting era is coming for people interested in understanding the mechanisms and evolution of these associations.

However, one should keep in mind that beside the molecular approaches driven by advances in technology, the characterization of the phenotypic effects of these symbionts remains a priority. The aim, in the end, is a whole organism biology. Each year, new phenotypes are described, and the emerging field of protective symbiosis is probably one of the best examples. The number of cases of symbiont-mediated protection that have been discovered in just the last few years is impressive. What new revolutions might appear in the coming years if sufficient attention is paid to this problem?

Beside the fundamental interest of these systems, it is now clear that this knowledge can also be used for applied purposes. Regulation of arthropod populations through CI, manipulation of symbiotic content to alter the ability of arthropods to transmit pathogens, endosymbionts in conservation biology, evolution of minimal genomes, and synthetic biology are just some of the subjects that are currently being investigated, and many others are foreseen.

REFERENCES

Allen, J.M., J.E. Light, M.A. Perotti, H.R. Braig, and D.L. Reed. 2009. Mutational meltdown in primary endosymbionts: selection limits Muller's ratchet. *PLoS ONE* 4:e4969.

Anbutsu, H., and T. Fukatsu. 2010. Evasion, suppression and tolerance of *Drosophila* innate immunity by a male-killing *Spiroplasma* endosymbiont. *Insect Molecular Biology* 19:481–488.

Anselme, C., V. Perez-Brocal, A. Vallier, C. Vincent-Monegat, D. Charif, A. Latorre, A. Moya, and A. Heddi. 2008. Identification of the weevil immune genes and their expression in the bacteriome tissue. *BMC Biology* 6:1–13.

Attardo, G.M., C. Lohs, A. Heddi, U.H. Alam, S. Yildirim, and S. Aksoy. 2008. Analysis of milk gland structure and function in *Glossina morsitans*: milk protein production, symbiont populations and fecundity. *Journal of Insect Physiology* 54:1236–1242.

Bai, X., W. Zhang, L. Orantes, T.-H. Jun, O. Mittapalli, M.A.R. Mian, and A.P. Michel. 2010. Combining next-generation sequencing strategies for rapid molecular resource development from an invasive aphid species, *Aphis glycines*. *PLoS ONE* 5:e11370.

Belda, E., A. Moya, S. Bentley, and F.J. Silva. 2010. Mobile genetic element proliferation and gene inactivation impact over the genome structure and metabolic capabilities of *Sodalis glossinidius*, the secondary endosymbiont of tsetse flies. *BMC Genomics* 11:449.

Blochmann, F. 1887. Über das regelmäßige Vorkommen von bakterienähnlichen Gebilden in den Geweben und Eiern verschiedener Insekten [On the regular occurrence of bacteria-like entities in the tissues and eggs of various insects]. *Zeitschrift für Biologie* 24(N. S.):1–16.

Blochmann, F. 1892. Über das Vorkommen von bakterienähnlichen Gebilden in den Geweben und Eiern verschiedener Insekten [On the occurrence of bacteria-like entities in the tissues and eggs of various insects]. *Zentralblatt für Bakteriologie* 11:234–240.

Bouchon, D., R. Cordaux, and P. Grève. 2009. Feminizing *Wolbachia* and the evolution of sex determination in isopods. In *Insect symbiosis*, K. Bourtzis and T.A. Miller, editors. Vol. 3. CRC Press, Boca Raton, FL, 273–294.

Bourtzis, K., and T.A. Miller. 2003. Insect symbiosis. In *Contemporary topics in entomology series*, T.A. Miller, editor. Vol. 1. CRC Press, Boca Raton, FL, 347.

Bourtzis, K., and T.A. Miller. 2006. Insect symbiosis. In *Contemporary topics in entomology series*, T.A. Miller, editor. Vol. 2. CRC Press, Boca Raton, FL, 276.

Bourtzis, K., and T.A. Miller. 2009. Insect symbiosis. In *Contemporary topics in entomology series*, T.A. Miller, editor. Vol. 3. CRC Press, Boca Raton, FL, 408.

Braig, H.R., B.D. Turner, B.B. Normark, and R. Stouthamer. 2002. Microorganism-induced parthenogenesis. In *Progress in asexual reproduction*, R.N. Hughes, editor. Vol. 11. John Wiley & Sons, Chichester, UK, 1–62.

Braig, H.R., B.D. Turner, and M.A. Perotti. 2009. Symbiotic *Rickettsia*. In *Insect symbiosis 3*, K. Bourtzis and T.A. Miller, editors. Taylor & Francis, Boca Raton, FL, 221–252.

Bressan, A., O. Semetey, J. Arneodo, J. Lherminier, and E. Boudon-Padieu. 2009. Vector transmission of a plant-pathogenic bacterium in the *Arsenophonus* clade sharing ecological traits with facultative insect endosymbionts. *Phytopathology* 99:1289–1296.

Bright, M., and S. Bulgheresi. 2010. A complex journey: transmission of microbial symbionts. *Nature Reviews Microbiology* 8:218–230.

Briones, A.M., J. Shililu, J. Githure, R. Novak, and L. Raskin. 2008. *Thorsellia anophelis* is the dominant bacterium in a Kenyan population of adult *Anopheles gaimbiae* mosquitoes. *ISME Journal* 2:74–82.

Brownlie, J.C., and K.N. Johnson. 2009. Symbiont-mediated protection in insect hosts. *Trends in Microbiology* 17:348–354.

Buchner, P. 1965. *Endosymbiosis of animals with plant microorganisms*. Intersciences Publishers, New York.

Capuzzo, C., G. Firrao, L. Mazzon, A. Squartini, and V. Girolami. 2005. '*Candidatus* Erwinia dacicola', a coevolved symbiotic bacterium of the olive fly *Bactrocera oleae* (Gmelin). *International Journal of Systematic and Evolutionary Microbiology* 55:1641–1647.

Clark, E.L., A.J. Karley, and S.F. Hubbard. 2010. Insect endosymbionts: manipulators of insect herbivore trophic interactions? *Protoplasma* 244:25–51.

Clay, K., O. Klyachko, N. Grindle, D. Civitello, D. Oleske, and C. Fuqua. 2008. Microbial communities and interactions in the lone star tick, *Amblyomma americanum*. *Molecular Ecology* 17:4371–4381.

Committee on Taxonomy. 1937. Supplement to the report of the twelfth annual meeting of the American Society of Parasitologists: report of the Committee on Taxonomy. *Journal of Parasitology* 23:325–329.

Cordaux, R., S. Pichon, A. Ling, P. Perez, C. Delaunay, F. Vavre, D. Bouchon, and P. Greve. 2008. Intense transpositional activity of insertion sequences in an ancient obligate endosymbiont. *Molecular Biology and Evolution* 25:1889–1896.

de Bary, A. 1879. Die Erscheinung der Symbiose [Manifestations of symbiosis]. In *Vortrag, gehalten auf der Versammlung der Deutschen Naturforscher und Ärzte zu Cassel*. Verlag von Karl J. Trübner, Strassburg, 1–30.

Degnan, P.H., T.E. Leonardo, B.N. Cass, B. Hurwitz, D. Stern, R.A. Gibbs, S. Richards, and N.A. Moran. 2010. Dynamics of genome evolution in facultative symbionts of aphids. *Environmental Microbiology* 12:2060–2069.

Duron, O., T.E. Wilkes, and G.D.D. Hurst. 2010. Interspecific transmission of a male-killing bacterium on an ecological timescale. *Ecology Letters* 13:1139–1148.

Ellison, D.W., T.R. Clark, D.E. Sturdevant, K. Virtaneva, S.F. Porcella, and T. Hackstadt. 2008. Genomic comparison of virulent *Rickettsia rickettsii* Sheila Smith and avirulent *Rickettsia rickettsii* Iowa. *Infection and Immunity* 76:542–550.

Engelstaedter, J., and G.D.D. Hurst. 2009. The ecology and evolution of microbes that manipulate host reproduction. *Annual Review of Ecology Evolution and Systematics* 40:127–149.

Estes, A.M., D.J. Hearn, J.L. Bronstein, and E.A. Pierson. 2009. The olive fly endosymbiont, "*Candidatus* Erwinia dacicola," switches from an intracellular existence to an extracellular existence during host insect development. *Applied and Environmental Microbiology* 75:7097–7106.

Everett, K.D.E., M.L. Thao, M. Horn, G.E. Dyszynski, and P. Baumann. 2005. Novel chlamydiae in whiteflies and scale insects: endosymbionts '*Candidatus* Fritschea bemisiae' strain Falk and '*Candidatus* Fritschea eriococci' strain Elm. *International Journal of Systematic and Evolutionary Microbiology* 55:1581–1587.

Frank, A.B. 1877. Über die biologischen Verhähltnisse des Thallus einiger Krustenflechten [About the biological relationships of the thallus of some crustose lichens]. *Beiträge zur Biologie der Pflanzen* 2:123–200.

Geiger, A., M.-L. Fardeau, E. Falsen, B. Ollivier, and G. Cuny. 2010. *Serratia glossinae* sp. nov., isolated from the midgut of the tsetse fly *Glossina palpalis gambiensis*. *International Journal of Systematic and Evolutionary Microbiology* 60:1261–1265.

Geiger, A., M.-L. Fardeau, P. Grebaut, G. Vatunga, T. Josenando, S. Herder, G. Cuny, P. Truc, and B. Ollivier. 2009. First isolation of *Enterobacter*, *Enterococcus*, and *Acinetobacter* spp. as inhabitants of the tsetse fly (*Glossina palpalis palpalis*) midgut. *Infection Genetics and Evolution* 9:SI1364–SI1370.

Gomez-Valero, L., M. Soriano-Navarro, V. Perez-Brocal, A. Heddi, A. Moya, J.M. Garcia-Verdugo, and A. Latorre. 2004. Coexistence of *Wolbachia* with *Buchnera aphidicola* and a secondary symbiont in the aphid *Cinara cedri*. *Journal of Bacteriology* 186:6626–6633.

Gosalbes, M.J., A. Latorre, A. Lamelas, and A. Moya. 2010. Genomics of intracellular symbionts in insects. *International Journal of Medical Microbiology* 300:271–278.

Gottlieb, Y., E. Zchori-Fein, N. Mozes-Daube, S. Kontsedalov, M. Skaljac, M. Brumin, I. Sobol, H. Czosnek, F. Vavre, F. Fleury, and M. Ghanim. 2010. The transmission efficiency of tomato yellow leaf curl virus by the whitefly *Bemisia tabaci* is correlated with the presence of a specific symbiotic bacterium species. *Journal of Virology* 84:9310–9317.

Gross, R., F. Vavre, A. Heddi, G.D.D. Hurst, E. Zchori-Fein, and K. Bourtzis. 2009. Immunity and symbiosis. *Molecular Microbiology* 73:751–759.

Gruenwald, S., M. Pilhofer, and W. Hoell. 2010. Microbial associations in gut systems of wood- and bark-inhabiting longhorned beetles [Coleoptera: Cerambycidae]. *Systematic and Applied Microbiology* 33:25–34.

Gusmao, D.S., A.V. Santos, D.C. Marini, M. Bacci, Jr., M.A. Berbert-Molina, and F.J.A. Lemos. 2010. Culture-dependent and culture-independent characterization of microorganisms associated with *Aedes aegypti* (Diptera: Culicidae) (L.) and dynamics of bacterial colonization in the midgut. *Acta Tropica* 115:275–281.

Harris, H.L., L.J. Brennan, B.A. Keddie, and H.R. Braig. 2010. Bacterial symbionts in insects: balancing life and death. *Symbiosis* 51:37–53.

Heise, S.R., M.S. Elshahed, and S.E. Little. 2010. Bacterial diversity in *Amblyomma americanum* (Acari: Ixodidae) with a focus on members of the genus *Rickettsia*. *Journal of Medical Entomology* 47:258–268.

Hosokawa, T., Y. Kikuchi, N. Nikoh, M. Shimada, and T. Fukatsu. 2006. Strict host-symbiont cospeciation and reductive genome evolution in insect gut bacteria. *PLoS Biology* 4:1841–1851.

Jaenike, J., J.K. Stahlhut, L.M. Boelio, and R.L. Unckless. 2010a. Association between *Wolbachia* and *Spiroplasma* within *Drosophila neotestacea*: an emerging symbiotic mutualism. *Molecular Ecology* 19:414–425.

Jaenike, J., R. Unckless, S.N. Cockburn, L.M. Boelio, and S.J. Perlman. 2010b. Adaptation via symbiosis: recent spread of a *Drosophila* defensive symbiont. *Science* 329:212–215.

Jones, E.O., A. White, and M. Boots. 2010. The evolutionary implications of conflict between parasites with different transmission modes *Evolution* 64:2408–2416.

Kaiwa, N., T. Hosokawa, Y. Kikuchi, N. Nikoh, X.Y. Meng, N. Kimura, M. Ito, and T. Fukatsu. 2010. Primary gut symbiont and secondary, *Sodalis*-allied symbiont of the scutellerid stinkbug *Cantao ocellatus*. *Applied and Environmental Microbiology* 76:3486–3494.

Kaltenpoth, M. 2009. Actinobacteria as mutualists: general healthcare for insects? *Trends in Microbiology* 17:529–535.

Kaltenpoth, M., T. Schmitt, C. Polidori, D. Koedam, and E. Strohm. 2010. Symbiotic streptomycetes in antennal glands of the South American digger wasp genus *Trachypus* (Hymenoptera, Crabronidae). *Physiological Entomology* 35:196–200.

Kikuchi, Y., T. Hosokawa, and T. Fukatsu. 2007. Insect-microbe mutualism without vertical transmission: a stinkbug acquires a beneficial gut symbiont from the environment every generation. *Applied and Environmental Microbiology* 73:4308–4316.

Kikuchi, Y., T. Hosokawa, N. Nikoh, X.-Y. Meng, Y. Kamagata, and T. Fukatsu. 2009. Host-symbiont co-speciation and reductive genome evolution in gut symbiotic bacteria of acanthosomatid stinkbugs. *BMC Biology* 7:2.

Kirkness, E.F., B.J. Haas, W. Sun, H.R. Braig, M.A. Perotti, J.M. Clark, S.H. Lee, H.M. Robertson, R.C. Kennedy, E. Elhaik, D. Gerlach, E.V. Kriventseva, C.G. Elsik, D. Graur, C.A. Hill, J.A. Veenstra, B. Walenz, J.M.C. Tubío, J.M.C. Ribeiro, J. Rozas, J.S. Johnston, J.T. Reese, A. Popadic, Y. Tomoyasu, M. Tojo, D. Raoult, D.L. Reed, E. Krause, O. Mittapalli, V.M. Margam, H.-M. Li, J.M. Meyer, R.M. Johnson, J. Romero-Severson, J.P. Van Zee, D. Alvarez-Ponce, F.G. Vieira, M. Aguadé, S. Guirao-Rico, J.M. Anzola, K.S. Yoon, J.P. Strycharz, M.F. Unger, S. Christley, N.F. Lobo, M.J. Seufferheld, N. Wang, G.A. Dasch, C.J. Struchiner, G. Madey, L.I. Hannick, S. Bidwell, V. Joardar, E. Caler, R. Shao, S.C. Barker, S. Cameron, R.V. Bruggner, A. Regier, J. Johnson, L. Viswanathan, T.R. Utterback, G.G. Sutton, D. Lawson, R.M. Waterhouse, J.C. Venter, R.L. Strausberg, M. Berenbaum, F.H. Collins, E.M. Zdobnov, and B.R. Pittendrigh. 2010. Genome sequences of the human body louse and its primary endosymbiont provide insights into the permanent parasitic lifestyle. *Proceedings of the National Academy of Sciences of the United States of America* 107:12168–12173.

Koivisto, R.K.K., and H.R. Braig. 2003. Microorganisms and parthenogenesis. *Biological Journal of the Linnean Society* 79:43–58.

Kremer, N., D. Charif, H. Henri, M. Bataille, G. Prevost, K. Kraaijeveld, and F. Vavre. 2009. A new case of *Wolbachia* dependence in the genus *Asobara*: evidence for parthenogenesis induction in *Asobara japonica Heredity* 103:248–256.

Kremer, N., F. Dedeine, D. Charif, C. Finet, R. Allemand, and F. Vavre. 2010. Do variable compensatory mechanisms explain the polymorphism of the dependence phenotype in the *Asobara tabida-Wolbachia* association? *Evolution* 64:2969–2979.

Küchler, S.M., K. Dettner, and S. Kehl. 2010. Molecular characterization and localization of the obligate endosymbiotic bacterium in the birch catkin bug *Kleidocerys resedae* (Heteroptera: Lygaeidae, Ischnorhynchinae). *FEMS Microbiology and Ecology* 73:408–418.

Lindh, J.M., A.-K. Borg-Karlson, and I. Faye. 2008. Transstadial and horizontal transfer of bacteria within a colony of *Anopheles gambiae* (Diptera: Culicidae) and oviposition response to bacteria-containing water. *Acta Tropica* 107:242–250.

Lopez-Sanchez, M.J., A. Neef, J. Pereto, R. Patino-Navarrete, M. Pignatelli, A. Latorre, and A. Moya. 2009. Evolutionary convergence and nitrogen metabolism in *Blattabacterium* strain Bge, primary endosymbiont of the cockroach *Blattella germanica*. *PLoS Genetics* 5:e1000721.

Majerus, T.M.O., and M.E.N. Majerus. 2010. Intergenomic arms races: detection of a nuclear rescue gene of male-killing in a ladybird. *PLoS Pathogens* 6:e1000987.

Mazzon, L., I. Martinez-Sanudo, M. Simonato, A. Squartini, C. Savio, and V. Girolami. 2010. Phylogenetic relationships between flies of the Tephritinae subfamily (Diptera, Tephritidae) and their symbiotic bacteria. *Molecular Phylogenetics and Evolution* 56:312–326.

Mazzon, L., A. Piscedda, M. Simonato, I. Martinez-Sanudo, A. Squartin, and V. Girolami. 2008. Presence of specific symbiotic bacteria in flies of the subfamily Tephritinae (Diptera Tephritidae) and their phylogenetic relationships: proposal of 'Candidatus Stammerula tephritidis'. *International Journal of Systematic and Evolutionary Microbiology* 58:1277–1287.

McFall-Ngai, M. 2008. Are biologists in 'future shock'? Symbiosis integrates biology across domains. *Nature Reviews Microbiology* 6:789–792.

Mediannikov, O., Z. Sekeyova, M.-L. Birg, and D. Raoult. 2010. A novel obligate intracellular gamma-Proteobacterium associated with ixodid ticks, *Diplorickettsia massiliensis*, gen. nov., sp nov. *PLoS ONE* 5:e11478.

Moran, N.A., J.A. Russell, R. Koga, and T. Fukatsu. 2005. Evolutionary relationships of three new species of Enterobacteriaceae living as symbionts of aphids and other insects. *Applied and Environmental Microbiology* 71:3302–3310.

Moya, A., J. Peretó, R. Gil, and A. Latorre. 2008. Learning how to live together: genomic insights into prokaryote-animal symbioses. *Nature Reviews Genetics* 9:218–229.

OED. 2009. Oxford English dictionary online. Oxford University Press.

Oliver, K.M., P.H. Degnan, M.S. Hunter, and N.A. Moran. 2009. Bacteriophages encode factors required for protection in a symbiotic mutualism. *Science* 325:992–994.

Pannebakker, B.A., B. Loppin, C.P.H. Elemans, L. Humblot, and F. Vavre. 2007. Parasitic inhibition of cell death facilitates symbiosis. *Proceedings of the National Academy of Sciences of the United States of America* 104:213–215.

Panteleev, D.Y., I.I. Goryacheva, B.V. Andrianov, N.L. Reznik, O.E. Lazebny, and A.M. Kulikov. 2007. The endosymbiotic bacterium *Wolbachia* enhances the nonspecific resistance to insect pathogens and alters behavior of *Drosophila melanogaster*. *Russian Journal of Genetics* 43:1066–1069.

Perotti, M.A., J.M. Allen, D.L. Reed, and H.R. Braig. 2007. Host-symbiont interactions of the primary endosymbiont of human head and body lice. *FASEB Journal* 21:1058–1066.

Perotti, M.A., H.K. Clarke, B.D. Turner, and H.R. Braig. 2006. *Rickettsia* as obligate and mycetomic bacteria. *FASEB Journal* 20:2372–2374, E1646–E1656.

Perotti, M.A., E.F. Kirkness, D.L. Reed, and H.R. Braig. 2009. Endosymbionts of lice. In *Insect Symbiosis 3*, K. Bourtzis and T.A. Miller, editors. Taylor & Francis, Boca Raton, FL, 205–220.

Perru, O. 2007a. Les insectes et leurs endosymbiotes, leur découverte de Blochmann à Buchner (1880–1930)—deuxième partie [Insects and their endosymbionts, their discovery from Blochmann to Buchner (1880–1930)—second part]. *Bulletin Mensuel de la Société Linnéenne de Lyon.*76:27–37.

Perru, O. 2007b. Les insectes et leurs endosymbiotes, leur découverte de Blochmann à Buchner (1880–1930)—première partie [Insects and their endosymbionts, their discovery from Blochmann to Buchner (1880–1930)—first part]. *Bulletin Mensuel de la Société Linnéenne de Lyon* 76:11–18.

Pierantoni, U. 1909. L'origine di alcuni organi d'*Icerya purchasi* e la simbiosi ereditaria. Nota preliminare [The origin of some organs of *Icerya purchasi* and hereditary symbiosis. Preliminary note]. *Bollettino della Società dei Naturalisti in Napoli* 23:147–150.

Pierantoni, U. 1951. Die physiologische Symbiose der Termiten mit Flagellaten und Bakterien [The physiological symbiosis of termites with flagellates and bacteria]. *Naturwissenschaften* 38:346–348.

Ramirez-Puebla, S.T., M. Rosenblueth, C.K. Chavez-Moreno, M.C. Catanho Pereira de Lyra, A. Tecante, and E. Martinez-Romero. 2010. Molecular phylogeny of the genus *Dactylopius* (Hemiptera: Dactylopiidae) and identification of the symbiotic bacteria. *Environmental Entomology* 39:1178–1183.

Ran, L., J. Larsson, T. Vigil-Stenman, J.A.A. Nylander, K. Ininbergs, W.-W. Zheng, A. Lapidus, S. Lowry, R. Haselkorn, and B. Bergman. 2010. Genome erosion in a nitrogen-fixing vertically transmitted endosymbiotic multicellular cyanobacterium. *PLoS ONE.* 5:e11486.

Rani, A., A. Sharma, R. Rajagopal, T. Adak, and R.K. Bhatnagar. 2009. Bacterial diversity analysis of larvae and adult midgut microflora using culture-dependent and culture-independent methods in lab-reared and field-collected *Anopheles stephensi*—an Asian malarial vector. *BMC Microbiology* 9:96.

Ryu, J.-H., E.-M. Ha, and W.-J. Lee. 2010. Innate immunity and gut-microbe mutualism in *Drosophila*. *Developmental and Comparative Immunology* 34:369–376.

Salar, P., O. Semetey, J.-L. Danet, E. Boudon-Padieu, and X. Foissac. 2010. 'Candidatus Phlomobacter fragariae' and the proteobacterium associated with the low sugar content syndrome of sugar beet are related to bacteria of the arsenophonus clade detected in hemipteran insects. *European Journal of Plant Pathology* 126:123–127.

Sapp, J. 1994. *Evolution by association, a history of symbiosis*. Oxford University Press, New York.

Schwemmler, W. 1989. *Symbiogenesis, a macro-mechanism of evolution—progress towards a unified theory of evolution based on studies in cell biology.* Walter de Gruyter, Berlin.

Sharples, G.J. 2009. For absent friends: life without recombination in mutualistic gamma-proteobacteria. *Trends in Microbiology* 17:233–242.

Shi, W., R. Syrenne, J.-Z. Sun, and J.S. Yuan. 2010. Molecular approaches to study the insect gut symbiotic microbiota at the 'omics' age. *Insect Science* 17:199–219.

Sicard, M., F. Chevalier, M. De Vlechouver, D. Bouchon, P. Greve, and C. Braquart-Varnier. 2010. Variations of immune parameters in terrestrial isopods: a matter of gender, aging and *Wolbachia*. *Naturwissenschaften* 97:819–826.

Snyder, A.K., J.W. Deberry, L. Runyen-Janecky, and R.V.M. Rio. 2010. Nutrient provisioning facilitates homeostasis between tsetse fly (Diptera: Glossinidae) symbionts. *Proceedings of the Royal Society of London: Series B: Biological Sciences* 277:2389–2397.

Sonthayanon, P., S.J. Peacock, W. Chierakul, V. Wuthiekanun, S.D. Blacksell, M.T.G. Holden, S.D. Bentley, E.J. Feil, and N.P.J. Day. 2010. High rates of homologous recombination in the mite endosymbiont and opportunistic human pathogen *Orientia tsutsugamushi*. *PLoS Neglected Tropical Diseases* 4:e752.

Starr, D.J., and T.W. Cline. 2002. A host-parasite interaction rescues *Drosophila* oogenesis defects. *Nature.* 418:76–79.

Stouthamer, R., J.E. Russell, F. Vavre, and L. Nunney. 2010. Intragenomic conflict in populations infected by parthenogenesis inducing *Wolbachia* ends with irreversible loss of sexual reproduction. *BMC Evolutionary Biology* 10:229.

Strassert, J.F.H., M.S. Desai, R. Radek, and A. Brune. 2010. Identification and localization of the multiple bacterial symbionts of the termite gut flagellate *Joenia annectens*. *Microbiology* 156:2068–2079.

Šulc, K. 1906. *Kermincola kermesina* n. gen. n. sp. und *physokermina* n. sp., neue Mikroendosymbiontiker der Cocciden [*Kermincola kermesina* n. gen. n. sp. and *physokermina* n. sp., new micro-endosymbionts of coccids]. *Sitzungsberichte der Königlichen Böhmischen Gesellschaft der Wissenschaften in Prag* Art. 19:1–6.

Šulc, K. 1910. "Pseudovitellus" und ähnliche Gewebe der Homopteren sind Wohnstätten symbiontischer Saccharomyceten ["Pseudovitellus" and similar tissues of Homoptera are living quarters of symbiotic Saccharomycetes]. *Sitzungsberichte der Königlichen Böhmischen Gesellschaft der Wissenschaften in Prag* Art. 3:1–39.

Šulc, K. 1931. O symbiose [The symbiosis]. *Biologické Listy* 16:209–214.

Thome, O.W. 1869. *Lehrbuch der Botanik für Gymnasien, Realschulen, forst- und landwirtschaftliche Lehranstalten, pharmaceutische Institute etc. sowie zum Selbstunterrichte.* F. Vieweg und Sohn, Braunschweig.

Thome, O.W. 1877. *Text-book of structural and physiological botany.* Longmans, Green & Co. London.

Thome, O.W. 1885. *Text-book of structural and physiological botany.* Longmans, Green & Co. London.

Toft, C., and S.G.E. Andersson. 2010. Evolutionary microbial genomics: insights into bacterial host adaptation *Nature Reviews Genetics* 11:465–475.

Toju, H., T. Hosokawa, R. Koga, N. Nikoh, X.Y. Meng, N. Kimura, and T. Fukatsu. 2010. "*Candidatus* Curculioniphilus buchneri," a novel clade of bacterial endocellular symbionts from weevils of the genus Curculio. *Applied and Environmental Microbiology* 76:275–282.

Vannini, C., S. Lucchesi, and G. Rosati. 2007a. *Polynucleobacter*: symbiotic bacteria in ciliates compensate for a genetic disorder in glycogenolysis. *Symbiosis* 44:85–91.

Vannini, C., M. Poeckl, G. Petroni, Q.L. Wu, E. Lang, E. Stackebrandt, M. Schrallhammer, P.M. Richardson, and M.W. Hahn. 2007b. Endosymbiosis in statu nascendi: close phylogenetic relationship between obligately endosymbiotic and obligately free-living *Polynucleobacter* strains (*Betaproteobacteria*). *Environmental Microbiology* 9:347–359.

Vautrin, E., and F. Vavre. 2009. Interactions between vertically transmitted symbionts: cooperation or conflict? *Trends in Microbiology.* 17:95–99.

Vavre, F., N. Kremer, B.A. Pannebakker, B. Loppin, and P. Mavingui. 2009a. Is symbiosis evolution influenced by the pleiotropic role of programmed cell death in immunity and development? In *Insect symbiosis,* K. Bourtzis and T.A. Miller, editors. Vol. 3. CRC Press, Boca Raton, FL, 57–75.

Vavre, F., L. Mouton, and B.A. Pannebakker. 2009b. *Drosophila*-parasitoid communities as model systems for host-*Wolbachia* interactions. *Advances in Parasitology* 70:299–331.

Wang, J., Y. Wu, G. Yang, and S. Aksoy. 2009. Interactions between mutualist *Wigglesworthia* and tsetse peptidoglycan recognition protein (PGRP-LB) influence trypanosome transmission. *Proceedings of the National Academy of Sciences of the United States of America* 106:12133–12138.

Weiss, B.L., Y. Wu, J.J. Schwank, N.S. Tolwinski, and S. Aksoy. 2008. An insect symbiosis is influenced by bacterium-specific polymorphisms in outer-membrane protein A. *Proceedings of the National Academy of Sciences of the United States of America* 105:15088–15093.

Wernegreen, J.J., S.N. Kauppinen, S.G. Brady, and P.S. Ward. 2009. One nutritional symbiosis begat another: phylogenetic evidence that the ant tribe Camponotini acquired *Blochmannia* by tending sap-feeding insects. *BMC Evolutionary Biology* 9:292.

Williams, K.P., J.J. Gillespie, B.W.S. Sobral, E.K. Nordberg, E.E. Snyder, J.M. Shallom, and A.W. Dickerman. 2010. Phylogeny of gamma-proteobacteria. *Journal of Bacteriology* 192:2305–2314.

Zhang, F., H. Guo, H. Zheng, T. Zhou, Y. Zhou, S. Wang, R. Fang, W. Qian, and X. Chen. 2010. Massively parallel pyrosequencing-based transcriptome analyses of small brown planthopper (*Laodelphax striatellus*), a vector insect transmitting rice stripe virus (RSV). *BMC Genomics* 11:303.

Zouache, K., D. Voronin, V. Tran-Van, L. Mousson, A.-B. Failloux, and P. Mavingui. 2009. Persistent *Wolbachia* and cultivable bacteria infection in the reproductive and somatic tissues of the mosquito vector *Aedes albopictus. PLoS ONE.* 4:e6388.

1 Proteobacteria as Primary Endosymbionts of Arthropods

Abdelaziz Heddi and Roy Gross

CONTENTS

INTRODUCTION

Arthropods are known to house a huge variety of microorganisms, and members of the Proteobacteria are among the most frequently found partners (Buchner 1965). One of the major peculiarities of arthropod-Proteobacteria associations is their ability to establish long-term intracellular symbioses, ranging from parasitism to mutualism and occurring at different levels of organism complexity and biotope. Among the Proteobacteria, mutualism most often involves the γ-proteobacterial

1

subgroup, while some α- and β-Proteobacteria can also share mutualistic relation-ships. However, to really confirm a dominance of the Proteobacteria in symbiotic interactions with arthropods, many more symbiotic systems must be investigated in the future, to get a real glimpse about the symbiotic universe of these animals. Here, we briefly review current knowledge about several well-characterized primary endo-symbionts of arthropods belonging to the Proteobacteria (Table 1.1 and Figure 1.1). Primary endosymbionts of other phylogenetic groups, such as the *Bacteroidetes*, will be reviewed in Chapter 2.

SPECIFIC SYMBIOTIC SYSTEMS

SAP-SUCKING ARTHROPODS

Aphids and *Buchnera aphidicola*

The symbiosis of aphids with the primary endosymbiont *Buchnera aphidicola* is clearly the best investigated primary symbiotic bacteria-insect association and became the paradigm example, since basic features found in this symbiosis are also valid for many other symbiotic systems in arthropods. Like many other insect symbionts, this bacterium belongs to the γ-Proteobacteria. *B. aphidicola* resides in host-derived vacuoles (symbiosome) within bacteriocytes forming a bacteriome associated with the midgut (Figure 1.2). The *Buchnera*-aphid association is very old and was established approximately 200 MYA ago, and during this long evolu-tionary period, strict host-bacteria cospeciation has occurred (Munson et al. 1991; Baumann et al. 1995). The bacteria are vertically transmitted to the progeny, either to embryos in the viviparous morph or to eggs in the oviparous morph (Wilkinson et al. 2003). The long-lasting confinement within a eukaryotic host cell providing a stable environment has had severe consequences for the genome of the endosym-bionts, which has experienced a drastic reduction of its size (Charles and Ishikawa 1999). The genomes of *Buchnera* isolates from different aphid hosts vary between approximately 425 and 650 kb in size (Shigenobu et al. 2000; Gil et al. 2002; Pérez-Brocal et al. 2006). Compared to the genome of their free-living relatives (such as *E. coli*), the genome of *Buchnera* is reduced about six- to eightfold. A hallmark of the *Buchnera* genome is its extremely high AT content of around 75%, which was then found to be a characteristic feature of most other endosymbiont genomes (see below), with the notable exceptions of *Sitophilus oryzae* primary endosymbiont (Heddi et al. 1998), *Candidatus* Hodgkinia cicadicola (McCutcheon et al. 2009a), and *Candidatus* Tremblaya princeps (Baumann et al. 2002) (see below). Moreover, genome sequence analysis of endosymbionts hosted by four aphids that diverged 60 MYA ago (i.e., *Acyrthosiphon pisum*, *Schizaphis graminum*, *Baizongia pista-ciae*, and *Cinara cedri*) has shown a remarkable gene order conservation (Tamas et al. 2002; van Ham et al. 2003; Pérez-Brocal et al. 2006), suggesting that symbi-ont genome reduction would occur early at the initial phases of symbiosis (Gómez-Valero et al. 2004a; Gil et al. 2008). The genome reduction process has eliminated all genes becoming needless for the association, e.g., because they are redundant in the symbiotic system, including many metabolic genes involved, for example, in the synthesis of nonessential amino acids (Shigenobu et al. 2000), but also some

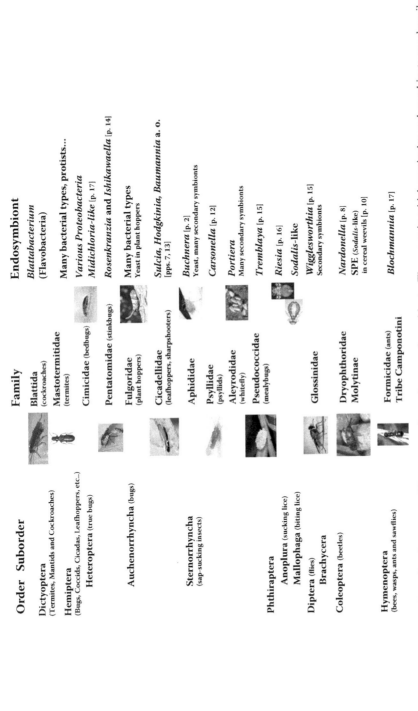

FIGURE 1.1 An overview of major groups of insects carrying primary endosymbionts. The pages in which particular endosymbionts are described are indicated in parentheses.

TABLE 1.1

Overview of the Nomenclature of the Primary Endosymbionts Discussed in the Chapter in Alphabetical Order

Symbiont[a]	Origin of Genus Name	Why	Origin of Species Name	Why	Reference
Baumannia cicadellinicola	After Linda and Paul Baumann	Introduced molecular tools to study insect symbiosis	Refers to the *Cicadellinae* subfamily of leafhoppers	Sharpshooters are the insect host	Moran et al. 2003
Blochmannia floridanus	After Friedrich Blochmann	First description of the symbiosis in 1887	From the ant *Camponotus floridanus*	*C. floridanus* is the insect host	Sauer et al. 2000
Buchnera aphidicola	After Paul Buchner	The founding father of insect symbiosis field	From aphid	Aphids are the host insect	Munson et al. 1991
Carsonella ruddii	After Rachel Carson	An American naturalist and author of *Silent Spring*	After Robert L. Rudd	An American naturalist and author of *Pesticides and the Living Landscape*	Thao et al. 2000
Hodgkinia cicadicola	After Dorothy Crowfoot Hodgkin	A British biochemist receiving the Nobel Prize for chemistry for her work on vitamin B12	From cicadas	Cicadas are the insect host	McCutcheon et al. 2009
Ishikawaella capsulata	After Haijme Ishikawa	Pioneering molecular studies in insect symbiosis	From *capsule*	Referring to the capsules encasing the symbionts	Hosokawa et al. 2006
Midichloria mitochondrii	Midichloria is derived from the midichlorians, organisms within the *Star Wars* epos of George Lucas	In the *Star Wars* epos the midichlorians are endosymbionts of cells that communicate with the Force	From mitochondria	The endosymbionts are found within mitochondria	Sassera et al. 2006

Name	Named after	Important contributions to the field	Name meaning	Host/note	Reference
Nardonella	After Paul Nardon	Important contributions to the field of weevil endosymbiosis			Lefevre et al. 2004
Riesia pediculicola	After Erich Ries	First comprehensive analysis of the symbiosis	Refers to the association of the bacteria with lice	Lice are the insect host	Sasaki-Fukatsu et al. 2006
Rosenkranzia clausaccus	After Werner Rosenkranz	First description of the symbiotic system of acanthosomatid stinkbugs	*Clausus* = "closed" *Saccus* = "bag"	The specific name refers to the completely isolated midgut crypts harboring the symbionts	Kikuchi et al. 2009
Tremblaya princeps	After Ermenegildo Tremblay	An Italian entomologist with significant contributions to the field of endosymbiosis in plant sap-sucking insects	Princeps means "first in rank"		Thao et al. 2002
Wigglesworthia glossinidia	After Vincent B. Wigglesworth	An entomologist with great impact on insect physiology	From the tsetse fly *Glossina*	Tsetse flies are the host insect	Aksoy 1995

a Systematically, all endosymbionts have the *Candidatus* state, since they cannot be cultivated *in vitro* yet and their phylogenetic position is derived mainly from very few characteristics, e.g., the sequence of the 16S rDNA (Murray and Schleifer 1994).

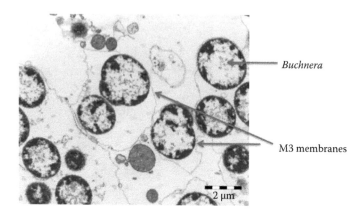

FIGURE 1.2 Electron microscopical image of bacteriocytes dissected from aphid *Acyrthosiphon pisum* adults. Shown are *Buchnera* bacteria surrounded with the M3 membrane forming the symbiosome structure. The M3 membrane is presumably of host origin, and it may include more than one bacterium. (Image kindly supplied by Yvan Rahbé [BF2I Lab and CTμ Center.])

bacteria-specific genes encoding bacterial cell wall structures, and genes involved in DNA repair mechanism, recombination, signal transduction, and gene regulation (Dale et al. 2003). The lack or the strong reduction of DNA repair and recombination factors and frequent population bottlenecks due to a strict vertical transmission may be responsible for the higher mutation rates observed in these bacteria (see below) (Moran 1996). In contrast, genes involved in the synthesis of essential amino acids and some vitamins have been kept by the symbiont. The main biological function of *Buchnera* is therefore the production of essential amino acids and other nutrients that are missing in the sole nutriment of the animals, phloem sap, which is particularly poor in these compounds (Liadouze et al. 1996; Douglas 1998; Zientz et al. 2004). Artificial elimination of this bacterium from its hosts by antibiotic treatment, thus producing so-called aposymbiotic animals, results in host metabolic perturbation and strongly affects aphid survival and reproduction.

Recently, by sequencing of the genomes of *Buchnera* strains from seven pea aphid hosts (*A. pisum*) diverging only about 135 years ago, the nucleotide substitution rate was found to be around 10^{-7} per site per year (Moran et al. 2009). However, the interpretation of such data is difficult, and this rate may be overestimated, as discussed recently by Ho et al. (2005).

Another hallmark of *B. aphidicola* also seen in other primary endosymbionts is the strong constitutive expression of stress factors, and in particular of GroEL. Due to its strong overexpression, GroEL was early recognized as a factor probably relevant in symbiosis and originally was termed symbionin (Ishikawa 1985; Aksoy 1995; Charles et al. 1995; Stoll et al. 2009a). This may indicate that the obligate intracellular lifestyle may also confer significant environmental stress to the bacteria. In addition, massive GroEL overexpression may also be a means to counteract an increasing number of slightly deleterious mutations accumulating in the endosymbiont proteins, thus conserving the function of these proteins (Fares et al. 2002).

In addition to *Buchnera*, some aphid populations also harbor one or more secondary endosymbionts, including three γ-Proteobacteria (*Serratia*, *Hamiltonella*, and *Regiella*) and two α-Proteobacteria (*Rickettsia* and *Wolbachia*) (Moran et al. 2008). Although the precise functions of these endosymbionts still remain unclear, recent works have provided strong evidence that secondary endosymbionts may benefit insects under heat stress (Montllor et al. 2002), confer resistance to parasitic wasps (Moran et al. 2005; Oliver et al. 2003, 2009), or broaden food plant range (Tsuchida et al. 2004). Secondary endosymbionts may also take over functions of the primary endosymbiont, for example, if the primary endosymbiont, due to ongoing genome erosion, is not able anymore to carry out all functions required by the host. As an example, in *Cinara cedri* carrying the *Buchnera* isolate with the smallest genome of about 420 kb, the symbiont has lost the capacity to synthesize trypthophan and riboflavin, both essential to the host. However, the secondary endosymbiont *Serratia symbiotica* is present in all *Cinara cedri* isolates, and apparently substitutes for the lacking metabolic activity of the primary endosymbiont (Pérez-Brocal et al. 2006).

Sharpshooters and *Candidatus* Baumannia

Leafhoppers are a species-rich group of insects and for a long time have been known to harbor symbiotic bacteria. They are agricultural pest animals themselves, or they can transmit plant pest agents; e.g., sharpshooters may be vectors for the plant pathogen *Xylella fastidiosa*. A phylogenetic analysis of five sharpshooter species of the subfamily Cicadellinae revealed that their bacteriocyte endosymbionts form a well-defined clade within the γ-Proteobacteria and show a coevolution with their host animals due to a strict vertical transmission of the endosymbionts. For these bacteria a new genus and species name, *Candidatus* Baumannia cicadellinicola, was proposed recently (Moran et al. 2003). Based on an extensive analysis of the 16S rDNA sequences of endosymbionts from 29 leafhopper species, it was discovered that another primary endosymbiont belonging to the *Bacteroidetes* group (*Candidatus* Sulcia muelleri; see Chapter 2) is present in all animals tested, while *Candidatus* Baumannia was found only in certain sharpshooter tribes in addition to *Candidatus* Sulcia. The phylogenetic analysis of the host animals harboring both primary endosymbionts revealed a congruent evolutionary history between the sharpshooters and their two companions, and accordingly, the term *coprimary symbionts* was coined for these bacteria (Takiya et al. 2006). *Candidatus* Baumannia cicadellinicola of the glassy-winged sharpshooter (*Homalodisca coagulata*) was further investigated, and its genome sequence was recently published (Wu et al. 2006). As in the case of most primary endosymbionts, the genome of *Candidatus* Baumannia was shown to be small, with a size of only about 686 kb, and it has a low GC content of 33.2%. Many metabolic pathways encoded by *Candidatus* Baumannia are involved in the biosynthesis of vitamins, cofactors, and prosthetic groups, such as riboflavin, folate, pyridoxal 5'-phosphate, and thiamine. There are also several genes of other incomplete biosynthetic pathways, for which this bacterium requires external sources to feed the specific intermediates into its own pathways; e.g., *Candidatus* Baumannia would need prophobilinogen and protoheme to complete the respective pathways, but it also requires certain amino acids, such as tyrosine, alanine, and aspartate.

This endosymbiont has retained very few functions involved in amino acid biosynthesis. Since the xylem sap taken up by the host is extremely scarce, in particular in essential amino acids, both the host and *Candidatus* Baumannia need these metabolites from another source (Wu et al. 2006). The recent establishment of the genome sequence of its coprimary symbiont *Candidatus* Sulcia muelleri revealed that this endosymbiont very well matches these nutritional requirements, since it encodes the required amino acid biosynthesis pathways (McCutcheon and Moran 2007). Metabolic complementarity of the two endosymbionts matches not only the host requirements, but in several cases also the specific requirements between the symbionts, which need to exchange the respective components between each other. For example, *Candidatus* Baumannia encodes the complete pathway to synthesize histidine, while this pathway is missing in *Candidatus* Sulcia, thus suggesting production of histidine not only for the host, but also for the coprimary symbiont. Overall, there is little redundancy between the metabolic capacities of the two endosymbionts, demonstrating a striking complementarity in their biological functions. Thus, the main function of *Candidatus* Baumannia is likely to produce vitamins and cofactors for the host and, in this respect, resembles *Wigglesworthia* in tsetse flies, while the major task of *Candidatus* Sulcia appears to resemble that of *Buchnera* in aphids in mainly supplying amino acids.

Weevils and *Candidatus* Nardonella

Weevils are probably the most species-rich animals (more than 60,000 species), and many are relevant agricultural pests. Weevils of the family Dryophthoridae are found virtually everywhere in the world in very diverse habitats, ranging from seashores to mountaintops and from deserts to rain forests. They also have succeeded to thrive on a large panel of nutriments, including roots, leafs, stipes, flowerheads, fruits, woods, and seeds of both angiosperms and gymnosperms. These variations in habitat offer the possibility to explore the extent and diversity of nutritional endosymbiosis in the insect world, and to examine the impact of the ecodiversity on symbiosis dynamics during evolution.

So far, 19 Dryophthoridae species have been studied histologically and molecularly. Except for *Sitophilus linearis* that thrives on Tamarin, a relatively balanced diet, all the other species were found to harbor γ-proteobacterial symbiotic bacteria permanently in the oocytes and in bacteriocytes forming a large bacteriome lining the gut tissue. Phylogenetic studies based on the 16 rDNA symbiont sequences have shown that symbioses have evolved several times in this family (Figure 1.3). While the insect phylogenetic tree showed that the Dryophthoridae are grouped in one monophyletic clade, endosymbiont phylogenetic trees exhibited as many as three clades characterized by different GC contents and evolutionary rates (Lefèvre et al. 2004). The ancestral lineage has an age of approximately 100 MYA, and recently, these bacteria were classified in the candidate genus *Candidatus* Nardonella. These symbionts show much higher relative rates of nucleotide substitution than both the Enterobacteriaceae and other insect endosymbionts. Furthermore, they exhibit a high AT bias and contain AT-rich insertions (40.5% GC in the 16S rDNA sequences). Conversely, the two other clades, including the clade of the cereal weevils *Sitophilus*, did not show such an AT bias and significant high rates of evolution. Interestingly, the

Bacterial phylogeny

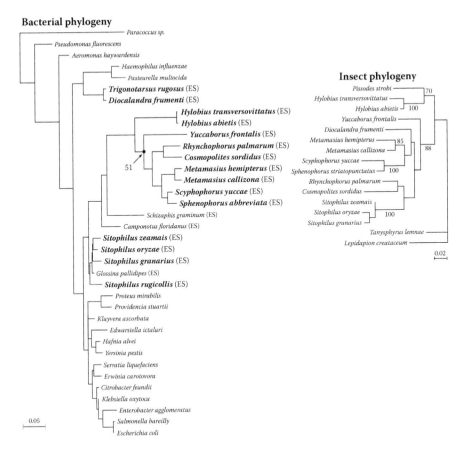

FIGURE 1.3 Maximum likelihood phylogenetic tree of Molytinae (*Hylobius genus*) and Dryophthoridae endosymbionts (left) and maximum likelihood phylogram of host insects based on COI amino acid sequences (right). ES indicates endosymbiotic bacteria; other taxa are free-living bacteria. The marked node (number 51) corresponds to the most likely position of the *Hylobius* clade on the tree topology. Numbers on the branches of the insect hosts' tree are bootstrap values (1,000 replicates). (Adapted from Conord, C., et al., *Mol. Biol. Evol.*, 25, 859–868, 2008.)

Sitophilus clade endosymbionts have been associated with their hosts more recently, probably less than 25 MYA ago (see below).

Endosymbiont replacement hypothesis was postulated in aphids (Moran and Baumann 1994), and recently, a potential interplay between old primary and recent secondary endosymbionts was reported by several works that have noted a beneficial role of secondary endosymbionts for the host metabolism and fitness in some aphids (Chen et al. 2000; Sabater et al. 2001; Gómez-Valero et al. 2004b; Gosalbes et al. 2008; Pérez-Brocal et al. 2006). In the Dryophthoridae, the more recent *Sitophilus* endosymbiont association is concomitant with the shift from stem feeding, the ancestral habit in these weevils, to seed feeding in *Sitophilus*. One hypothesis would be that symbiont replacement in *Sitophilus* is functionally correlated with this habitat

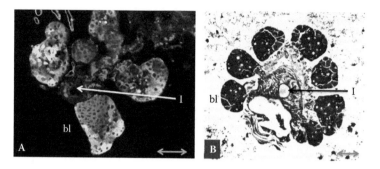

FIGURE 1.4 See color insert. Bacteriome morphology of *Hylobius abietis* (A) and *Metamasius hemipterus* (B). Shown is the structure of isolated larval bacteriome lobelets from the foregut-midgut junction. Slide A was hybridized with a specific probe designed on *Hylobius* 16S rDNA and stained with DAPI. (Modified from Conord, C., et al., *Mol. Biol. Evol.*, 25, 859–868, 2008.) Slide B was stained with Tolouid blue. bl, bacteriome lobelets; I, intestine. Scale bar: 50 μm.

shift, thus generating a situation where *Candidatus* Nardonella came into competition with the ancestor of *Sitophilus* primary endosymbiont (SPE), in the newly established symbiotic entity, and in interaction with the new environment (see below) (Lefèvre et al. 2004).

Recently, the hypothesis of symbiont replacement in the Dryophthoridae was strongly supported through the analysis of endosymbiosis in the Molytinae family, the members of which would have diverged from the Dryophthoridae approximately 125 MYA ago. In contrast to Dryophthoridae, Molytinae members are widely distributed in the northern and southern hemispheres. Some of them, such as *Hylobius abietis*, are restricted to conifer trees and breed in the roots of dying or recently dead conifers. Two *Hylobius* species (*H. abietis* and *H. transversovittatus*) were shown to harbor pleiomorphic endosymbiotic bacteria in a conserved bacteriome structure similar to that of the Dryophthoridae (Figure 1.4) (Conord et al. 2008). The phylogenetic trees of the host insects (Molytinae and Dryophthoridae) and their endosymbionts have revealed a strong concordance of the trees of the endosymbionts and their host animals. The phylogenetic analysis also revealed that Molytinae endosymbionts are placed among *Candidatus* Nardonella endosymbionts, which are well separated from the endosymbionts of the *Sitophilus* clade (Figure 1.3) (Conord et al. 2008). In conclusion, endosymbiosis seems to be very widespread in the Curculionoidea superfamily, older than previously estimated on the Dryophthoridae insects, and symbiont displacement may have occurred several times within this large insect superfamily.

Sitophilus Weevils and Their Primary Endosymbionts (SPE)

The cereal weevils *Sitophilus* spp. are the best investigated insects among the Coleoptera order, because of their agronomical importance, but also because of some particular biological and evolutionary features, on which we will focus in this part. Discovered by Pierantoni in 1927, symbiosis in *Sitophilus* was intensively studied by Nardon, who provided the first comparative data when he succeeded in obtaining aposymbiotic insects from a wild-type symbiotic strain (Nardon 1973). This system is unique in insect symbioses since a mild heat treatment (35°C) and 90% relative humidity for 1 month results in symbiont elimination from both somatic and germ cells, and the resulting aposymbiotic insects are viable and could be maintained for many generations. However, the resulting aposymbiotic strain is less fertile than the wild type, it develops slowly during the larval stages and is not able to fly. Several biochemical aspects were tested on *Sitophilus oryzae* to understand how *Sitophilus oryzae* primary endosymbiont (SOPE) enhances the host fitness. SOPE was shown to provide the insect with many components that are poorly represented in the wheat grains or completely absent from the albumen part of the grain on which *Sitophilus* spp. larvae feed. These include pantothenic acid, biotin, riboflavin (Wicker 1983), and amino acids, particularly phenylalanine and tyrosine (Wicker and Nardon 1982). SOPE also interferes with insect metabolism either directly through the modification of some products or indirectly by improving some insect enzymatic pathways (Gasnier-Fauchet and Nardon 1986; Gasnier-Fauchet et al. 1986; Heddi et al. 1993). Finally, SOPE was shown to increase mitochondrial activities and mitochondrial ATP production by supplying the host with vitamins, such as pantothenic acid and riboflavin, which are involved in the synthesis of mitochondrial coenzymes such as coenzyme A, NAD, and FAD (Heddi and Nardon 1993; Heddi et al. 1993, 1999). Moreover, similar experiments have shown that flight ability of aposymbiotic adults is partly restored when insects are supplemented with these vitamins (Grenier et al. 1994). This finding links mitochondrial oxidative phosphorylation with flight consuming a lot of ATP, and indicates that SOPE not only interacts with physiological traits but also impacts the behavior and invasive power of weevils.

Molecularly, SOPE is distinct from the other insect primary endosymbionts as well as from the primary endosymbionts of weevils, *Candidatus* Nardonella, in several ways. First, it has a quite large genome of about 3 Mb (Charles et al. 1997), and second, it has retained a high GC content of about 54% (Heddi et al. 1998). Such a large genome size is in agreement with the assumption of a relatively recent replacement of an ancient primary endosymbiont by SOPE. The relatively young association of SOPE with its host may explain that genome degeneration is apparently quite mild in this bacterium. Similarly, the secondary endosymbiont of tsetse flies *S. glossinidius* (*Sodalis*) is also thought to be at an early stage of symbiosis with its host flies, since, in contrast to SOPE, it can be found intra- and extracellularly and can be cultivated *in vitro*. A striking feature underlining the relatively young symbiotic association and an early state of symbiotic genome evolution is the fact that the SOPE genome harbors a large number of insertion sequences (ISs) (Gil et al. 2008), which usually are not found in the streamlined genomes of other primary endosymbionts. The large number of such ISs (more than 25% of the genome) may be indicative of an early stage of genome erosion since these elements may massively promote genome rearrangements, gene inactivation, and finally gene loss (Gil et al. 2008). Since many genes of bacteria that establish a symbiotic relationship with a eukaryotic host are not essential anymore under the relatively stable conditions within their host organisms, such genes are preferentially lost without detrimental effects for the bacteria, thus leading to reduced genomes. The reason for the increase in the number of IS elements could also be interpreted the same way. As many genes become needless during the first steps of symbiosis, evolutionary pressure is relaxed on these genes, which may favor IS multiplication and integration at different sites of the genome. On the other hand, the decrease in genome size probably increases the evolutionary constraints on genes absolutely required for the association. This would exclude the activity and multiplication of ISs at a certain point, as they may increasingly be a danger for the bacteria and the symbiotic association itself.

By use of degenerated polymerase chain reaction (PCR) primers based on the sequences available for *Sodalis*, several genes of *Sitophilus zeamais* primary endosymbiont (SZPE), the primary endosymbiont of *Sitophilus zeamais*, homologous to genes (*inv/spa* genes) encoding factors of a type III secretion system (TTSS), were identified (Dale et al. 2001, 2002). Obviously, this TTSS was retained in the genomes of both the primary (SZPE) and secondary (*Sodalis*) endosymbionts, since it is essential for establishment and maintenance of a chronic infection. It is noteworthy that SZPE may require the invasion of new host cells during the weevil's nymphal metamorphosis, and that the TTSS genes were found to be upregulated in particular in this phase of insect development (Anselme et al. 2006; Dale et al. 2002). Interestingly, several host immune genes were shown to be induced parallel to TTSS gene expression at the nymphal stage, probably in response to bacterial invasion, which indicates a tight interplay between the host immune system and bacterial TTSS. Moreover, a functional quorum-sensing system highly related to the respective *S. glossinidius* system was identified in SOPE (Pontes et al. 2008). Quorum sensing appears to be implicated in protection from oxidative stress in the bacteriocytes, where recently high metabolic activity, along with the upregulation of many genes involved in cellular stress, was documented (Heddi et al. 2005). Comparison of SPE and *Sodalis* shows that mutualistic endosymbionts exhibit similar features with pathogenic microorganisms at the first steps of symbiogenesis. These features are apparently still required for SPE and *Sodalis* to sustain mutualistic relationships at this symbiotic stage.

Psyllids and *Candidatus* Carsonella

Candidatus Carsonella ruddii is the primary bacteriocyte endosymbiont of phloem sap-feeding psyllids. It belongs to the γ-Proteobacteria and is present in all psyllid species investigated so far. The genome sequence of *Candidatus* Carsonella of the hackberry petiole gall psyllid *Pachypsylla venusta* resulted in the discovery of an unprecedented extremely reduced genome of only about 160 kb, with the most extreme GC content, about 16.5%, known so far (Nakabachi et al. 2006). Another striking feature is the extremely high coding density of 97.3% due to a large number of overlapping genes, an almost complete lack of intergenic regions, and possibly, a lack of characteristic promoter sequences, suggesting translational coupling and the synthesis of long mRNAs as basic features of gene expression in this symbiont (Clark et al. 2001). The genome encodes only 182 open reading frames (ORFs), and comparison with the ORFs of other sequenced endosymbionts revealed that the *Candidatus* Carsonella ORFs are around 18% shorter on average. The genome reduction or degeneration concerns many functions thought to be essential not only for the symbiotic interaction with the host, but also for the survival of a bacterial cell (Moya et al. 2009). For example, many genes involved in DNA replication, transcription, and translation are lacking entirely, or the shortening of the ORFs has led to loss of domains believed to be essential for protein function. Moreover, even several pathways required for biosynthesis of amino acids, which are not present in the phloem sap but are essential for the host and the symbiont, are missing (Tamames et al. 2007). It remains enigmatic how the host can thrive on its nutrient-restricted diet assisted only by *Candidatus* Carsonella, as so far no secondary endosymbiont that could supplement this deficiency was found in *Pachypsylla venusta*, contrary to other systems, such as *Buchnera* of the aphid *Cinara cedri* (see above) (Pérez-Brocal et al. 2006). To explain these puzzling features, it is discussed whether genes of *Candidatus* Carsonella may have been transferred to the host nucleus, or whether the mitochondrial machinery encoded by the host genome may be involved in maintenance of the endosymbiont. An important step to understand this symbiosis will therefore be the sequencing of the host genome (Nakabachi et al. 2009).

Cicadas and *Candidatus* Hodgkinia

The molecular characterization of the primary endosymbiont *Candidatus* Hodgkinia cicadicola of cicadas led to a great surprise since it differs substantially from the other endosymbionts described so far. *Candidatus* Hodgkinia belongs to the α-Proteobacteria and is most closely related to the Rhizobiales, while other known arthropod symbionts within the α-Proteobacteria belong to the Rickettsiales. In the cicada *Diceroprocta semicincta*, this endosymbiont was found to coexist with a second primary endosymbiont (much like in the sharpshooter symbiosis described above) in the same bacteriocytes. With only 144 kb, it has the smallest genome size so far detected among the primary endosymbionts, even smaller than that of *Carsonella* (McCutcheon et al. 2009a). Similar to other primary endosymbionts with reduced genome sizes, it exhibits a high coding density and apparently shortened open reading frames. The genomic analysis has identified putative stop codons in many ORFs, which should lead to premature termination of protein translation of these ORFs. However, similar to mycoplasmas and several mitochondrial lineages, a recoding of the UGA stop codon to code for tryptophane has apparently occurred in this bacterium. This is substantiated by the fact that this microbe has lost the gene encoding the translational release factor TF2 that specifically recognizes UGA as a stop codon. However, it still encodes the release factor TF1 recognizing the other two stop codons. Mycoplasmas using the same modified genetic code also lack the TF2 encoding gene. In contrast to the other sequenced endosymbionts, the genome is characterized by a very high GC content of 58.4%. The high GC content of this endosymbiont challenges previous thoughts about the evolution of the extreme AT-rich genomes of the other endosymbionts exhibiting a strongly reduced genome size. Environmental cues as driving forces for the changes in the GC content can be excluded since *Sulcia* coexisting in the same bacteriocytes as *Hodgkinia* has the characteristic low GC content of most other reduced genomes (22.6%). Therefore, it was suggested that in *Candidatus* Hodgkinia the replication process or the mutagenic environment differs from that of other endosymbionts. In fact, astonishingly, this reduced genome encodes only two genes involved in replication, *dnaE* and *dnaQ*, encoding the α- and ε-subunits of DNA polymerase III. The lack of all known DNA repair enzymes in *Candidatus* Hodgkinia implies that the lack of repair functions does not necessarily lead to an extreme AT bias, as previously suggested as a possible general explanation for the AT bias in reduced genomes that typically encode only very few repair enzymes (McCutcheon et al. 2009a). The gene content of *Candidatus* Hodgkinia strikingly resembles that of *B. cicadellinicola* of sharpshooters, which is also living in close association with *Sulcia* strains (see above), since both *Candidatus* Hodgkinia and *Candidatus* Baumannia encode similar genes required for biosynthesis of some essential amino acids. However, a unique feature of *Candidatus* Hodgkinia among primary endosymbionts of insects is the presence of a gene set required for biosynthesis of cobalamin (vitamin B_{12}) (McCutcheon et al. 2009b).

Stinkbugs and *Candidatus* Rosenkranzia and Ishikawaella

Many stinkbugs are plant-sucking insects and are endowed with endosymbiotic bacteria. Endosymbiont-bearing stinkbugs have special crypts in the terminal region of their midgut that are filled with extracellular symbiotic bacteria. These bacteria must play an important role for the animals since without the endosymbionts, the stinkbugs show retarded growth and increased mortality and sterility. Transmission of the endosymbionts occurs vertically. In stinkbugs of the family Plataspidae the γ-proteobacterial endosymbionts classified in the genus *Candidatus* Ishikawaella capsulata are transferred to the progeny by the so-called capsule transmission process, which is characterized by the deposition of bacteria-filled so-called symbiont capsules together with the eggs. These symbiont capsules are formed in the gut, excreted, and finally ingested by the hatchlings (Hosokawa et al. 2006). Acanthosomatid stinkbugs also carry extracellular γ-Proteobacteria classified in the genus *Candidatus* Rosenkranzia clausaccus in midgut crypts, which, in contrast to other stinkbugs, are sealed off from the lumen of the midgut. Therefore, the bacteria cannot be excreted from the anus, a prerequisite for egg surface contamination in other stinkbugs. These animals have evolved a special organ, the so-called lubricating organ, harboring these bacteria in tubulets. The lubricating organs are closely associated with the female ovipositor and guarantee endosymbiont transmission to the eggs. Despite living in an extracellular niche in the midgut of the animals, *Candidatus* Ishikawaella and *Candidatus* Rosenkranzia exhibit properties previously thought to be confined to obligate intracellular bacteria, namely, a strongly reduced genome size, an AT-biased genome, and a congruent coevolution with their hosts. In fact, the genome sizes of *Candidatus* Ishikawaella and *Candidatus* Rosenkranzia were determined to range from about 820 to 830 kb and 900 to 960 kb, respectively. The AT contents of the 16S rDNA sequences were found to be in the range of 61.5 to 64%, which is similar to obligate intracellular symbionts, but significantly higher than the values of free-living Enterobacteriaceae, which have an AT content of around 45% (Kikuchi et al. 2009). These findings strongly suggest that similar evolutionary forces experienced by intracellular endosymbionts may also shape the genomes of extracellular endosymbionts. In the future, it will be very interesting to obtain genome sequences from these bacteria to see whether there are basic differences in the gene content of the extra- vs. intracellular endosymbionts, particularly in relation to genes and pathways involved in secretion and bacterial cell wall biosynthesis.

Mealybugs and *Candidatus* Tremblaya

Mealybugs carry a spectacular endosymbiotic community within their bacteri-oytes. In the citrus mealybug *Planococcus citri* bacteriocytes two endosymbionts were detected by PCR, a β-Proteobacterium classified as *Candidatus* Tremblaya princeps, and a γ-Proteobacterium. A combination of electron microscopy, con-focal microscopy, and fluorescence *in situ* hybridization (FISH) revealed that the γ-Proteobacterium surprisingly is localized intracellularly within *Tremblaya* cells (von Dohlen et al. 2001), thus demonstrating the first and only intracellular prokary-ote-prokaryote endosymbiosis known so far. While *Tremblaya* shows a strict pattern of coevolution with different mealybug hosts, the phylogeny of the intrabacterial γ-Proteobacterium does not point to a common ancestor. Instead, the comparison with the phylogeny of the γ-proteobacterial companion suggests multiple infections of *Tremblaya* by this bacterium, but also suggests a cotransmission of both symbi-onts after successful establishment of the bacteria-bacteria symbiosis (von Dohlen et al. 2001; Thao et al. 2002). The genomic features of *Candidatus* Tremblaya differ from those of other primary endosymbionts. In particular, *Candidatus* Tremblaya has a high GC content of 57.1%, and this GC content is similar in intergenic spaces as well as in the coding regions. Moreover, highly and poorly conserved proteins are encoded by genes with a similar GC content, which is usually not the case in the other endosymbionts (Baumann et al. 2002).

BLOOD-SUCKING ARTHROPODS

Tsetse Flies and *Candidatus* Wigglesworthia

The tsetse fly is the vector for the human African trypanosomiasis (sleeping sick-ness) caused by the protozoan *Trypanosoma brucei*. These flies feed exclusively on vertebrate blood, which is a very imbalanced food resource, since several essen-tial vitamins are lacking. The flies carry the primary endosymbiont *Candidatus* Wigglesworthia glossinidia and may be infected by the secondary facultative endo-symbiont *Sodalis glossinidius*, which is phylogenetically closely related to the pri-mary endosymbiont of *Sitophilus* (see above) (Lefèvre et al. 2004). *Candidatus* Wigglesworthia is located free in the cytosol of bacteriocytes forming a bacteriome close to the anterior midgut and, astonishingly, extracellulary in the lumen of milk glands, permitting transmission to the offspring of these viviparous animals (Attardo et al. 2008; Pais et al. 2008). *S. glossinidius* is primarily found in the midgut in extra- and intracellular locations, but it is also present in the hemolymph and other host tissues, as well as in the lumen of milk glands (Attardo et al. 2008). Phylogenetic analysis led to the estimation of an ancient association of the primary endosymbiont with the flies (50 to 80 MYA years) and shows a cospeciation of these bacteria with their host, while no strict cospeciation with the host is obvious in the case of *S. glos-sinidius* (Aksoy et al. 1997). Heat elimination of *Candidatus* Wigglesworthia from the flies led to sterility of females. Fertility could be partially rescued by feeding of the aposymbiotic animals with a vitamin cocktail (Nogge 1976). Thus, it was antici-pated that *Candidatus* Wigglesworthia may upgrade the unbalanced diet of the host with vitamins. Recently, by use of antibiotics, a selective elimination of *Candidatus*

Wigglesworthia not affecting survival of the secondary endosymbiont was achieved, which clearly confirmed the primary endosymbiont-dependent loss of fertility in females but not in males. Both aposymbiotic sexes were impaired in their longevity and were compromised in their ability to digest their blood meal. The genome of *Candidatus* Wigglesworthia glossinidiae was established shortly after the first genome of *Buchnera aphidicola* and revealed a strongly reduced genome of 698 kb with a high AT bias (22% GC content) (Akman et al. 2002). The reconstruction of the metabolic capacity of *Candidatus* Wigglesworthia revealed a nearly complete lack of amino acid biosynthesis pathways, while many pathways for vitamin and cofactor biosynthesis are present. Thus, the genome sequence strikingly confirms the previous experiments that this endosymbiont enriches the vitamin-poor diet of the host with vitamins and cofactors, in particular, vitamins of the B-group, thus allowing the host to thrive on vertebrate blood as the sole nutrient source (Zientz et al. 2004). Interestingly, the presence of *Candidatus* Wigglesworthia also has an effect on the parasite load of the flies, since older flies, which usually are quite resistant to parasite infection, devoid of the primary endosymbiont, were more susceptible to trypanosome infection. This suggests that the primary endosymbiont in some way is involved in the control of host functions regulating parasite performance (Pais et al. 2008).

Lice and *Candidatus* Riesia

Sucking lice are ectoparasites of mammals that exclusively feed on blood. As mentioned above, mammalian blood lacks several nutrients essential to the animals, such as certain vitamins. Thus, the blood-feeding lice, such as the head louse *Pediculus capitis* living in the hair of man, are known for a long time to live in close association with endosymbiotic bacteria that are believed to contribute to host nutrition. Historically, in a human louse (presumably the body louse) for the first time mycetomes/bacteriomes were seen by Robert Hooke in 1664, and shortly thereafter described by Jan Swammerdam in 1669 (Perotti et al. 2007). These early observations were confirmed at the beginning of the twentieth century, when Sikora (1919) described that the nymphs of human lice have large aggregates of bacteriocytes (also called stomach discs) in their midguts harboring rod-like endosymbionts. These bacteria are transmitted vertically and are able to infiltrate the developing oocytes of females. In fact, in adult females the bacteria are exclusively found in the lateral oviducts and the posterior pole of the oocytes in the ovarioles, while in adult males the bacteria are still found in the characteristic stomach discs. Phylogenetic analysis of the endosymbionts by sequencing of their 16S rDNA genes revealed that the endosymbionts of the head louse constitute a basal lineage within the γ-Proteobacteria, and that they cannot be differentiated significantly from the symbionts of the closely related body lice. In line with the genomic properties of most other endosymbionts, the 16S rDNA genes exhibit a significant AT bias. This bacterium was found in all human lice investigated so far, underlining its important role for the host animals. Based on these data, the new genus *Candidatus* Riesia pediculicola was proposed for the bacterial companions of human lice (Sasaki-Fukatsu et al. 2006). Highly related endosymbionts were also recently reported from lice of anthropoid primates (Allen et al. 2007). However, based on 16S rDNA and *groEL* gene sequencing of their endosymbionts, lice of the genus *Pedicinus* parasitizing old world monkeys were shown to carry phylogenetically distant companions compared to lice of humans or anthropoid primates. This suggests that the bacterial companions of old world monkey lice evolved independently from *Candidatus* Riesia present in the lice of anthropoid primates. For this lineage of endosymbionts the name *Candidatus* Puchtella pedicinophila was proposed recently (Fukatsu et al. 2009).

Ticks and *Candidatus* Midichloria

Ticks harbor a large variety of bacterial endosymbionts, but they are also relevant vectors for several human pathogens (Reisen 2010). Recently, an endosymbiont with a striking intracellular location was identified in the ovaries of females of the hard tick *Ixodes ricinus*. This bacterium was found free in the cytosol of various cell types in the ovaries or, occasionally, included in a host-derived membrane. Strikingly, the bacteria were also found in the intermembrane space between the two membranes of mitochondria in luminal cells and oocytes, where they can also multiply (Sassera et al. 2006). This endosymbiont appears to have an extremely high prevalence in female ticks, while they are quite rare in males. In accordance with their presence in the ovaries, these bacteria are transmitted vertically. For these endosymbionts the novel genus *Candidatus* Midichloria mitochondrii was proposed recently. The biological function of this endosymbiont is unknown so far, but it was recently shown that its multiplication in the host correlates with the blood meal by the ticks (Sassera et al. 2008). A PCR screen of other ixodid ticks for the presence of *Candidatus* Midichloria in seven species from five different genera revealed the presence of this bacterium, and intracellular or intramitochondrial location could be confirmed for the tick *Rhipicephalus bursa*. The phylogenetic analysis of the 16S rDNA sequences revealed that the bacteria form a monophylogenetic group within the Rickettsiales clade of the α-Proteobacteria; however, there are some discrepancies of the phylogenies of the endosymbionts and their hosts, indicating occasional horizontal transmission (Epis et al. 2008). PCR screens in other invertebrates, such as bedbugs of the genus *Cimex* (Richard et al. 2009) or horse flies of the genus *Tabanus* (Hornok et al. 2008), also revealed the presence of 16S rDNA sequences very closely related to *Candidatus* Midichloria, indicating that this endosymbiont is much more widespread than initially thought.

Omnivorous Arthropods

Carpenter Ants and *Candidatus* Blochmannia

The endosymbiosis of carpenter ants (Hymenoptera: Formicidae) with intracellular bacteria was the first animal-bacteria symbiosis ever described (Blochmann 1887). The fact that they carry bacteriocyte endosymbionts on the first view is enigmatic, because these animals were not known to live on a nutrient-imbalanced diet and are considered to be omnivorous (Koch 1960). The endosymbionts classified in the novel genus *Candidatus* Blochmannia have been found in all carpenter ant (*Camponotus*) species investigated so far, and are also found in the related *Camponotini* genera *Calomyrmex*, *Colobopsis*, *Echinopla*, *Opisthopsis*, and *Polyrhachis*, thus pointing to an important function of these endosymbionts for their host animals (Wernegreen et al. 2009). *Candidatus* Blochmannia is classified within the γ-Proteobacteria, and a strict cospeciation between endosymbionts and their hosts is obvious (Sauer et al. 2000; Schröder et al. 1996, Wernegreen et al. 2009). Based on fossil records, this endosymbiosis was estimated to have an age of between 50 and 70 MYA years. The bacteria reside in the cytosol of bacteriocytes, which, in contrast to most other bacteriocyte endosymbioses, do not form bacteriomes. Instead, the bacteriocytes are intercalated between midgut epithelial cells and tend to disappear in older animals. Moreover, in line with their vertical transmission, the bacteria are present in the oocytes (Sauer et al. 2002). So far, apart from *Wolbachia*, no secondary endosymbionts could be found in any of the ant species tested (Zientz et al. 2005). The genome sequences of two *Camponotus* species, *C. floridanus* and *C. pennsylvanicus*, which diverged approximately 25 MYA years ago, are available. With 705 and 795 kb, respectively, once more these genomes are reduced in size and have a similar low GC content of between 27.4 and 29.6% (Degnan et al. 2005; Gil et al. 2003). As in the case of *Buchnera aphidicola*, the two genomes show a complete stasis in architecture, indicating that the reductive genome evolution is mainly based on large deletions and point mutations, and that the reductive process occurs early after symbiosis establishment and before symbiont cospeciation with the host lineage (Degnan et al. 2005). The analysis of the metabolic capability of *Candidatus* Blochmannia revealed that most pathways required for biosynthesis of essential amino acids and several vitamin and cofactor biosynthesis pathways are present in both sequenced organisms, thus pointing again to a nutritional basis of this endosymbiosis. Moreover, both strains code for a potentially functional urease. The somewhat larger genome of *Candidatus* Blochmannia pennsylvanicus encodes several additional, mainly biosynthetic functions, e.g., required for isoprenoid biosynthesis, but overall the metabolic capabilities of both sequenced strains are very similar. In *C. floridanus*, by feeding experiments with N^{15}-labeled urea or a defined chemical diet including or devoid of essential amino acids, in combination with antibiotics to reduce the symbiont load of *Candidatus* Blochmannia, evidence was provided that the animals in fact provide essential amino acids to the host and have a functional urease possibly involved in nitrogen recycling (Feldhaar et al. 2007). *Candidatus* Blochmannia encodes only four putative dedicated transcriptional factors, and a whole genome transcriptome analysis through the different life stages of the animal (larvae, pupae, imagines) revealed, in fact, very mild transcriptional changes throughout the life span of the animals. However, a significant differential pattern of gene expression could be observed, which indicated a slightly stronger expression of certain metabolic genes, in particular in the pupal stage of the animals (Stoll et al. 2009a), thus pointing to a particular relevance of this symbiosis during metamorphosis when the animals do not take up nutrients. In line with this, the bacterial load of the animals is highest in the pupal phase, and also, the polyploidy of the symbionts with about 80 genomes per cell is maximal (Stoll et al. 2009a). How the differential expression of symbiosis relevant factors during pupal stages is achieved is not known. However, it could be shown that despite the very high AT content of the genome, the RNA polymerase still specifically recognizes distinct promoter sites that are highly conserved with the consensus promoter sequence of *E. coli* (Stoll et al. 2009b). The fact that *Candidatus* Blochmannia provides essential nutrients to the host is somewhat surprising given the omnivorous lifestyle of the host. However, it is known that many *Camponotus* species may suffer from strong variations in food supply during the year, or they live in habitats that in principle are quite nitrogen poor, such as rain forest canopies, where the animals mainly rely on sweet secretions from plants or other insects, such as aphids. Thus, many *Camponotus* species can be considered secondary herbivores, and thus may require their endosymbionts to thrive in such habitats. The endosymbiosis of carpenter ants with *Candidatus* Blochmannia may therefore be of importance for the biological success of this extremely successful ant genus, which comprises more than 1,000 species, since the symbiosis enabled the animals to colonize habitats that would have been out of their reach without their bacterial companions.

INTERACTIONS OF ENDOSYMBIONTS AND THEIR HOSTS

Each living organism, independent of its lifestyle (free living, mutualistic, or pathogenic), has to interact with its environment and usually other creatures. However, since the postulates of Koch and the work of Pasteur, bacterial interactions with eukaryotic cells or organisms have mainly been viewed as a permanent conflict between the interacting organisms, and because of their ethical, social, and economical impact, an anthropocentric view of the microbial world was prevalent. Accordingly, most attention has been paid to bacterial pathogens of man and livestock, and cure or prevention of infectious diseases. Mutualistic associations with bacteria were much less investigated in general, but in particular with regard to microbial interaction with host defense systems, although these associations are by far much more frequent in nature, and despite the fact that they are one of the most powerful forces driving the diversity of life and evolution. For instance, the human symbiocosm may consist of thousands of bacterial species living permanently in commensalism or mutualism with the human host, and in particular in the digestive tract, while less than a hundred different bacteria appear to be relevant human pathogens. Accordingly, the basic features of why some bacteria-host interactions end up with mutualism or commensalism and others with pathogenesis are still poorly understood.

How do arthropods manage to limit these symbionts to specific tissues, and how do they control their proliferation? Such questions have been investigated by symbiosis researchers for decades, and despite most of the early investigations remaining on the descriptive level, they provided important basic knowledge. For example, Nardon et al. (1998) reported with a simple experiment that weevil's endosymbiont (SPE) may divide seven times over 21 days, from the embryo to the last larval stage. Counting the bacteria was used to show that symbiont density is stable within a given weevil population, but changes from one population to the another. Genetic crosses and back-crosses between populations with high and low bacterial density have demonstrated that symbionts must be genetically controlled by the host. Moreover, endosymbiont location was examined in almost all symbiotic arthropod models. Regardless of the manner of their transmission to the offspring, the majority of arthropod endosymbionts infiltrate the bacterioytes early during embryogenesis. From the first larval stage, primary endosymbionts fail to proliferate outside these "symbiotic cells." Such data have suggested that symbiont survival is usually not permitted outside the bacteriocytes, and that the bacteriocytes may possess appropriate immune responses to control excessive bacterial replication. Currently, several groups are assessing these questions on the molecular level, in order to uncover genes involved in the specific bacteriocyte immune response and to decipher their function and impact on symbiosis (Gerardo et al. 2009; Gross et al. 2009; Heddi et al. 2005; Nakabachi et al. 2005; Wang et al. 2009).

The increasing interest in innate immunity and its role in pathogen defense and maintenance of symbiotic interactions is mainly due to the establishment of *Drosophila melanogaster* as a model system. This fly can easily be manipulated genetically, and it was the first insect to be entirely genome sequenced. Briefly, invertebrate innate immunity against microorganisms relies on two principal ways: cellular and humoral defense mechanisms. The former consists mainly of phagocytosis by plasmatocytes/hemocytes (macrophage-like cells) and encapsulation by lamellocytes. The latter involves proteolytic cascades that result in local production of melanin and synthesis and secretion in the hemolymph of antimicrobial effectors (Lemaitre and Hoffmann 2007). The humoral system relies on inducible synthesis of antimicrobial peptides (AMPs), and the type of molecules may change according to the microbe infecting the insect (i.e., Gram-negative bacteria, Gram-positive bacteria, fungi). This specific recognition is due to pattern recognition receptors (PRRs) that are associated with microbial-associated molecular patterns (MAMPs). Two receptor types have been identified: the Gram-negative bacteria-binding proteins (GNBPs) that are specific to invertebrates and bind to lipopolysaccharide (LPS) and bacterial lipoteichoic acids, as well as to β-1,3-glucans (Kim et al. 2000), and peptidoglycan recognition proteins (PGRPs) that recognize both bacterial peptidoglycan (DAP-type or Lys-type peptidoglycan) and LPS (Kurata 2004; Werner et al. 2003).

Only recently, was the first evidence provided that the innate immune system of insects is also involved in endosymbiont perception and control. For example, two immune genes, a PGRP and a lysozyme, were shown to be highly expressed in insect bacteriocytes (Heddi et al. 2005; Nakabachi et al. 2005). Along with the discovery of TTSS in SZPE and *Sodalis* (Dale et al. 2001, 2002), this shows that mutualistic relationships rely on bacterial and immune genes similar to those coming into action in pathogenic relationships, and that innate immunity may function to protect or control the endosymbionts at certain conditions. Recently, Anselme and coworkers compared the expression of a large number of weevil immune genes in the bacteriocytes and in larvae as a response to the endosymbiont SZPE injected in the hemolymph. As a result, expression of AMPs and lysozymes was highly induced, attesting that the humoral response recognizes the symbiont as an intruder when "infecting" the hemolymph (Anselme et al. 2008). Fascinatingly, however, the expression pattern in the bacteriome was different. All investigated genes encoding immune functions (including AMPs) were downregulated, with the notable exception of three genes: one AMP (*coleoptericin*), one PGRP (*wpgrp1*), and one putative signaling pathway regulator (*Tollip*). The failure to express most AMPs indicates that immune sanctions are relaxed inside the bacteriome, which may help bacterial survival and symbiosis maintenance (Reynolds and Rolff 2008). However, additional work is necessary to evaluate the functional significance of the weevil genes that are expressed at high levels in the bacteriome. Another example for the role of immunity in symbiosis was recently described in the tsetse fly. The tsetse PRR GmPGRP-LB, a homologue of the *D. melanogaster* PGRP-LB protein, known to be an inhibitor of the Imd immune signaling pathway, was also shown to be highly expressed in the bacteriome of the *Candidatus* Wigglesworthia-harboring tsetse fly (Wang et al. 2009). This gene appears to play an immunomodulatory role in the bacteriocytes of the tsetse flies, since its silencing by RNAi leads to a reduction in the number of *Candidatus* Wigglesworthia, and an increase in the expression of the antimicrobial effector attacin (Wang et al. 2009). Lack of the endosymbiont also reduced the expression of GmPGRP-LB and, interestingly, led to an increase of susceptibility of the tsetse toward *Trypanosoma* infections (Pais et al. 2008). This indicates, on the one hand, that host immune responses are highly adapted in the bacteriocytes to protect and control the endosymbionts, and on the other hand, symbiosis-driven host responses may directly influence parasite resistance traits (Wang et al. 2009).

The development of exciting novel genetic and genomic molecular tools, including RNAi (Jaubert-Possamai et al. 2007; Vallier et al. 2009), has dramatically improved our experimental repertoire to understand host-microbe interactions in the future. These technical improvements allow a detailed investigation of the host side, in particular of immune functions of the host, and enable us to broaden our research focus away from only a few model organisms to a large variety of different, nonconventional host organisms. This is of particular importance for the symbiosis field, which in the past decades has mainly investigated the microbe side and, apart from mainly descriptive research, has largely neglected to consider the host side due to experimental limitations.

ACKNOWLEDGMENTS

We thank Dagmar Beier for critically reading the manuscript and Yvan Rahbé (BF2I and CTμ Center) for providing the symbiosome picture. We thank the COST Action FA0701, *Arthropod Symbioses: From Fundamental Studies to Pest and Disease Management*, for support.

REFERENCES

Akman, L., Yamashita, A., Watanabe, H., et al. 2002. Genome sequence of the endocellular obligate symbiont of tsetse flies, *Wigglesworthia glossinidia*. *Nat. Genet.* 32: 402–407.

Aksoy, S. 1995. Molecular characteristics of the endosymbionts of tsetse flies: 16S rDNA locus and over-expression of chaperonins. *Insect Mol. Biol.* 4: 23–29.

Aksoy, S., Chen, X., and Hypsa, V. 1997. Phylogeny and potential transmission routes of midgut-associated endosymbionts of tsetse (Diptera: Glossinidae). *Insect Mol. Biol.* 6: 183–190.

Allen, J.M., Reed, D.L., Perotti, M.A., and Braig, H.R. 2007. Evolutionary relationships of "*Candidatus* Riesia spp.," endosymbiotic enterobacteriaceae living within hematophagous primate lice. *Appl. Environ. Microbiol.* 73: 1659–1664.

Anselme, C., Pérez-Brocal, V., Vallier, A., et al. 2008. Identification of the weevil immune genes and their expression in the bacteriome tissue. *BMC Biol.* 6: 43.

Anselme, C., Vallier, A., Balmand, S., Fauvarque, M.O., and Heddi, A. 2006. Host PGRP gene expression and bacterial release in endosymbiosis of the weevil *Sitophilus zeamais*. *Appl. Environ. Microbiol.* 72: 6766–6772.

Attardo, G.M., Lohs, C., Heddi, A., Alam, U.H., Yildirim, S., and Aksoy, S. 2008. Analysis of milk gland structure and function in *Glossina morsitans*: milk protein production, symbiont populations and fecundity. *J. Insect Physiol.* 54: 1236–1242.

Baumann, P., Baumann, L., Lai, C.-Y., Rouhbakhsh, D., Moran, N.A., and Clark, M.A. 1995. Genetics, physiology, and evolutionary relationships of the genus *Buchnera*: intracellular symbionts of aphids. *Ann. Rev. Microbiol.* 49: 55–94.

Baumann, L., Thao, M.L., Hess, J.M., Johnson, M.W., and Baumann, P. 2002. The genetic properties of the primary endosymbionts of mealybugs differ from those of other endosymbionts of plant sap-sucking insects. *Appl. Environ. Microbiol.* 68: 3198–3205.

Blochmann, F. 1887. Über das Vorkommen bakterienähnlicher Gebilde in den Geweben und Eiern verschiedener Insekten. *Zbl. Bakt.* 11, 234–240.

Buchner, P. 1965. *Endosmybiosis of animals with plant microorganisms.* Interscience Publishers, New York.

Charles, H., Heddi, A., Guillaud, J., Nardon, C., and Nardon, P. 1997. A molecular aspect of symbiotic interactions between the weevil *Sitophilus oryzae* and its endosymbiotic bacteria: overexpression of a chaperonin. *Biochem. Biophys. Res. Commun.* 239: 769–774.

Charles, H., and Ishikawa, H. 1999. Physical and genetic map of the genome of *Buchnera*, the primary endosymbiont of the pea aphid *Acyrthosiphon pisum*. *J. Mol. Evol.* 48(2): 142–50.

Charles, H., Ishikawa, H., and Nardon, P. 1995. Presence of a protein specific of endocytobiosis (symbionin) in the weevil *Sitophilus*. *C. R. Acad. Sci. Paris* 318: 35–41.

Chen, D.Q., Clytia, B., Montllor, C.B., and Pucell, A.H. 2000. Fitness effects of two facultative endosymbiotic bacteria on the pea aphid, *Acyrthosiphon pisum*, and the blue alfalfa aphid, *A. kondoi*. *Ent. Exp. Appl.* 95: 315–323.

Clark, M.A., Baumann, L., Thao, M.L., Moran, N.A., and Baumann, P. 2001. Degenerative minimalism in the genome of a psyllid endosymbiont. *J. Bacteriol.* 183: 1853–1861.

Conord, C., Despres, L., Vallier, A., et al. 2008. Long-term evolutionary stability of bacterial endosymbiosis in Curculionoidea: additional evidence of symbiont replacement in the Dryophthoridae family. *Mol. Biol. Evol.* 25: 859–868.

Dale, C., Plague, G.R., Wang, B., Ochman, H., and Moran, N.A. 2002. Type III secretion systems and the evolution of mutualistic endosymbiosis. *Proc. Natl. Acad. Sci. USA* 99: 12397–12402.

Dale, C., Wang, B., Moran, N.A., and Ochman, H. 2003 Loss of DNA recombinational repair enzymes in the initial stages of genome degeneration. *Mol. Biol. Evol.* 20: 1188–1194.

Dale, C., Young, S.A., Haydon, D.T., and Welburn, S.C. 2001. The insect endosymbiont *Sodalis glossinidius* utilizes a type III secretion for cell invasion. *Proc. Natl. Acad. Sci. USA* 98: 1883–1888.

Degnan, P.H., Lazarus, A.B., and Wernegreen, J.J. 2005. Genome sequence of *Blochmannia pennsylvanicus* indicates parallel evolutionary trends among bacterial mutualists of insects. *Genome Res.* 15: 1023–33.

Douglas, A.E. 1998. Host benefit and the evolution of specialization in symbiosis. *Heredity* 81: 599–603.

Epis, S., Sassera, D., Beninati, T., et al. 2008. *Midichloria mitochondrii* is widespread in hard ticks (Ixodidae) and resides in the mitochondria of phylogenetically diverse species. *Parasitology* 135: 485–494.

Fares, M.A., Ruiz-González, M.X., Moya, A., Elena, S.F., and Barrio, E. 2002. Endosymbiotic bacteria: *groEL* buffers against deleterious mutations. *Nature* 417: 398.

Feldhaar, H., Straka, J., Krischke, M., et al. 2007. Nutritional upgrading for omnivorous carpenter ants by the endosymbiont *Blochmannia*. *BMC Biol.* 5: 48.

Fukatsu, T., Hosokawa, T., Koga, R., et al. 2009. Intestinal endocellular symbiotic bacterium of the macaque louse *Pedicinus obtusus*: distinct endosymbiont origins in anthropoid primate lice and the old world monkey louse. *Appl. Environ. Microbiol.* 75: 3796–3769.

Gasnier-Fauchet, F., Gharib, A., and Nardon, P. 1986. Comparison of methionine metabolism in symbiotic and aposymbiotic larvae of *Sitophilus oryzae* L. (Coleoptera: Curculionidae). 1. Evidence for a glycine N-methyltransferase-like activity in the aposymbiotic larvae. *Comp. Biochem. Physiol.* 85B: 245–250.

Gasnier-Fauchet, F., and Nardon, P. 1986. Comparison of methionine metabolism in symbiotic and aposymbiotic larvae of *Sitophilus oryzae* L. (Coleoptera: Curculionidae). 2. Involvement of the symbiotic bacteria in the oxidation of methionine. *Comp. Biochem. Physiol.* 85B: 251–254.

Gerardo, N., Altincicek, B., Anselme, C., Atamian, H., Barribeau, S., de Vos, M., Evans, J.D., Gabaldón, T., Ghanim, M., Heddi, A., Kaloshian, I., Latorre, A., Monegat, C., Moya, A., Nakabachi, A., Parker, B.J., Pérez-Brocal, V., Pignatelli, M., Rahbé, Y., Ramsey, J., Spragg, C., Tamames, J.S., Tamarit, D., Tamborindeguy, C., and Vilcinskas, A. 2009. Immunity and defense in pea aphids, *Acyrthosiphon pisum. Genome Biol.* 11: R21.

Gil, R., Belda, E., Gosalbes, M.J., et al. 2008. Massive presence of insertion sequences in the genome of SOPE, the primary endosymbiont of the rice weevil *Sitophilus oryzae. Int. Microbiol.* 11: 41–48.

Gil, R., Sabater-Muñoz, B., Latorre, A., Silva, F.J., and Moya, A. 2002. Extreme genome reduction in *Buchnera* spp.: toward the minimal genome needed for symbiotic life. *Proc. Natl. Acad. Sci. USA* 99: 4454–4458.

Gil, R., Silva, F.J., Zientz, E., et al. 2003. The genome sequence of *Blochmannia floridanus*: comparative analysis of reduced genomes. *Proc. Natl. Acad. Sci. USA* 100: 9388–9393.

Gómez-Valero, L., Latorre, A., and Silva, F.J. 2004a. The evolutionary fate of nonfunctional DNA in the bacterial endosymbiont *Buchnera aphidicola. Mol. Biol. Evol.* 21: 2172–2181.

Gómez-Valero, L., Soriano-Navarro, M., Pérez-Brocal, V., et al. 2004b. Coexistence of *Wolbachia* with *Buchnera aphidicola* and a secondary symbiont in the aphid *Cinara cedri. J. Bacteriol.* 186: 6626–6633.

Gosalbes, M.J., Lamelas, A., Moya, A., and Latorre, A. 2008. The striking case of tryptophan provision in the cedar aphid *Cinara cedri. J. Bacteriol.* 190: 6026–6029.

Grenier, A.M., Nardon, C., and Nardon, P. 1994. The role of symbiotes in flight activity of *Sitophilus* weevils. *Entomol. Exp. Appl.* 70: 201–208.

Gross, R., Vavre, F., Heddi, A., Hurst, G.D.D., Zchori-Fein, E., and Bourtzis, K. 2009. Immunity and symbiosis. *Mol. Microbiol.* 73: 751–759.

Heddi, A., Charles, H., Khatchadourian, C., Bonnot, G., and Nardon, P. 1998. Molecular characterization of the principal symbiotic bacteria of the weevil *Sitophilus oryzae*: a peculiar G-C content of an endocytobiotic DNA. *J. Mol. Evol.* 47: 52–61.

Heddi, A., Grenier, A.M., Khatchadouria, C., Charles, H., and Nardon, P. 1999. Four intracellular genomes direct weevil biology: nuclear, mitochondrial, principal endosymbionts, and *Wolbachia. Proc. Natl. Acad. Sci. USA* 96: 6814–6819.

Heddi, A., Lefebvre, F., and Nardon, P. 1991. The influence of symbiosis on the respiratory control ratio (RCR) and the ADP/O ratio in the adult weevil *Sitophilus oryzae* (Coleoptera, Curculionidae). *Endocytobiosis Cell Res.* 8: 61–73.

Heddi, A., Lefebvre, F., and Nardon, P. 1993. Effect of endocytobiotic bacteria on mitochondrial enzymatic activities in the weevil *Sitophilus oryzae* (Coleoptera, Curculionidae). *Insect Biochem. Mol. Biol.* 23: 403–411.

Heddi, A., and Nardon, P. 1993. Mitochondrial DNA expression in symbiotic and aposymbiotic strains of *Sitophilus oryzae. J. Stored Prod. Res.* 29: 243–252.

Heddi, A., Vallier, A., Anselme, C., Xin, H., Rahbé, Y., and Wäckers F. 2005. Molecular and cellular profiles of insect bacteriocytes: mutualism and harm at the initial evolutionary step of symbiogenesis. *Cell. Microbiol.* 7: 293–305.

Ho, S.Y., Phillips, M.J., Cooper, A., and Drummond, A.J. 2005. Time dependency of molecular rate estimates and systematic overestimation of recent divergence times. *Mol. Biol. Evol.* 22: 1561–1568.

Hornok, S., Földvári, G., Elek, V., Naranjo, V., Farkas, R., and de la Fuente, J. 2008. Molecular identification of *Anaplasma marginale* and rickettsial endosymbionts in blood-sucking flies (Diptera: Tabanidae, Muscidae) and hard ticks (Acari: Ixodidae). *Vet. Parasitol.* 154: 354–359.

Hosokawa, T., Kikuchi, Y., Nikoh, N., Shimada, M., and Fukatsu, T. 2006. Strict host-symbiont cospeciation and reductive genome evolution in insect gut bacteria. *PLoS Biol.* 4: e337.

Ishikawa, H. 1985. Symbionin, an aphid endosymbiot-specific protein. II. Diminution of symbionin during post embryonic development of aposymbiotic insects. *Insect Biochem.* 15: 165–174.

Jaubert-Possamai, S., Le Trionnaire, G., Bonhomme, J., Christophides, G.K., Rispe, C., and Tagu, D. 2007. Gene knockdown by RNAi in the pea aphid *Acyrthosiphon pisum*. *BMC Biotechnol.* 7: 63.

Kikuchi, Y., Hosokawa, T., Nikoh, N., Meng, X.Y., Kamagata, Y., and Fukatsu, T. 2009. Host-symbiont co-speciation and reductive genome evolution in gut symbiotic bacteria of acanthosomatid stinkbugs. *BMC Biol.* 7: 2.

Kim, Y.S., Ryu, J.H., Han et al. 2000. Gram-negative bacteria-binding protein, a pattern recognition receptor for lipopolysaccharide and beta-1,3-glucan that mediates the signaling for the induction of innate immune genes in *Drosophila melanogaster* cells. *J. Biol. Chem.* 275: 32721–32727.

Koch, A. 1960. Intracellular symbiosis in insects. *Annu. Rev. Microbiol.* 14: 121–140.

Kurata, S. 2004. Recognition of infectious non-self and activation of immune responses by peptidoglycan recognition protein (PGRP)-family members in *Drosophila*. *Dev. Comp. Immunol.* 28: 89–95.

Lefèvre, C., Charles, H., Vallier, A., Delobel, B., Farrell, B., and Heddi, A. 2004. Endosymbiont phylogenesis in the dryophthoridae weevils: evidence for bacterial replacement. *Mol. Biol. Evol.* 21: 965–973.

Lemaitre, B., and Hoffmann, J. 2007. The host defense of *Drosophila melanogaster*. *Annu. Rev. Immunol.* 25: 697–743.

Liadouze, I., Febvay, G., Guillaud, J., and Bonnot, G. 1996. Metabolic fate of energetic amino acids in the aposymbiotic pea aphid *Acyrthosiphon pisum* (Harris) (Homoptera: Aphididae). *Symbiosis* 21: 115–127.

McCutcheon, J.P., McDonald, B.R., and Moran, N.A. 2009. Origin of an alternative genetic code in the extremely small and GC-rich genome of a bacterial symbiont. *PLoS Genet.* 5: e1000565.

McCutcheon, J.P., McDonald, B.R., and Moran, N.A. 2009b. Convergent evolution of metabolic roles in bacterial co-symbionts of insects. *Proc. Natl. Acad. Sci. USA* 106: 15394–15399.

McCutcheon, J.P, and Moran, N.A. 2007. Parallel genomic evolution and metabolic interdependence in an ancient symbiosis. *Proc. Natl. Acad. Sci. USA* 104: 19392–19397.

Montllor, C.B., Maxmen, A., and Purcell, A.H. 2002. Facultative bacterial endosymbionts benefit pea aphids *Acyrthosiphon pisum* under heat stress. *Ecol. Entomol.* 27: 189–195.

Moran, N.A. 1996. Accelerated evolution and Muller's rachet in endosymbiotic bacteria. *Proc. Natl. Acad. Sci. USA* 93: 2873–2878.

Moran, N., and Baumann, L. 1994. Phylogenetics of cytoplasmically inherited microorganisms of arthropods. *Trends Ecol. Evol.* 9: 15–20.

Moran, N.A., Dale, C., Dunbar, H., Smith, W.A., and Ochman, H. 2003. Intracellular symbionts of sharpshooters (Insecta: Hemiptera: Cicadellinae) form a distinct clade with a small genome. *Environ. Microbiol.* 5: 116–126.

Moran, N.A., Degnan, P.H., Santos, S.R., Dunbar, H.E., and Ochman, H. 2005. The players in a mutualistic symbiosis: insects, bacteria, viruses, and virulence genes. *Proc. Natl. Acad. Sci. USA* 102: 16919–16926.

Moran, N.A., McCutcheon, J.P., and Nakabachi, A. 2008. Genomics and evolution of heritable bacterial symbionts. *Annu. Rev. Genet.* 42: 165–90.

Moran, N.A., McLaughlin, H.J., and Sorek, R. 2009. The dynamics and time scale of ongoing genomic erosion in symbiotic bacteria. *Science* 323: 379–382.

Moya, A., Gil, R., Latorre, A., Peretó, J., Pilar Garcillán-Barcia, M., and de la Cruz, F. 2009. Toward minimal bacterial cells: evolution vs. design. *FEMS Microbiol. Rev.* 33: 225–235.

Munson, M.A., Baumann, P., Clark, M. A., et al. 1991. Evidence for the establishment of aphid eubacterium endosymbiosis in an ancestor of four aphid families. *J. Bacteriol.* 173: 6321–6324.

Murray, R.G., and Schleifer, K.H. 1994. Taxonomic notes: a proposal for recording the properties of putative taxa of procaryotes. *Int. J. Syst. Bacteriol.* 44: 174–176.

Nakabachi, A., Koshikawa, S., Miura, T., and Miyagishima, S. 2009. Genome size of *Pachypsylla venusta* (Hemiptera: Psyllidae) and the ploidy of its bacteriocyte, the symbiotic host cell that harbors intracellular mutualistic bacteria with the smallest cellular genome. *Bull. Entomol. Res.* 23: 1–7.

Nakabachi, A., Shigenobu, S., Sakazume, N., et al. 2005. Transcriptome analysis of the aphid bacteriocyte, the symbiotic host cell that harbors an endocellular mutualistic bacterium, *Buchnera. Proc. Natl. Acad. Sci. USA* 102: 5477–5482.

Nakabachi, A., Yamashita, A., Toh, H., et al. 2006. The 160-kilobase genome of the bacterial endosymbiont *Carsonella. Science* 314: 267.

Nardon, P. 1973. Obtention d'une souche asymbiotique chez le charançon Sitophilus sasakii Tak: différentes méthodes d'obtention et comparaison avec la souche symbiotique d'origine. *C. R. Acad. Sci. Paris* 277D: 981–984.

Nardon, P., Grenier, A.M., and Heddi, A. 1998. Endocytobiote control by the host in the weevil *Sitophilus oryzae* (Coleoptera: Curculionidae). *Symbiosis* 25: 237–250.

Nogge, G. 1976. Sterility in tsetse flies (*Glossina morsitans* Westwood) caused by loss of symbionts. *Experientia* 32: 995–996.

Oliver, K.M., Degnan, P.H., Burke, G.R., and Moran, N.A. 2010. Facultative symbionts of aphids and the horizontal transfer of ecologically important traits. *Annu. Rev. Entomol.* 55: 247–266.

Oliver, K.M., Russell, J.A., Moran, N.A., and Hunter, M.S. 2003. Facultative bacterial symbionts in aphids confer resistance to parasitic wasps. *Proc. Natl. Acad. Sci. USA* 100: 1803–1807.

Pais, R., Lohs, C., Wu, Y., Wang, J., and Aksoy, S. 2008. The obligate mutualist *Wigglesworthia glossinidia* influences reproduction, digestion, and immunity processes of its host, the tsetse fly. *Appl. Environ. Microbiol.* 74: 5965–5974.

Pérez-Brocal, V., Gil, R., Ramos, S., et al. 2006. A small microbial genome: the end of a long symbiotic relationship? *Science* 314: 312–313.

Perotti, M.A., Allen, J.M., Reed, D.L., and Braig, H.R. 2007. Host-symbiont interactions of the primary endosymbiont of human head and body lice. *FASEB J.* 21: 1058–1066.

Pierantoni, U. 1927. L'organo simbiotico nello sviluppo di *Calandra oryzae. Rend. Reale Acad. Sci. Fis. Mat Napoli* 35: 244–250.

Pontes, M.H., Babst, M., Lochhead, R., Oakeson, K., Smith, K., and Dale, C. 2008. Quorum sensing primes the oxidative stress response in the insect endosymbiont, *Sodalis glossinidius. PLoS One* 3: e3541.

Reisen, W.K. 2010. Landscape epidemiology of vector-borne diseases. *Annu. Rev. Entomol.* 55: 461–483.

Richard, S., Seng, P., Parola, P., Raoult, D., Davoust, B., and Brouqui, P. 2009. Detection of a new bacterium related to '*andidatus* Midichloria mitochondrii' in bed bugs. *Clin. Microbiol. Infect.* Suppl. 2: 85–85.

Sabater, B., van Ham, R., Martinez Torres, D., Silva, F.J., Latorre, A., and Moya, A. 2001. Molecular evolution of aphids and their primary (*Buchnera* sp.) and secondary endosymbionts: implications for the role of symbiosis in insect evolution. *Interciencia* 26: 508–512.

Sasaki-Fukatsu, K., Koga, R., Nikoh, N., et al. 2006. Symbiotic bacteria associated with stomach discs of human lice. *Appl. Environ. Microbiol.* 72: 7349–7352.

Sassera, D., Beninati, T., Bandi, C., et al. 2006. '*Candidatus* Midichloria mitochondrii', an endosymbiont of the tick *Ixodes ricinus* with a unique intramitochondrial lifestyle. *Int. J. Syst. Evol. Microbiol.* 56: 2535–2540.

Sassera, D., Lo, N., Bouman, E.A., Epis, S., Mortarino, M., and Bandi, C. 2008. "*Candidatus* Midichloria" endosymbionts bloom after the blood meal of the host, the hard tick *Ixodes ricinus*. *Appl. Environ. Microbiol.* 74: 6138–6140.

Sauer, C., Dudaczek, D., Hölldobler, B., and Gross, R. 2002. Tissue localization of the endosymbiotic bacterium "*Candidatus* Blochmannia floridanus" in adults and larvae of the carpenter ant *Camponotus floridanus*. *Appl. Environ. Microbiol.* 68: 4187–4193.

Sauer, C., Stackebrandt, E., Gadau, J., Hölldobler, B., and Gross, R. 2000. Systematic relationships and cospeciation of bacterial endosymbionts and their carpenter ant host species: proposal of the new taxon *Candidatus* Blochmannia gen. nov. *Int. J. Syst. Evol. Microbiol.* 50: 1877–1886.

Schröder, D., Deppisch, H., Obermayer, M., et al. 1996. Intracellular endosymbiotic bacteria of *Camponotus* species (carpenter ants): systematics, evolution and ultrastructural characterization. *Mol. Microbiol.* 21: 479–489.

Shigenobu, S., Watanabe, H., Hattori, M., Sakaki, Y., and Ishikawa, H. 2000. Genome sequence of the endocellular bacterial symbiont of aphids *Buchnera* sp. APS. *Nature* 407: 81–86.

Sikora, H. 1919. Vorläufige Mitteilungen über Mycetome bei Pediculiden. *Biol. Zentralbl.* 39: 287–288.

Stoll, S., Feldhaar, H., and Gross, R. 2009a. Transcriptional profiling of the endosymbiont *Blochmannia floridanus* during different developmental stages of its holometabolous ant host. *Environ. Microbiol.* 11: 877–888.

Stoll, S., Feldhaar, H., and Gross, R. 2009b. Promoter characterization in the AT-rich genome of the obligate endosymbiont "*Candidatus* Blochmannia floridanus." *J. Bacteriol.* 191: 3747–3751.

Takiya, D.M., Tran, P.L., Dietrich, C.H., and Moran, N.A. 2006. Co-cladogenesis spanning three phyla: leafhoppers (Insecta: Hemiptera: Cicadellidae) and their dual bacterial symbionts. *Mol. Ecol.* 15: 4175–4191.

Tamames, J., Gil, R., Latorre, A., Peretó, J., Silva, F.J., and Moya, A. 2007. The frontier between cell and organelle: genome analysis of *Candidatus* Carsonella ruddii. *BMC Evol. Biol.* 7: 181.

Tamas, I., Klasson, L., Canbäck, B., et al. 2002. 50 million years of genomic stasis in endosymbiotic bacteria. *Science* 296: 2376–2379.

Thao, M.L., Gullan, P.J., and Baumann, P. 2002. Secondary (γ-Proteobacteria) endosymbionts infect the primary (γ-Proteobacteria) endosymbionts of mealybugs multiple times and coevolve with their hosts. *Appl. Environ. Microbiol.* 68: 3190–3197.

Thao, M.L., Moran, N.A., Abbot, P., Brennan, E.B., Burckhardt, D.H., and Baumann, P. 2000. Cospeciation of psyllids and their primary prokaryotic endosymbionts. *Appl. Environ. Microbiol.* 66: 2898–2905.

Tsuchida, T., Koga, R., and Fukatsu, T. 2004. Host plant specialization governed by facultative symbiont. *Science* 303: 1989.

Vallier, A., Vincent-Monégat, C., Laurençon, A., and Heddi, A. 2009. RNAi in the cereal weevil *Sitophilus* spp.: systemic RNAi mediated gene knockdown in the bacteriome tissue. *BMC Biotechnol.* 9: 44.

van Ham, R.C., Kamerbeek, J., Palacios, C., et al. 2003. Reductive genome evolution in *Buchnera aphidicola*. *Proc. Natl. Acad. Sci. USA* 100: 581–586.

von Dohlen, C.D., Kohler, S., Alsop, S.T., and McManus, W.R. 2001. Mealybug β-proteobacterial endosymbionts contain γ-proteobacterial symbionts. *Nature* 412: 433–436.

Wang, J., Wu, Y., Yang, G., and Aksoy, S. 2009. Interactions between mutualist *Wigglesworthia* and tsetse peptidoglycan recognition protein (PGRP-LB) influence trypanosome transmission. *Proc. Natl. Acad. Sci. USA* 106: 12133–12138.

Wernegreen, J.J., Kauppinen, S.N., Brady, S.G., and Ward, P.S. 2009. One nutritional symbiosis begat another: phylogenetic evidence that the ant tribe *Camponotini* acquired *Blochmannia* by tending sap-feeding insects. *BMC Evol. Biol.* 9: 292.

Werner, T., Borge-Renberg, K., Mellroth, P., Steiner, H., and Hultmark, D. 2003. Functional diversity of the *Drosophila* PGRP-LC gene cluster in the response to lipopolysaccharide and peptidoglycan. *J. Biol. Chem.* 278: 26319–22.

Wicker, C. 1983. Differential vitamin and choline requirements of symbiotic and aposymbiotic *S. oryzae* (Coleoptera: Curculionidae). *Comp. Biochem. Physiol.* 76A: 177–182.

Wicker, C., and Nardon, P. 1982. Development responses of symbiotic and aposymbiotic weevil *Sitophilus oryzae* L. (Coleoptera, Curculionidae) to a diet supplemented with aromatic amino acids. *J. Insect Physiol.* 28: 1021–1024.

Wilkinson, T.L., Fukatsu, T., and Ishikawa, H. 2003. Transmission of symbiotic bacteria *Buchnera* to parthenogenetic embryos in the aphid *Acyrthosiphon pisum* (Hemiptera: Aphidoidea). *Arthropod Struct. Dev.* 32: 241–245.

Wu, D., Daugherty, S.C., van Aken, S.E., Pai, G.H., et al. 2006. Metabolic complementarity and genomics of the dual bacterial symbiosis of sharpshooters. *PLoS Biol.* 4: e188.

Zientz, E., Dandekar, T., and Gross, R. 2004. Metabolic interdependence of obligate intracellular bacteria and their insect hosts. *Microbiol. Mol. Biol. Rev.* 70: 4096–4102.

Zientz, E., Feldhaar, H., Stoll, S., and Gross, R. 2005. Insights into the microbial world associated with ants. *Arch. Microbiol.* 184: 199–206.

2 The Bacteroidetes *Blattabacterium* and *Sulcia* as Primary Endosymbionts of Arthropods

Matteo Montagna, Luciano Sacchi, Nathan Lo, Emanuela Clementi, Daniele Daffonchio, Alberto Alma, Davide Sassera, and Claudio Bandi

CONTENTS

Genus	*Sulcia*
Species	*muelleri*
Family	Not determined
Order	Flavobacterales
Description year	2005
Origin of name	The genus name refers to Karel Sulc, while the species name refers to H. J. Müller. Karel Sulc (1872–1952), a Moravian embryologist at the University of Brno, was one of the first scientists to observe symbiotic microorganisms in cicadas (Müller 1910).
	A student of Buchner, H. J. Müller (1890–1967) studied symbionts of auchenorrhynchans and recognized them as bacteria (Buchner 1940).
Description	
Reference	Moran et al. (2005)

INTRODUCTION

THE EARLY DISCOVERY OF BACTEROIDETES AS SYMBIONTS

The study of intracellular microorganisms harbored by arthropods, particularly by insects, began in the second half of the nineteenth century, following the wider use of light microscopy. It was only at the end of the century, however, that Blochmann (1887) observed peculiar structures in cells of the fat body of cockroaches, cells that are now known as bacteriocytes (Sacchi et al. 1988). Bacteriocytes are cells specialized to harbor obligatory beneficial symbionts in insects. At the beginning of the twentieth century, Karel Sulc (1910, 1924) described aggregations of bacteriocytes (bacteriomes) in the body cavity of cicads, while Carlo Jucci (1932) observed bacteriocytes in the fat body of the termite *Mastotermes darwiniensis*. Jucci recognized the similarity between this type of cell in *M. darwiniensis* and those present in the fat body of cockroaches. These three scientists (Blochmann, Sulc, and Jucci) thus reported the first observations of the bacterial symbionts that are now known to belong to the Bacteroidetes phylum. The work of Blochmann, Sulc, and Jucci is reviewed in detail in Buchner (1965).

THE SYMBIONT ROLE

Intracellular symbiotic microorganisms are harbored by a variety of eukaryotes. A very active research area in this field is focused on the symbioses of insects. There is evidence that symbionts played a central role in insect evolution. They may have influenced radiation of insect groups, contributing to their plasticity and adaptability to different environments. It is well known that plant materials, like phloem, xylem sap, or leaf parenchymal tissue, have low amounts of nitrogen and organic carbon, and are generally insufficient to support the metabolic requirements of metazoans. Thus, in addition to insect self-made adaptations, such as the one evolved by aphids that filter phloem sap concentrating useful substances, several insects evolved the intriguing solution of relying on symbiotic bacteria capable of synthesizing the essential compounds needed by the insect itself.

Symbiosis

It is possible to consider symbionts from different points of view. There are a variety of body areas in which hosts could harbor symbionts, such as body cavities (gut and salivary glands), fluids (hemolymph), or intracellular spaces. Intracellular symbionts can reside in either normal or specialized cells, the latter being called bacteriocytes (or mycetocytes when the symbionts are yeasts or fungi). Symbionts can play different roles, for example, supplementing important metabolic functions, or determining a variety of alterations, from manipulations of host reproduction to the protection against pathogens. The types of interactions between hosts and symbionts are thus difficult to classify. The classic division into beneficial, commensal, and parasitic symbionts can be regarded as a first interpretative framework. Of course, the three concepts are not always easily applicable. For example, a *Wolbachia* that induces cytoplasmic incompatibility (CI) can be seen as a parasite at the population or species level, but could actually confer an advantage at the individual level. Similarly, *Midichloria mitochondrii* is likely a facultative, beneficial symbiont for the individual tick (Sassera et al. 2006), but behaves as a parasite toward the mitochondrion (if we consider this organelle a type of microorganism).

Another classification divides insect-associated symbionts into primary and secondary symbionts (Dale and Moran 2006; Moya et al. 2008). Primary symbionts (hereafter indicated as p-symbionts) are obligate mutualistic bacteria, generally anciently acquired and vertically transmitted from mother to offspring (Bressan et al. 2009; Dale and Moran 2006). Examples of Bacteroidetes p-symbionts are *Blattabacterium* spp., *Candidatus* Sulcia muelleri (hereafter *Sulcia*), and *Candidatus* Uzinura diaspidicola (hereafter *Uzinura*) of Diaspididae (Gruwell et al. 2007). Secondary symbionts (hereafter indicated as s-symbionts) are generally facultative symbionts and can be deleterious or beneficial; in the latter case they can confer some advantage on host survival or reproductive rates. S-symbionts are primarily vertically transmitted but can also colonize hosts through horizontal transfer (Baumann 2006; Dale and Moran 2006). This chapter will be focused only on p-symbionts belonging to the phylum Bacteroidetes. Based on the information so far acquired, p-symbionts increase the host fitness through some form of nutritional support, e.g., through the provision of amino acids and cofactors, or they can contribute to the management of host metabolic wastes (Moran and Telang 1998).

Co-Cladogenesis of Hosts and Symbionts

Comparative studies of the cladogenesis of hosts and symbionts can be used to study the history of a symbiotic association, and to indirectly infer the primary route of transmission of a symbiont. Observation of the overall pattern of host-symbiont phylogenies may also offer clues on the type of interaction. In cases where symbionts are strictly vertically transmitted, host and symbiont phylogenies are expected to overlap, showing a pattern of co-cladogenesis (i.e., a parallel evolutionary history).

P-symbionts normally show a pattern of coevolution with the host, while co-cladogenesis is typically less stringent for S-symbionts. Where symbionts are commensals, parasites, or manipulators of reproduction, horizontal transfer can be expected to be more frequent, and host-symbiont phylogenies could show several discrepancies (Dale and Moran 2006). The first study of the coevolution of hosts and symbionts is probably the work performed on the aphid symbiont *Buchnera*; in 1993 this work showed the perfect matching of the host and symbiont phylogenies (Moran et al. 1993). Soon after, Bandi et al. (1994) published the identification of the cockroach symbiont *Blattabacterium* (now assigned to the phylum Bacteroidetes), and the comparison of its phylogeny with the host taxonomy provided further evidence for the coevolution of a group of hosts with their p-symbionts.

THE BACTEROIDETES PHYLUM

Bacteroidetes is a phylum of Bacteria (Gherna and Woese 1992; Garrity and Holt 2001; Gupta 2004), previously also known as Cytophaga-Flexibacter-Bacteroides (CFB). This phylum encompasses bacteria with different oxygen affinities (aerobic, anaerobic, and microaerophilic), adapted to different environments, including animal tissues, plants, soil, and seawater (Hunter and Zchori-Fein 2006; Gupta 2004). They are organized into three large classes: Bacteroidia (Bacteroidaceae, Rikenellaceae, Porphyromonadaceae, Prevotellaceae), Sphingobacteria (Sphingobacteriaceae, Saprospiraceae, Flexibacteraceae, Flammeoviraceae, Crenotrichaceae), and Flavobacteria (Flavobacteriaceae, Myroidaceae, Blattabacteriaceae) (Garrity and Holt 2001; Hunter and Zchori-Fein 2006). The various families/classes of the phylum Bacteroidetes display a limited number of shared, derived phenotypic traits (apomorphies) that allow unambiguous classification. Indeed, the description and classification of this phylum and its classes and families required the application of molecular approaches.

There are currently three genera of Bacteroidetes that are known to play the role of p-symbionts in insects: *Blattabacterium*, found in all but one cockroach species (Dictyoptera: Blattaria) and in the termite *Mastotermes darwiniensis* (Dictyoptera: Isoptera); *Sulcia*, which has so far been detected in over 20 species of cicadellids of the suborder Auchenorryncha (Takiya et al. 2006); and *Uzinura*, which infects Diaspididae (Andersen et al. 2010). Another well-known insect-associated Bacteroides is *Cardinium*, which does not, however, appear to behave as a p-symbiont (this bacterium is indeed known as a manipulator of reproduction; Weeks et al. 2001; Zchori-Fein et al. 2001). In addition, bacteria causing the male-killing phenotype in lady beetles (*Adonia variegata* and *Coleomegilla maculata*) have been shown to belong to the Bacteroidetes (Hurst et al. 1996, 1997, 1999). This chapter will focus on *Blattabacterium* and *Sulcia*, which are the most well-characterized Bacteroidetes P-symbionts of insects.

BLATTABACTERIUM CUENOTI (FLAVOBACTERALES, BLATTABACTERIACEAE)

INTRODUCTION

All cockroach species thus far examined and the termite *M. darwiniensis* (Mastotermitidae) live in symbiosis with intracellular P-symbionts of the genus *Blattabacterium* (Dasch et al. 1984). The only exception is the cave cockroach *Nocticola* spp., which has recently been shown not to harbor these bacteria (Lo et al. 2007). It is well known that cockroaches concentrate and store excess nitrogen as uric acid in their fat body, and this has been postulated to provide a supplementary source of nitrogen in cases of nutritional deficit. In the fat body of cockroaches and *M. darwiniensis*, *Blattabacterium* spp. are located in bacteriocytes, and are transovarially transmitted to the progeny (Bandi et al. 1994, 1995; Bandi and Sacchi 2000).

The association of *Blattabacterium* with the cockroach is well known to be essential for the growth, reproduction, and long-term survival of the host, and several studies in the past decades have indicated a role for the symbionts in supplementing the host with essential compounds (reviewed in Douglas 1989; Sacchi and Grigolo 1989). Only in 2009, after the sequencing of the entire genomes of *Blattabacterium* from two species of cockroaches (*Blattella germanica*, *Periplaneta americana*), have old clues on the role of *Blattabacterium* in nitrogen metabolism and the provision of amino acids finally been confirmed (Sabree et al. 2009; López-Sánchez et al. 2009).

HISTORY

Bacterial symbionts had been observed in the cockroach fat body since the end of the nineteenth century (Blochmann 1887). In 1906, the bacteria harbored in the fat body of *Blatta orientalis* had been named *Bacillus cuenoti* by Mercier (1906); later, the name was changed to *Blattabacterium cuenoti* (Hollande and Favre 1931). The bacteria in the cockroach fat body are easily observed by light microscopy. A few decades after its first description, it became clear that *Blattabacterium* was widespread in cockroaches, all of the species examined being infected (Jucci 1932, 1952), with the exception of *Nocticola* spp. (Lo et al. 2007).

The first studies of intracellular symbiosis in termites were carried out by Jucci (1924), when he discovered the presence of bacteria in the fat body cells of the Australian termite *M. darwiniensis* (reviewed in Bandi and Sacchi 2000). Jucci emphasized that the cells harboring bacteria in the fat body of *M. darwiniensis* resembled those in the cockroach fat body. After a survey for intracellular bacteria of termites, he formulated the hypothesis that the symbiosis was established before the evolutionary divergence of cockroaches and termites, and that only the lineage of the Mastotermitidae retained the symbionts that every cockroach possesses (Jucci 1932, 1952). Koch (1938), in agreement with Jucci, supported the hypothesis that cockroaches and *M. darwiniensis* share the same intracellular symbiotic association and showed that the bacteria of *M. darwiniensis* are transovarially transmitted to the offspring by a mechanism similar to the one adopted by *Blattabacterium* in cockroaches. Additional specimens examined by Jucci and Koch included representatives of the main termite lineages (Termopsidae and Termitidae), but none contained bacteriocytes with the exception of the fat body of *M. darwiniensis*. They thus concluded that the symbiotic association was established in a common ancestor of cockroaches and termites, but it was then lost in termite lineages, with the exception of the one that gave rise to *M. darwiniensis*. The loss of *Blattabacterium* in the main termite lineages was supposed to be linked with the acquisition and evolution of the gut microbiota. It has indeed been supposed that some of the functions of fat body endosymbionts (e.g., in nitrogen recycling) became superfluous during the coevolution of termites with their gut bacteria (Buchner 1965; Bandi and Sacchi 2000; Lo et al. 2003). According to this hypothesis, the gut microbiota of termites could be seen as a community of symbionts that reduces the need for *Blattabacterium*. This hypothesis is also supported by the fact that termite gut bacteria are involved in nitrogen fixation and nitrogen recycling from uric acid (Breznak et al. 1973; French et al. 1976; Slaytor and Chappell 1994; Breznak 2000; Hunter and Zchori-Fein 2006). Termites, during their evolution, established social strategies to recycle nitrogen at the colony level, developing behaviors like cannibalism and consumption of exuviae (i.e., exoskeleton after molt) (Nalepa 1994).

PHYLOGENY AND HOST-SYMBIONT COEVOLUTION

Molecular phylogenetic analyses have allowed us to address three key questions about the evolution of cockroaches and *Blattabacterium*: (1) What is the phylogenetic affiliation of *Blattabacterium*? (2) Are the bacteriocyte endosymbionts of *M. darwiniensis* phylogenetically related with those of cockroaches? (3) Did *Blattabacterium* coevolve with the insect hosts?

As already emphasized, the phylogenetic positioning of *Blattabacterium* as a relative of the genera Cytophaga-Flexibacter-Bacteroides (CFB) was published in 1994 (Bandi et al.). At that time, the phylum Bacteroidetes had not yet been described. The description of this bacterial phylum led to a formal systematic placement of the genus *Blattabacterium* as a member of this phylum, within the class Flavobacteria (Gupta 2004). Until the identification of *Sulcia* (Moran et al. 2005), *Blattabacterium* was the sole P-symbiont assigned to the CFB. It was thus a unique symbiont, in that most of the P-symbionts described during the last decade of the past century had been assigned to the Gammaproteobacteria.

Phylogenetic analysis of the 16S rRNA genes of *Blattabacterium* from *M. darwiniensis* and from seven species of cockroaches (*Blattella germanica*, *Periplaneta americana*, *P. australasiae*, *Nauphoeta cinerea*, *Pycnoscelus surinamensis*, *Cryptocercus punctulatus*, and *Blaberus craniifer*) placed the endosymbiont of *M. darwiniensis* as the sister group of cockroach endosymbionts (Bandi et al. 1994). In addition, a comparison of the phylogenies of the endosymbionts with the taxonomy of cockroaches provided the evidence for the co-cladogenesis of hosts and symbionts. The work also confirmed the scenarios proposed by Jucci and Koch, positioning the establishment of symbiosis with *Blattabacterium* in a common ancestor of cockroaches and termites, and the loss of the endosymbiont along one or more termite lineages, after the split of the one that gave rise to *M. darwiniensis* (Bandi et al. 1995).

The understanding of the evolution of symbiosis in cockroaches and termites had a major advance after the publication of the first robust phylogenetic tree of the Dictyoptera, the insect order including cockroaches (Blattaria), termites (Isoptera), and mantids (Mantodea) (Lo et al. 2000). This work confirmed the close phylogenetic relationship of cockroaches and termites, showing that cockroaches, as commonly understood, are actually paraphyletic (i.e., a group that does not contain all of the descendants of a common ancestor) with wood roaches of the genus *Cryptocercus* sister group to termites. As a side product of this study, gene sequences from cockroaches and *M. darwiniensis* became available for formal analysis of the co-clado-genesis with *Blattabacterium* symbionts (Lo et al. 2003). It was thus shown that the phylogeny of cockroaches and *M. darwiniensis* matches that of *Blattabacterium*, in a pattern of coevolution similar to that already uncovered for aphids with their endosymbionts of the genus *Buchnera* (Moran et al. 1993).

More recently, three species of *Nocticola* (Nocticolidae), a genus from a divergent family of cockroaches, were screened with molecular and microscopical approaches for the presence of *Blattabacterium*. Surprisingly, *Nocticola* did to not harbor *Blattabacterium*, nor has it any cell resembling a bacteriocyte in the fat body cells (Lo et al. 2007). This could be explained by an early divergence of the Nocticolidae, before the establishment of the symbiosis with *Blattabacterium*. This would imply that the split of this cockroach family occurred before the split of cockroaches and termites (if we assume, as previously stated, that the common ancestor of these lineages harbored *Blattabacterium*). An alternative hypothesis would be that symbiosis with *Blattabacterium* was lost along the Nocticolidae lineage. These two possible hypotheses emphasize the importance of phylogenetic analysis for reconstructing scenarios on the evolution of symbiotic associations. It would be interesting to examine the environmental conditions that might explain the absence of *Blattabacterium* in termites and in the Nocticolidae.

GENOME SEQUENCES

The complete genome sequencing of the P-symbionts of *Periplaneta americana* (*Blattabacterium cuenoti* strain BPLAN; hereafter BPLAN) and of *Blattella germanica* (*Blattabacterium cuenoti* strain Bge; hereafter Bge) confirms their phylogenetic closeness with *Sulcia muelleri* (López-Sánchez et al. 2009; Sabree et al. 2009) and sheds light on the metabolic interactions between host and symbiont. The sizes of the genomes of BPLAN and Bge have been shown to be almost identical, being circles of 636,994 and 636,850 bp, respectively (Sabree et al. 2009; López-Sánchez et al. 2009). The two genomes are also very similar in terms of the number of protein coding genes (581 CDS (coding sequence) in BPLAN, 586 CDS in Bge). Approximately 50% of the genes of both genomes code for proteins involved in amino acid metabolism.

Comparison of the genome sizes of free-living Bacteroidetes (from about 2.5 to 6 Mb) with those of endosymbionts like BPLAN and Bge (about 0.64 Mb) and *S. muelleri* (about 0.25 Mb) highlights the tendency of P-symbionts toward a reduction of the genome size (Tokuda et al. 2008; McCutcheon and Moran 2007; López-Sánchez et al. 2009; Sabree et al. 2009). Furthermore, comparison of the genome sizes in different p-symbionts seems to suggest that there is a sort of "molecular clock" in genome

size reduction. The association between *Blattabacterium* and the Dyctioptera is estimated to have occurred over 140 MYA (Bandi et al. 1995; Lo et al. 2003), while
establishment of the symbiosis of *Sulcia* (smaller genome size) with its hemipteran
hosts could be traced back to over 250 MYA (Moran et al. 2005).

However, if we consider Blattabacterium only, we must remember that *B. germanica* and *P. americana* belong to different cockroach families (Blattellidae and
Blattidae) that diverged during the late Jurassic or early Cretaceous (about 140 MYA).
The P-symbionts of these cockroaches, BPLAN and Bge, are thus separated by a
long evolutionary period. As already emphasized, the genomes of the two symbionts
are almost identical in terms of both size and number of genes. We can thus conclude that these two genomes experienced a long evolutionary stasis after the split
of the host cockroaches. We could thus hypothesize that a dramatic reduction of the
symbiont genome occurred in the ancestor of the Blattidae and Blattellidae, or even
in the common ancestor of cockroaches and termites. If we accept this hypothesis,
the phase of the establishment of the symbiosis with the cockroach ancestor could be
seen as an evolutionary "punctuation," characterized by a dramatic reorganization
and size reduction of the symbiont genome, which could have occurred in a relatively
short time (in geological terms). The subsequent evolutionary time could then have
proceeded with a sort of stasis in terms of genome size and overall metabolic capacities of the symbionts.

METABOLIC RELATIONS IN THE GENOMIC ERA

Experiments with aposymbiotic (i.e., lacking the symbionts) cockroaches, obtained
by feeding diets supplemented with antibiotics or lysozyme, were performed some
decades ago with the aim of uncovering the role of the symbionts, and the metabolic relations with the host. In cockroaches, *Blattabacterium* has been shown to
be essential for the growth and reproduction of the host, and has been proposed to
be involved in recycling of nitrogen waste, mobilizing uric acid in the fat body and
providing the host with essential amino acids (Henry 1962; Malke and Schwartz
1966; Valovage and Brooks 1979; Cochran 1985; Wren and Cochran 1987; Bandi and
Sacchi 2000). A generally accepted idea was that *Blattabacterium* was capable of
mobilizing the urates from the urate cells of the fat body, and that the resulting nitrogen compounds were then used for the synthesis of amino acids essential for the host
cockroach. Based on this view, biochemical approaches have been used to detect
enzymes involved in uricolysis produced by the bacteria, with apparent evidence for
a xanthine dehydrogenase activity associated with the symbionts (Cochran 1985).

The capacity of symbionts to recycle nitrogen waste could explain the ecological
success of cockroaches. In addition, the cytological characteristics of the fat body
of cockroaches fit with the idea that symbionts participate in nitrogen recycling.
Differently from any other insect, the cockroach fat body is formed by three cell
types (Figure 2.1): bacteriocytes, harboring the bacterial symbionts; trophocytes,
which store the lipids; and urocytes, which contain crystals of urates (Sacchi et al.
1998). The cell membranes of bacteriocytes and urocytes have been shown to adhere
tightly to each other, even having a junctional complex, which might allow an easy
exchange of molecules (Sacchi et al. 1998).

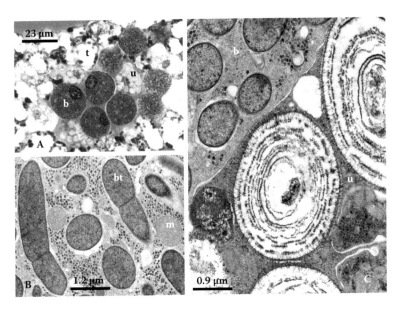

FIGURE 2.1 The fat body of the cockroach *Blattella germanica*. (A) Micrograph showing three cell types: bacteriocytes (b), trophocytes (t), and urocytes (u). (B) Transmission electron microscopy (TEM) micrograph of a bacteriocyte harboring bacterial symbionts (bt) in the cytoplasm; m = mitochondrion. (C) TEM micrograph of the interface between a bacteriocyte (b) and a urocyte (u) that contains crystals of urates.

In recent years, genome analysis allowed researchers to obtain a clearer picture of the metabolic role of *Blattabacterium* in cockroaches, partly confirming previous hypotheses, but also disproving some generally accepted ideas.

Both Bge and BPLAN genomes code for the gluconeogenesis pathway, TCA cycle, and phosphogluconate pathway, by which symbionts generate essential carbon precursors, and also generate ATP aerobically, using a peculiar type of complex IV, cytochrome cbb3 (López-Sánchez et al. 2009; Sabree et al. 2009). Interestingly BPLAN and Bge have been shown to be able to synthesize most amino acids, including those essential for the host. However, the pools of amino acids synthesized by Bge and BPLAN are not completely overlapping. A very important discovery that resulted from the sequencing of the genome is that *Blattabacterium* possesses a complete urea cycle and a urease, which makes this bacterium capable of synthesizing the amino acids starting from urea, ammonia, and glutamate (López-Sánchez et al. 2009).

The hypothesis that *Blattabacterium* could recycle nitrogenous waste to provide amino acids is supported by the fact that *Blattabacterium* encodes urease and glutamate dehydrogenase, and that it uses urea and ammonia to synthesize glutamate (Sabree et al. 2009). The urease genes *ureAB* and *ureC* are present in the genome of *Blattabacterium* Bge, and urease activity was demonstrated by enzymatic assay on endosymbiont crude extracts (López-Sánchez et al. 2009).

As already emphasized, bacteriocytes harboring symbionts are physically close to urocytes (Bandi and Sacchi 2000), and this suggested that *Blattabacterium* could be involved in recycling the nitrogen stored as uric acid. However, no genes coding for urea-degrading enzymes have been found in both BPLAN and Bge. This suggests other interesting hypotheses, e.g., that the host supplies the uricolytic enzymes, or that uric acid is transported to the gut and catabolized by the cockroach gut microbiota (Sabree et al. 2009; López-Sánchez et al. 2009). *Blattabacterium* could thus use urea and ammonia, the degradation products of uricolysis, to synthesize amino acids. As for the carbon source, *Blattabacterium* possesses an ABC-type transport system, and probably uses imported glycerol, since it lacks the glycolysis pathway (López-Sánchez et al. 2009).

The detection of the complete urea cycle and urease in *Blattabacterium* finally explains the ammoniotelic condition of cockroaches. This makes cockroaches the only insects able to discard ammonia.

Genome studies on the P-symbiont of the aphid *Buchnera aphidicola* showed that these bacteria have lost the capacity to synthesize lipopolysaccharides (LPSs), which are an important component of the bacterial cell envelope (Shigenobu et al. 2000). This loss of the LPS was probably facilitated by the protected intracellular environment, which relaxes the structural requirements of the cell envelope. Some years ago a study on *Blattabacterium* from *P. americana* and *B. germanica* provided no evidence for the presence of LPSs in these symbionts (Scudo et al. 1996). The authors of this study proposed the interesting idea that LPS synthetic capacity was lost following the transition to the obligatory symbiotic state.

CANDIDATUS SULCIA MUELLERI

INTRODUCTION

Interest in symbiosis in cicadoids arose at the beginning of the twentieth century, when Karel Sulc first observed symbiotic microorganisms in these insects, and indicated these symbionts as saccharomycetes. It was only in the middle of the last century, after morphological studies by Resühr (1938) and Müller (1940), that these symbionts were recognized as bacteria, described as "rich in fluid." Buchner (1965) found these "sausage-shaped" bacteria in well-defined pairs of bacteriocytes in a number of species of the jassids, membracids, euscelids, and cicads. It was only half a century later that Moran and colleagues (2005), with a 16S rRNA polymerase chain reaction (PCR)-based approach, identified the symbiont of leafhoppers, treehoppers, cicadas, spittlebugs, and planthoppers as a member of the Bacteroidetes phylum. This symbiont was then named *Candidatus* Sulcia muelleri (hereafter *Sulcia*) in honor of the two scientists, K. Sulc and H. J. Müller (Moran et al. 2005).

In the following years, studies focused on symbionts of sap-feeding insects using 16S rRNA as a marker and detected the presence of *Sulcia* in a number of species belonging to the Hemiptera order. The current picture indicates the presence of *Sulcia* in members of the superfamilies Fulgoroidea, Cicadoidea, Cercopoidea, and Membracoidea. Surprisingly, *Sulcia* is not described in the families Delphacidae, Flatidae, and Acanaloniidae (Coleorrhyncha) and in members of Heteroptera. (Moran et al. 2005; Takiya et al. 2006; Bressan et al. 2009).

PHYLOGENY AND HOST-SYMBIONT COEVOLUTION

The phylogeny of the Hemiptera is not fully resolved, thus making any hypothesis on the pattern of acquisitions and losses of *Sulcia* problematic. Currently the order is subdivided into five suborders: Coleorrhyncha, Heteroptera, Stenorrhyncha, Fulgoromorpha, and Cicadomorpha (the latter two had been for a long time joined into Auchenorrhyncha).

Buchner and Müller proposed that acquisitions and loss of *Sulcia* happened during Fulgoromorpha (planthoppers) cladogenesis, and in their opinion this symbiosis could explain the extant diversity of planthoppers (Müller 1940; Buchner 1965). The presence of *Sulcia* in members of Cicadomorpha and Fulgoromorpha and the absence in Heteroptera suggest a different hypothesis for the origin of symbiosis and host-symbiont coevolution.

Phylogenetic analyses on *Sulcia* support the hypothesis of host-symbiont coevolution and suggest that the ancestor of *Sulcia* was present in the ancestors of Auchenorrhyncha, Coleorrhyncha, and Heteroptera (250 MYA in Permian), but was subsequently lost in the Heteroptera and Coleorrhyncha suborders, as well as in some families of the suborders Fulgoromorpha and Cicadomorpha (Delphacidae, Flatidae, Acanalonidae, and Cicadellidae) (Moran et al. 2005). The acquisition of a more recent p-symbiont (*Baumannia cicadellicola*, belonging to the Gammaproteobacteria) happened during the Cretaceous period (70 to 100 MYA) (Moran et al. 2005). The co-cladogenesis between Cicadomorpha and their p-symbionts *Sulcia* and *Baumannia* was further supported by a second study on leafhoppers (Takiya et al. 2006). The analysis was conducted on 16S rRNA from 29 species belonging to four leafhopper subfamilies: Coelidiinae, Evacanthinae, Phereurhininae, and Cicadellinae (Hemiptera, Cicadomorpha). All 29 species examined were shown to harbor *Sulcia*, with a high percentage of identity in 16S rRNA sequences, while 25 species harbor *Baumannia*.

METABOLIC NETWORK AND COPRIMARY SYMBIONTS

PCR and fluorescence *in situ* hybridization (FISH) analyses were conducted to understand the presence and distribution of symbionts in the host bacteriome. Moran and colleagues (2005) showed with both approaches (PCR and FISH) that strapshaped cells of *Sulcia* harbored in *Clastoptera arizonana* are present only into the dark-red portion of the bacteriome, often coiled into structures encircled by a membrane (supposed to be of host origin). A second study, focused on *Pentastiridius leporinus* (Cixiidae family), using light microscopy, provided evidence for the presence of eight bacteriomes of two morphological types inside each dissected insect (six globular yellow types and two white types) (Bressan et al. 2009). In order to confirm what symbiont is present in each of the two bacteriome types, Bressan and colleagues designed two PCR protocols and primers, one specific for *Sulcia* and another specific for the Gammaproteobacterial symbiont. This approach detected *Sulcia* only inside the white bacteriomes and the novel Gammaproteobacterial symbiont only in the yellow ones, indicating the spatial isolation of the two symbionts. One hypothesis that could explain the coexistence of two P-symbionts in close structures involves their complementarity in biosynthetic capabilities (Moran et al. 2005; Takiya et al. 2006).

Wu and colleagues (2006), working on the genome of *Baumannia cicadellicola* and *Sulcia muelleri* of the sharpshooter *Homalodisca coagulata*, tested the hypothesis of metabolic codependency between host and the two bacteria, also confirming the presence of both symbionts in the cytosol of host bacteriocytes (sometimes within the same cells). In the genome of *Baumannia* they found 83 genes that encode for proteins involved in the synthesis of vitamins, cofactors, prosthetic groups, and related compounds and the presence of just histidine and methionine complete pathways (17 genes involved in amino acid synthesis), suggesting that both host and *Baumannia* receive essential amino acid from other sources. In the same study, partial genome sequencing of *Sulcia* (146 kb) revealed the presence of 33 genes involved in amino acid biosynthesis (in particular, essential amino acids: lysine, arginine, leucine, threonine, tryptophan, valine, and isoleucine) and only a few genes for cofactor and vitamin synthesis, strongly suggesting the possibility that both bacteria complement each other in this symbiotic system (e.g., *Sulcia* lacks the histidine pathway that is present in *Baumannia*, while the pathways for ubiquinone and menaquinone synthesis are absent in *Sulcia* but present in *Baumannia*) (Wu et al. 2006).

In recent years the genome of *Sulcia* from different hosts was completed: *Sulcia* GWSS from *Homalodisca coagulata* (Cicadomorpha) glassy-winged sharpshooter (GWSS) (McCutcheon and Moran 2007) and *Sulcia* from *Diceroprocta semicincta* (Cicadomorpha) (McCutcheon et al. 2009). These studies confirmed the presence of a mutualistic metabolic network between *Sulcia*, *Baumannia*, and their host, the glassy-winged sharpshooter. *Sulcia* retains pathways to synthesize eight essential amino acids not present in the *Baumannia* genome, which in turn possess genes for two other amino acids, vitamin and cofactor biosynthesis, that are absent in *Sulcia* (McCutcheon and Moran 2007). *Sulcia* GWSS seems to be able to generate reducing power in the form of NADH and transmit it to cbb-3 cytochrome to generate ATP (McCutcheon and Moran 2007). Another peculiar characteristic of the *Sulcia* genome is the presence of a probably functional pathway of Tat (twin-arginine translocation) and SecY, but the absence of genes coding for amino acid transporters (McCutcheon and Moran 2007).

To explain the insect-*Sulcia*-*Baumannia* system in light of the nitrogen cycle, we can propose different hypotheses on the roles of the three partners; for example, the ammonium host waste could be used by *Sulcia* for biosynthesis of amino acids, while nonessential amino acids in host diets supply *Sulcia* and *Baumannia* requirements for nitrogen, with the host providing enzymes that allow ammonium incorporation. In all of the hypotheses considered, *Sulcia* plays a central role in ammonium assimilation (McCutcheon and Moran 2007).

To further highlight the metabolic network relations in Cicadellidae, two systems were compared by McCutcheon and Moran (2009). The previously described network involving *Sulcia* GWSS and *Baumannia cicadellinicola* present in xylem-feeding glassy-winged sharpshooters was compared with *Diceroprocta semicincta* and its two p-symbionts *Sulcia muelleri* SMDSEM and *Hodgkinia cicadicola* (Alphaproteobacteria). The two hosts separated around 200 MYA (Shcherbakov and Popov 2002), and the comparison of the two genomes of *Sulcia muelleri*, from the two different hosts, shows a surprisingly conserved size and a similar gene pattern (276,984 bp, 278 genes, 242 of which are protein coding, in *S. muelleri* SMDSEM, while 245,530 bp, 263 genes, 227 of which are protein coding, in *S. muelleri* GWSS). The two *Sulcia* genomes both retain the genes encoding for 8 of the 10 essential amino acids, leaving to their respective coprimary symbionts the synthesis of the other 2 (McCutcheon et al. 2009). Thus, the coprimary symbionts, *Baumannia cicadellinicola* and *Hodgkinia cicadicola*, also show overlapping gene sets, supplying similar metabolic capabilities with respect to the insect host and the other p-symbiont *Sulcia*.

ACKNOWLEDGMENTS

Networking activity and missions of young scientists supported by COST Action FAO701. CB research supported by Ministero della Salute and ISPESLS and by a MIUR-PRIN grant to Claudio Genchi.

REFERENCES

Andersen, J. C., Wu, J., Gruwell, M. E., Gwiazdowski, R., Santana, S. E., Feliciano, N. M., Morse, G. E., and Normark, B. B. 2010. A phylogenetic analysis of armored scale insects (Hemiptera: Diaspididae), based upon nuclear, mitochondrial, and endosymbiont gene sequences. *Molecular Phylogenetics Evolution* 57: 992–1003.

Bandi, C., Damiani, G., Magrassi, L., Grigolo, A., Fani, R., and Sacchi, L. 1994. Flavobacteria as intracellular symbionts in cockroaches. *Proceedings of the Royal Society of London B* 257: 43–48.

Bandi, C., and Sacchi, L. 2000. Intracellular symbiosis in termites. In *Termites: evolution, sociality, symbioses, ecology*, ed. T. Abe, D. E. Bignell, and M. Higashi, 261–273. Netherlands: Kluwer Academic Publishers.

Bandi, C., Sironi, M., Damiani, G., et al. 1995. The establishment of intracellular symbiosis in an ancestor of cockroaches and termites. *Proceedings of the Royal Society of London B* 259: 293–299.

Baumann, P. 2006. Diversity of prokaryote-insect associations within the Stenorrhyncha (psyllids, whiteflies, aphids, mealybugs). In *Insect symbiosis*, ed. K. Bourtzis and T. A. Miller, 1–24. Vol. II. Boca Raton, FL: CRC Press.

Blochmann, F. 1887. Vorkommen bakterienaehnliche Koerpechen in den Geweben und Eiern verschiederen Insecten. *Biologisches Zentralblatt* 7: 606–608.

Bressan, A., Arneodo, J., Simonato, M., Haines, W. P., and Boudon-Padieu. 2009. Characterization and evolution of two bacteriome-inhabiting symbionts in cixiid planthoppers (Hemiptera: Fulgoromorpha: Pentastirini). *Environmental Microbiology* 11: 3265–3279.

Breznak, J. A. 2000. Ecology of prokaryotic microbes in the guts of wood- and litter-feeding termites. In *Termites: evolution, sociality, symbiosis, ecology*, ed. T. Abe, D. E. Bignell, and M. Higashi, 209–232. Netherlands: Kluwer Academic Publishers.

Breznak, J. A., Brill, W. J., Mertins, J. W., and Coppel H. C. 1973. Nitrogen fixation in termites. *Nature* 244: 577–580.

Buchner, P. 1965. *Endosymbiosis of animals with plant microorganism*. New York: John Wiley.

Cochran, D. G. 1985. Nitrogen excretion in cockroaches. *Annual Review of Entomology* 30: 29–49.

Dale, C., and Moran, N. A. 2006. Molecular interactions between bacterial symbionts and their hosts. *Cell* 126: 453–465.

Dasch, G. A., Weiss, E., and Chang, K. 1984. Endosymbionts in insects. In *Bergey's manual of systematic bacteriology*, ed. N. R. Krieg and J. G. Holt, 811–833. Vol. I. Baltimore: Williams and Wilkins.

Douglas, A. E. 1989. Mycetocyte symbiosis in insets. *Biological Reviews* 64: 409–434.

French, J. R. J., Turner, G. L., and Bradbury, J. F. 1976. Nitrogen fixation by bacteria from the hind gut of termites. *Journal of General Microbiology* 95: 202–206.

Garrity, G. M., and Holt, J. G. 2001. The road map to the manual. In *Bergey's manual of systematic bacteriology*, ed. D. R. Boone, R. W. Castenholz, and G. M. Garrity, 119–141. 2nd ed., vol. 1. New York: Springer.

Gherna, R., and Woese, C. R. 1992. A partial phylogenetic analysis of the "flavobacter-bacteroides" phylum: basis for taxonomic restructuring. *Systematic and Applied Microbiology* 15: 513–521.

Gruwell, M. E., Morse, G. E., and Normark, B. B. 2007. Phylogenetic congruence of armored scale insects (Hemiptera: Diaspididae) and their primary endosymbionts from the phylum Bacteroidetes. *Molecular Phylogenetics and Evolution* 44: 267–280.

Gupta, R. S. 2004. The phylogeny and signature sequences characteristics of *Fibrobacteres, Chlorobi*, and *Bacteroidetes*. *Critical Reviews in Microbiology* 30: 123–143.

Henry, S. M. 1962. The significance of microorganisms in the nutrition of insects. *Transactions of the New York Academy of Sciences* 24: 676–683.

Hollande, A. C., and Favre, R. 1931. La structure cytologique de *Blattabacterium cuenoti* (Mercier) N.G., symbiote du tissu adipeux des blattides. *Comptes Rendus de la Société de Biologie* 107: 752–754.

Hunter, M. S., and Zchori-Fein, E. 2006. Inherited Bacteroidetes symbionts in arthropods. In *Insect symbiosis*, ed. K. Bourtzis and T. A. Miller, 39–56. Vol. II. Boca Raton, FL: CRC Press.

Hurst, G. D., Hammarton, T. C, Obrycki, J. J., et al. 1996. Male-killing bacterium in a fifth ladybird beetle, *Coleomegilla maculate* (Coleoptera: Coccinellidae). *Heredity* 77: 177–185.

Hurst, G. D. D., Bandi, C., Sacchi, L., et al. 1999. *Adonia variegata* (Coleoptera: Coccinellidae) bears maternally inherited Flavobacteria that kill males only. *Parasitology* 18: 125–134.

Hurst, G. D. D., Hammarton, T. C., Bandi, C., Majerus, T. M. O., Bertrand, D., and Majerus, M. E. N. 1997. The male-killing agent of the ladybird beetle *Coleomegilla maculata* is related to *Blattabacterium*, the beneficial symbiont of cockroaches. *Genetics Research* 70: 1–6.

Jucci, C. 1924. La differenziazione de le caste ne la società dei termitidi. I. I Neotenici. *Reale Accademia Nazionale dei Lincei* 14: 198–205.

Jucci, C. 1932. Sulla presenza dei batteriociti nel tessuto adiposo dei termitidi. *Archivio Zoologico Italiano* 16: 1422–1429.

Jucci, C. 1952. Symbiosis and phylogenesis in the Isoptera. *Nature* 169: 1422–1429.

Koch, A. 1938. Symbiosestudien III: Die intrazellulare bakteriensymbiose von *Mastotermes darwiniensis* Frogratt (Isoptera). *Zeitschrift fur Morphologie und Okologie der Tiere* 34: 584–609.

Lo, N., Bandi, C., Watanabe, H., Nalepa, C., and Bennati, T. 2003. Evidence for cocladogenesis between diverse dictyopteran lineages and their intracellular endosymbionts. *Molecular Biology and Evolution* 20: 907–913.

Lo, N., Beninati, T., Stone, F., Walker, J., and Sacchi, L. 2007. Cockroaches that lack *Blattabacterium* endosymbionts: the phylogenetically divergent genus *Nocticola*. *Biology Letters* 3: 327–330.

Lo, N., Tokuda, G., Watanabe, H., et al. 2000. Evidence from multiple gene sequences indicates that termites evolved from wood-feeding cockroaches. *Current Biology* 10: 801–804.

López-Sánchez, M. J., Neef, A., Peretó, J., et al. 2009. Evolutionary convergence and nitrogen metabolism in *Blattabacterium* strain Bge, primary endosymbiont of the cockroach *Blattella germanica*. *PLoS Genetics* 5 (11): e1000721.

Malke, H., and Schwartz, W. 1966. Untersuchungen uber die symbiose von tieren mit pilzen und bakterien. XII. Die bedeutung der blattiden-symbiose. *Zeitschrift fur Allgemeine Mikrobiologie* 6: 34–68.

McCutcheon, J. P., McDonald, B. R., and Moran, N. A. 2009. Convergent evolution of metabolic roles in bacterial co-symbionts of insects. *Proceedings of the National Academy of Science of the United States of America* 106: 15394–15399.

McCutcheon, J. P., and Moran, N. A. 2007. Parallel genomic evolution and metabolic interdependence in an ancient symbiosis. *Proceedings of the National Academy of Science of the United States of America* 104: 19392–19397.

Mercier, L. 1906. Les corps bacteroides de la blatte (*Periplaneta orientalis*): *Bacillus cuenoti* (n. sp. L. Mercier). (Note preliminaire). *Comptes Rendus de la Société de Biologie* 61: 682–684.

Moran, N. A., Munson, M. A., Baumann, P., and Ishikawa, H. 1993. A molecular clock in endosymbiotic bacteria is calibrated using the insect host. *Proceedings of the Royal Society of London B* 253: 167–172.

Moran, N. A., and A. Telang 1998. Bacteriocyte-associated symbions of insects. *BioScience* 48: 295–304.

Moran, N. A., Tran, P., Gerardo, N. M. 2005. Symbiosis and insect diversification: an ancient symbiont of sap-feeding insects from the bacterial phylum Bacteroidetes. *Applied and Environmental Microbiology* 71: 8802–8810.

Moya, A., Pereto, J., Gil, R., and Latorre, A. 2008. Learning how to live together: genomic insights into prokaryote-animal symbioses. *Nature Reviews Genetics* 9: 218–229.

Müller, H. J. 1940. *Die symbiose der Fulgoroiden (Homoptera Cicadina)*. Stuttgart, Germany: E. Schweizerbart'sche-Verlagsbuchhandlung.

Nalepa, C. A. 1994. Nourishment and the origin of termite eusociality. In *Nourishment and evolution in insect societies*, ed. J. H. Hunt and C. A. Nalepa, 57–104. Boulder, CO: Westview Press.

Resühr, B. 1938. Zur morphologie und protoplasmatik der bakteroiden symbionten einiger homopteren (*Philaenus spumarius* L., *Cicadella viridis* und *Pseudococcus citri* Risso). *Archives Mikrobiology* 9: 31–79.

Sabree, Z. L., Kambhampati, S., and Moran N. A. 2009. Nitrogen recycling and nutritional provisioning by *Blattabacterium*, the cockroach endosymbiont. *Proceedings of the National Academy of Science of the United States of America* 106: 19521–19526.

Sacchi, L., and Grigolo, A. 1989. Endocytobiosis in *Blattella germanica* L. (Blattodea): recent acquisitions. *Endocytobiosis and Cell Research* 6: 121–147.

Sacchi, L., Grigolo, A., Mazzini, M., Bigliardi, E., Baccetti, B., and Laudani, U. 1988. Symbionts in the oocytes of *Blattella germanica* (L.) (Dyctyoptera: Blattellidae): their mode of transmission. *International Journal of Insect Morphology and Embryology* 17: 437–446.

Sacchi, L., Nalepa, C. A., Bigliardi, E., et al. 1998. Ultrastructural studies of the fat body and bacterial endosymbionts of *Cryptocercus punctulatus* Scudder (Blattaria: Cryptocercidae). *Symbiosis* 25: 252–269.

Sassera, D., Beninati, T., Bandi, C., et al. 2006. '*Candidatus* Midichloria motichondrii', an endosymbiont of the tick *Ixodes ricinus* with a unique intramitochondrial lifestyle. *International Journal of Systematic and Evolutionary Microbiology* 56: 2535–2540.

Scudo, F. M, Sacchi, L., Freudenberg, M. A., Gigolo, A., and Galanos, C. 1996. On the absence of lipopolysaccharides in the endocellular symbionts of cockroaches and its evolutionary implications. *Endocytobiosis and Cell Research* 11: 119–127.

Shcherbakov, D. E., and Popov, Y. A. 2002. Superorder Cimicidea Laicharting, 1781; Order Hemiptera Linné, 1758. The bugs, cicadas, plantlice, scale insects, etc. In *Hystory of Insects*, ed. A. P. Rasnitsyn and D. L. J. Quicke, 143–157. Dordrecht: Kluwer.

Shigenobu, S., Watanabe, H., Hattori, M., Sakaki, I., and Ishikawa, H. 2000. Genome sequence of the endocellular bacterial symbiont of aphids *Buchnera* sp. APS. *Nature* 407: 81–86.

Slaytor, M., and Chappell, D. J. 1994. Nitrogen metabolism in termites. *Comparative Biochemistry and Physiology* 107B: 1–10.

Sulc, K. 1910. Symbiontische Saccharomyceten der echten Cicaden. *Sitzungsberichte der Königlich Böhmischen Gesellschaft der Wissenschaften*, Prag.

Sulc, K. 1924. O intracellularni hereditarni symbiosa u Fulgorid (Homoptera). *Publ. Biol. Ecole haute etudes Veterin*, 3, Brno.

Takiya, D. M., Tran, P. L., Dietrich, C. H., and Moran, N. A. 2006. Co-cladogenesis spanning three phyla: leafhoppers (Insecta: Hemiptera: Cicadellidae) and their dual bacterial symbionts. *Molecular Ecology* 15: 4175–4191.

Tokuda, G., Lo, N., Takase, A., Yamada, A., Hayashi, Y., and Watanabe, H. 2008. Purification and partial genome characterization of the bacterial endosymbiont *Blattabacterium cuenoti* from the fat bodies of cockroaches. *BMC Research Notes* I: 118.

Valovage, W. D., and Brooks, M. A. 1979. Uric acid quantities in the fat body of normal and aposymbiotic German cockroaches, *Blattella germanica* (Blattodea). *Annals of the Entomological Society of America* 72: 687–689.

Weeks, A. R., Marec, F., and Breeuwer, J. A. J. 2001. A mite species that consists entirely of haploid females. *Science* 292: 2479–2482.

Wren, H. N., and Cochran, D. G. 1987. Xanthine dehydrogenase activity in the cockroach endosymbiont *Blattabacterium cuenoti* (Mercier 1906) Hollande and Favre 1931 and in the cockroach fat body. *Comparative Biochemistry and Physiology* 88B: 1023–1026.

Wu, D., Daugherty, S. C., Van Aken, S. E., et al. 2006. Metabolic complementarity and genomics of the dual bacterial symbiosis of sharpshooters. *PLoS Biology* 4: e188.

Zchori-Fein, E., Gottlieb, Y., Kelly, S. E., et al. 2001. A newly discovered bacterium associated with parthenogenesis and a change in host selection behavior in parasitoid wasps. *Proceedings of the National Academy of Science of the United States of America* 98: 12555–12560.

3 Secondary Symbionts of Insects
Acetic Acid Bacteria

Elena Crotti, Elena Gonella, Irene Ricci,
Emanuela Clementi, Mauro Mandrioli,
Luciano Sacchi, Guido Favia, Alberto Alma,
Kostas Bourtzis, Ameur Cherif,
Claudio Bandi, and Daniele Daffonchio

CONTENTS

Genus	*Asaia*
Species	*bogorensis, siamensis, krungthepensis, lannaensis*
Family	Acetobacteraceae
Order	Rhodospirillales
Description year	2000, with *A. bogorensis* as the first described species
Origin of name	*Asaia* derived from Toshinobu Asai, a Japanese bacteriologist who contributed to the systematics of acetic acid bacteria.
	Bogorensis derived from Bogor, Java, Indonesia, where most of the strains of *A. bogorensis* were isolated.
Description	
Reference	Yamada et al. (2000)

A picture of Professor Toshinobu Asai taken at his retirement from the University of Tokyo in 1962. (Kindly provided by Professor Yuzo Yamada and Professor Kazuo Komagata.)

Genus	*Acetobacter*
Species	*tropicalis*
Family	Acetobacteraceae
Order	Rhodospirillales
Description year	1898 for the genus *Acetobacter*
	2000 for the species *A. tropicalis*
Origin of name	*Acetobacter* derives from the combination of the words *acetum* ("vinegar") and *bacter* ("rod"). Thus, *Acetobacter* means vinegar rod.
	Tropicalis refers to the tropical region where the first strains of *A. tropicalis* were isolated.
Description	*Acetobacter* was described by Beijerinck.
	A. tropicalis was described by Lisdiyanti et al.
Reference	Beijerinck (1898), Lisdiyanti et al. (2000)

Martinus Willem Beijerinck (1851–1931). (Source: http://it.wikipedia.org/wiki/Martinus Willem_Beijerinck.)

INTRODUCTION

Insects host a large diversity of microorganisms, and during a long time of coevolution, different symbiotic interactions of mutualistic, commensal, and parasitic natures have been established. Despite that pathogenic interactions have been initially studied more intensively, like in the case of *Bacillus thuringiensis*, the recent rebirth of endosymbiont research moved the focus away from the pathogenic relationships, emphasizing the extent to which microbes have integrated their biology with that of insects.

Insects are the most diverse group of animals on Earth, and their evolutionary success under different ecological conditions has been attributed to their remarkable adaptability to a vast array of terrestrial habitats, including those that are strongly limited or imbalanced in nutrients. This success has to be seen in the context of the versatile roles played by the microbial partners. In many cases, the host insects can specialize in nutrient-imbalanced food resources by exploiting the microbial meta-bolic potential (Feldhaar and Gross, 2009). Insects that feed on nutritionally poor diets tend to possess bacterial endosymbionts. For instance, aphids that subsist only on sugar-rich phloem sap of higher plants harbor the primary vertically transmitted *Buchnera* symbionts, which provide essential amino acids to the hosts. Plant phloem sap contains carbohydrates, but only a very limited amount of organic nitrogen com-pounds. Blood-sucking arthropods such as ticks, lice, bedbugs, reduviid bugs, and tsetse flies harbor microbes able to synthesize vitamin B, since blood is deficient in this essential compound (Dale and Moran, 2006).

The role played by the arthropod-associated microorganisms is not restricted to nutritional aspects. Development, reproduction and speciation, the defense against natural enemies, and immunity can be positively influenced by endosymbionts. As examples, the production of indole derivatives and siderophores by symbiotic insect gut bacteria has been documented to antagonize pathogenic bacteria and fungi. The extracellular hydrolytic enzyme chitinase produced by symbiotic bacteria helps to maintain a physical property (thickness) of the peritrophic membrane, which is imperative to maintain an appropriate nutrient diffusion (Indiragandhi et al., 2007). In the widely studied model system of aphid interactions with their endosymbionts, the secondary symbiont *Hamiltonella defensa* appears to play a role in protection from parasitoid wasps (Oliver et al., 2003).

Hence, a comprehensive understanding of the biology of insects requires that they must be studied in an ecological context that considers microorganisms as an integral component of the system. Knowledge of the symbiotic-associated microbiota is therefore essential and obligatory (Dillon and Dillon, 2004).

Molecular biology techniques are increasingly used in microbial ecology studies of insects to assess the diversity, richness, and structure of their associated microbial populations. Culture-dependent methods have been complemented by the introduction of molecular tools that made possible a more rigorous whole community description of the microbial symbionts associated with insects and, more generally, arthropods. Such complementary approaches allowed the identification and characterization of new, previously unrecognized taxonomic groups associated with insects.

This chapter is focused on a particular group of prokaryotes, the acetic acid bacteria (AAB), which have been increasingly detected in insects in the last few years. In particular, insects with a sugar-based diet are a new niche from which AAB have been recovered. Bees, mosquitoes, fruit flies, and sugarcane mealybugs are the major hosts of the AAB reported so far, but it is anticipated that the number of insect species harboring AAB will increase in the future. A summary of the hitherto available information about AAB recovered from the above-mentioned insects is presented, and the emerging roles of these bacteria for arthropod metabolism and biology are discussed.

GENERAL CHARACTERISICS OF AAB

AAB are Gram-negative, ellipsoidal to rod-shaped bacteria belonging to the family of Acetobacteraceae within the subclass of α-Proteobacteria. Currently the members of this family encompass 12 genera: *Acetobacter, Gluconobacter, Gluconacetobacter, Acidomonas, Asaia, Kozakia* (Kersters et al., 2006), *Swaminathania* (Loganathan and Nair, 2004), *Saccharibacter* (Jojima et al., 2004), *Neoasaia* (Yukphan et al., 2005), *Granulibacter* (Greenberg et al., 2006a), and included very recently, *Tanticharoenia* (Yukphan et al., 2008) and *Ameyamaea* (Yukphan et al., 2009). A new genus, *Commensalibacter*, has just been proposed (Roh et al., 2008). Among these, nine monotypic genera are included: *Acidomonas, Kozakia, Swaminathania, Saccharibacter, Neoasaia, Granulibacter, Tanticharoenia, Ameyamaea,* and *Commensalibacter*. The genus *Asaia* contains four species (Malimas et al., 2008; Yukphan et al., 2004; Katsura et al., 2001; Yamada et al., 2000), whereas the remaining three genera, *Acetobacter, Gluconobacter,* and *Gluconacetobacter,* comprise many species that are, still nowadays, subject to rearrangements due to (1) the difficulty of the classification below genus level, and (2) the improvements in molecular tools that can lead to a better insight in the taxonomic relations between the species (Figure 3.1).

The main feature of AAB, which gave this group its name, is the remarkable ability to oxidize ethanol into acetic acid with the accumulation of the latter in the medium. This special primary metabolism at neutral and acidic (pH 4.5) conditions distinguishes the AAB from all other bacteria. Exceptions are *Asaia, Saccharibacter,* and *Granulibacter,* which produce no, or very little, amounts of acetic acid from ethanol. Moreover, after complete oxidation of ethanol, strains of *Acetobacter, Gluconacetobacter,* and *Acidomonas* are able to oxidize acetic acid further to CO_2

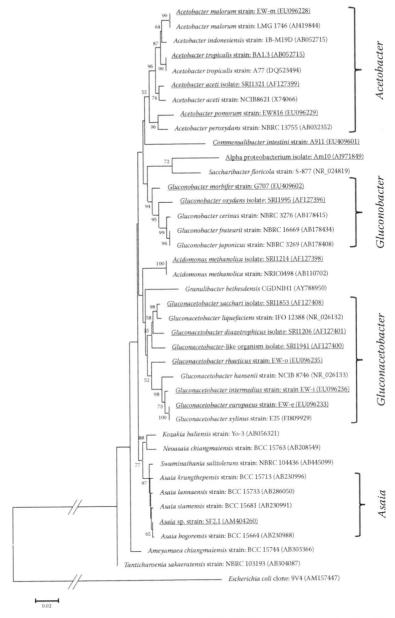

FIGURE 3.1 Neighbor-joining tree based on 16S rRNA gene sequences showing the phylogenetic positions of some representative insect-associated AAB among the most representative members of the family Acetobacteraceae. AAB recovered from insects are underlined. Accession numbers of reference sequences are between parentheses. *Escherichia coli* has been chosen as the outgroup. Numbers at nodes indicate bootstrap values as calculated on the basis of neighbor joining (Kimura two-parameter correction) expressed as percentages of 1,000 replications. Bootstrap values greater than 50% are shown at the branch points. Bar, 0.02 accumulated changes per nucleotide.

and H_2O, a process known as overoxidation of ethanol. This capability is weak in *Asaia*, *Kozakia*, *Swaminathania*, and *Granulibacter*, and absent in the other genera of the Acetobacteraceae.

AAB are characterized by a strictly respiratory metabolism in which oxygen is the final electron acceptor. Many of them can grow in the presence of 0.35% acetic acid, and gluconic acid is usually produced from glucose. They are able to oxidize a wide range of alcohols and sugars incompletely, leading to the accumulation of organic acids as end products and exhibiting resistance to high acetic acid concentrations and low pH (Kersters et al., 2006).

AAB are involved in the production of commercially important foods and chemical compounds (vinegar, kombucha tea, cocoa, sorbose, gluconic acid, etc.) due to their ability to oxidize different alcohols and sugars. However, these microbes not only play desirable roles, but also can be detrimental to foodstuffs and beverages. They often occur as spoilers of wine, beer, soft drinks, and fruits. Long before the recognition of AAB as the causative agent in vinegar fermentation, people benefited from their action, and today many new fields in biotechnology are exploiting the beneficial potential of AAB (Raspor and Goranovič, 2008).

Recently, the taxonomy of AAB has been fundamentally revised, as new techniques based on molecular tools have been introduced. The identification and classification of AAB based on the phenotypic features is time-consuming and not very accurate. The application of molecular methods offers a solution for quick and accurate identification of AAB (Cleenwerck and De Vos, 2008; Kersters et al., 2006).

Historically, the core genera of Acetobacteraceae were represented by *Acetobacter* and *Gluconobacter*. An important change in the classification of the AAB was the introduction of the genus *Gluconacetobacter* by Yamada et al. (1997, 1998) on the basis of partial sequence analysis of 16S rRNA and chemotaxonomic comparisons of ubiquinone systems. Before the middle of the 1990s, the study of AAB had mainly focused on the isolation and identification of strains from sources obtained from temperate regions (i.e., Europe and North America). The discovery of tropical sources for the isolation of novel species and genera of AAB led to the isolation and identification of a large number of AAB from fruits, flowers, and traditional fermented foods collected in tropical countries, such as Indonesia, Thailand, and the Philippines (Kersters et al., 2006). Following the investigations of new biotopes, the genera *Asaia* (Yamada et al., 2000), *Saccharibacter* (Jojima et al., 2004), *Kozakia* (Lisdiyanti et al., 2002), *Swaminathania* (Loganathan and Nair, 2004), and *Neoasaia* (Yukphan et al., 2005) have been introduced, while several new species have been described or old ones reclassified (Kersters et al., 2006). The most recent additions to the family have been the genera *Granulibacter*, isolated from patients with chronic granulomatous disease (Greenberg et al., 2006a); *Tanticharoenia*, isolated from soil collected in Thailand (Yukphan et al., 2008); *Commensalibacter*, isolated from the gut of *Drosophila melanogaster* (Roh et al., 2008); and *Ameyamaea*, isolated from flowers of red ginger collected in Thailand (Yukphan et al., 2009).

CLASSICAL ISOLATION SOURCES OF AAB
AND THE INSECT BODY NICHE

AAB are frequently found in nature and isolated from various plants, flowers, and fruits. Beer and wine, where they occur as spoilers or agents for the vinegar fermentation, are other classical niches for AAB isolation, but besides these, there are other less known niches, such as fermented food, tanning processes, and what will be the topic of this chapter, the insect body (Kersters et al., 2006; Figure 3.1).

According to our knowledge, the first insect from which members of Acetobacteraceae have been isolated is the honeybee *Apis mellifera* (White, 1921). Gluconobacters, environment sugar-loving microbes, have been isolated from this species since the beginning of the twentieth century. *Acetobacter* spp., *Gluconobacter* spp., and *Gluconacetobacter* spp. dominate *Drosophila* fruit flies microbiota, while *Anopheles* mosquitoes are able to support the growth of AAB belonging to the genus *Asaia* (Ryu et al., 2008; Cox and Gilmore, 2007; Favia et al., 2007). A sugar-rich diet is a common characteristic of the mentioned insects.

N$_2$-fixing bacteria and cellulose producers are also found in the family of Acetobacteraceae. These two metabolic capabilities are noteworthy because they are potentially important for the interaction with insect hosts. Several strains of AAB have been described as nitrogen fixing. The first one was *Acetobacter diazotrophicus* (Gillis et al., 1989; Cavalcante and Döbereiner, 1988), described in Brazil from sugarcane plant tissue, and subsequently renamed *Gluconacetobacter diazotrophicus* (Yamada et al., 1997). Later, two species associated with coffee plants in Mexico were recognized as diazotrophic bacteria and classified as *Ga. johannae* and *Ga. azotocaptans* (Fuentes-Ramírez et al., 2001). In 2007 *Ga. kombuchae* was proposed to be a novel nitrogen-fixing species in the *Gluconacetobacter* genus, able to produce cellulose even in a nitrogen-free broth (Dutta and Gachhui, 2007). Other members of Acetobacteraceae known to fix nitrogen are *Ac. peroxydans* (Muthukumarasamy et al., 2005), *Ac. nitrogenifigens* (Dutta and Gachhui, 2006), and *Swaminathania salitolerans* (Loganathan and Nair, 2004) associated with rice plants, Kombucha tea, and salt-tolerant, mangrove-associated wild rice, respectively.

The possible importance of nitrogen-fixing bacteria as symbionts of arthropods, for both the growth of the arthropods and their ecosystem, through processing of carbon and nitrogen, has been suggested recently (Behar et al., 2005; Nardi et al., 2002). Many arthropods can grow on diets with extremely high carbon-to-nitrogen (C:N) ratios, implying that they do not obtain sufficient nitrogen from the diets, and suggesting that they must obtain additional nitrogen from other source(s), i.e., atmospheric nitrogen fixed by symbiotic partners (Nardi et al., 2002).

Cellulose is a polymer of β-1,4-linked glucose; the most well-known bacterial cellulose producer is *Ga. xylinus*. Cellulose produced by *Ga. xylinus* has excellent properties, such as transparency, tensile strength, fiber-binding ability, adaptability to the living body, and biodegradability (Raspor and Goranovič, 2008). Other Acetobacteraceae cellulose producer species are the already mentioned *Ga. kombuchae* (Dutta and Gachhui, 2007), *Ga. swingsii*, *Ga. rhaeticus*, *Ga. nataicola*, and some strains of *Ga. hansenii*, *Ga. europeus*, and *Ga. oboediens* (Dutta and Gachhui, 2007; Lisdiyanti et al., 2006). Cellulose has been shown to have great importance for the interaction of the bacterial producers with other prokaryotes, as well as with eukaryotic cells such as the epithelial cells in the arthropod gut. In animal pathogens, cellulose has also been shown to participate in biofilm formation, multicellular behavior, adherence to animal cells, or stress tolerance (Barak et al., 2007).

INSECT GUT AND AAB

One of the major habitats in insects for microorganisms is the digestive system, in which nutrients are degraded by both host enzymes and metabolic activities of the microbial communities (Dillon and Dillon, 2004). Beneficial symbioses in invertebrates can be with complex or simple (<10 species) microbial consortia. For instance, a complex microbiota has been found in the gut of termites, while approximately 25 phylotypes, with just a few dominant bacterial species, have been shown to be associated with the gut of the fruit fly *D. melanogaster* (Cox and Gilmore, 2007).

The gut is typically influenced by structural and physiological factors and, in addition, by the quality of the ingested food. In view of the diversity of insects and the absence of detailed information about many arthropod genera, it is impossible to generalize the location and characteristics of the gut indigenous strains. Here, some properties of the guts of those insects from which AAB have been recovered will be illustrated.

Anatomically, the gastrointestinal (GI) tracts of fruit flies, bees, and mosquitoes have an overall comparable organization. These digestive systems possess a single alimentary canal beginning at the esophagus, connecting to a ventriculus (stomach), extending to the intestine, and then proceeding to the rectum and terminating at the anus. *Drosophila* possesses also an acidic crop, maintaining at the same moment a highly alkaline ventriculus and a neutral to acidic hindgut. In contrast, the mosquitoes have three diverticula that arise near the posterior end of the esophagus, one from the ventral and two from the dorsolateral wall of the gut, all of them surrounded by a thin impermeable cuticle. The ventral diverticulum, also called crop, is acidic and large and can extend into the abdomen. A sugar meal, such as floral nectar, is stored in the diverticula, from which it then passes slowly to the midgut, where the digestion occurs (Dapples and Lea, 1974). The bee gut is reported to be always acidic in comparison to the larval one that has been reported to vary from 4.8 to 7.0, depending on the pH of the pollen fed to the bee colonies. Moreover, the bee food source, as nectar and honey, has a low pH of approximately 4 (Mohr and Tebbe, 2006).

The gut of adult insects is more structured than the gut of larvae, and the anterior hindgut (ileum and colon) is in general the region most densely populated by microbes receiving the excretory products from the Malpighian tubules, as well as predigested food from the midgut (Nardi et al., 2002). The bacterial cell densities have been reported to be in the range of 10^8–10^9 cells per g of gut content in the case of bees (Mohr and Tebbe, 2006), while over 10^6 bacteria are typically associated with an old adult individual of *Drosophila* according to Ren et al. (2007).

Another feature to bear in mind is that the nature of the open circulatory system of these insects may account for the lack of anoxic environments in the GI tract (Cox and Gilmore, 2007; Bodenstein et al., 1950). Studies on the characterization of the microbiota associated with *Drosophila* or *Apis* spp. confirmed the absence of obligate anaerobic phylotypes (Cox and Gilmore, 2007; Mohr and Tebbe, 2006).

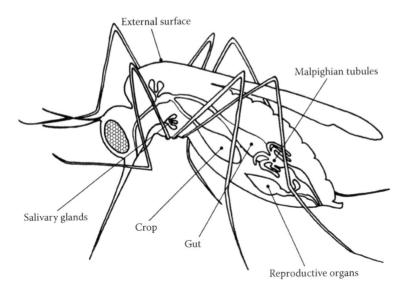

FIGURE 3.2 AAB symbionts have been shown to inhabit primarily the insect digestive system. Other insect tissues and organs, such as the salivary glands, the Malpighian tubule, and the reproductive organs, have been reported to host AAB. Moreover, AAB have been detected on the insect external surface.

The aerobic environment, acidic pH, and presence of diet-derived sugars in the insect gut are conditions that enable AAB to thrive in the insect gut niche, as demonstrated by the recovery of these bacteria from the digestive system of several insects (Figure 3.2). Bees, fruit flies, mosquitoes, and leafhoppers rely on a sugar-based diet, such as nectars, sugars derived from fruits, and phloematic sap.

INSECT SPECIES INHABITED BY AAB

To date, AAB have been found associated with insect species belonging to Hymenoptera, Diptera, and Hemiptera (Table 3.1).

Bees visit many environmental niches for feeding, thereby coming into intimate contact with many microorganisms. The microbiota associated with these hymenopterans, especially honeybees, as well as the microorganisms present within their food and the products obtained by their breeding, such as honey, wax, or royal jelly, has been the subject of several studies. Over 6,000 microbial strains associated with bees and their food were isolated and identified before 1997 (Gilliam, 1997), and nowadays, with the introduction of molecular tools, the number of the identified phylotypes is increasing.

Five phylogenetic groups contribute almost equally to the total diversity, i.e., the α-, β-, and γ-Proteobacteria, Firmicutes, and Bacteroidetes; in addition, yeasts and molds and some Actinobacteria have been found associated with these insects (Mohr and Tebbe, 2006). Among these, the sugar-loving and flower-associated gluconobacters are one of the predominant bacterial groups in bees. The first detection of AAB

TABLE 3.1
AAB Symbionts of Different Insect Species

In the table are listed insect species that host AAB, together with their taxonomic affiliation, the name(s) of AAB that they host, the developmental stage of the analyzed insects, the localization of AAB, and when available, the population abundance of AAB in the different insect samples.

Insect Species	Insect Taxonomic Affiliation	AAB	Developmental Stage[a]	Localization	Population Abundance[b]	References
Apis mellifera	Hymenoptera: Apidae	Acetobacter sp.	Lp	Whole tube-like gut		Mohr and Tebbe, 2006
			A	Intestine (midgut and hindgut)		Babendreier et al., 2007
		Gluconobacter sp.	A	Intestine (midgut and hindgut)		Babendreier et al., 2007
			L	Gut		Mohr and Tebbe, 2007
		Gluconacetobacter sp.	A, E	Abdomen (in the case of adults); pools of eggs		Jeyaprakash et al., 2003
			A	Hindgut		Mohr and Tebbe, 2006
			A	Intestine (midgut and hindgut)		Babendreier et al., 2007
		Saccharibacter floricola	L	Gut		Mohr and Tebbe, 2007
Marietta leopardiana	Hymenoptera: Aphelinidae	Asaia sp.	A	Whole insect		Matalon et al., 2007
Asobara tabida	Hymenoptera: Braconidae	Acetobacter sp.	A	Whole insect		Zouache et al., 2009

Host	Taxonomy	Symbiont	Stage	Tissue	Abundance	Reference
Drosophila melanogaster	Diptera: Drosophilidae	Acidomonas methanolica	A	Whole insect		Zouache et al., 2009
		Acetobacter sp.	A	Whole insect		Corby-Harris et al., 2007
			A	Whole insect	Clones in the libraries: 29%	Cox and Gilmore, 2007
			A	Whole insect		Ren et al., 2007
			A	Posterior midgut	A. pomorum: ~3.3 × 10^4 CFUs per gut	Ryu et al., 2008
		Commensalibacter intestini	A	Posterior midgut	~1.4 × 10^5 CFUs per gut	Ryu et al., 2008
		Gluconobacter sp.	A	Whole insect		Corby-Harris et al., 2007
			A	Whole insect	Clones in the libraries: ~5%	Cox and Gilmore, 2007
			A	Whole insect		Ren et al., 2007
			A	Posterior midgut	G. morbifer: ~800 CFUs per gut	Ryu et al., 2008
		Gluconacetobacter sp.	A	Whole insect		Corby-Harris et al., 2007
			A	Whole insect	Clones in the libraries: ~0.15%	Cox and Gilmore, 2007
			A	Posterior midgut		Ryu et al., 2008
Bactrocera oleae	Diptera: Tephritidae	Acetobacter sp.	A, P, L	Whole insect	A. tropicalis; clones in the libraries: 43.8% (males), 16.9% (females)	Kounatidis et al., 2009

(continued)

TABLE 3.1 (continued)
AAB Symbionts of Different Insect Species

Insect Species	Insect Taxonomic Affiliation	AAB	Developmental Stage[a]	Localization	Population Abundance[b]	References
Anopheles mosquitoes	Diptera: Culicidae	Asaia sp.	A	Gut, salivary glands, female reproductive organs, male reproductive organs (observed also in recolonization experiments)	ABR[c]: 41% (gut), 25% (salivary glands), 20% (female reproductive organs) Clones in the libraries: 90% (An. stephensi), 20% (An. maculipennis), 5% (An. gambiae) MPN counts: 9.8×10^5 CFUs per female, 9.8×10^4 CFUs per male	Favia et al., 2007
			A, L	Midgut, ventral diverticulum, salivary glands (observed in recolonization experiments)		Damiani et al., 2008
			A	Gut, salivary glands, female reproductive organs, male reproductive organs (observed in recolonization experiments)		Crotti et al., 2009

Insect	Order: Family	Symbiont	Stage[a]	Tissue		Reference
Aedes aegypti	Diptera: Culicidae		A, P, L, E	Gut, salivary glands, female reproductive organs, male reproductive organs		Damiani et al., 2010
		Asaia sp.	A, P, L	Whole insect, gut, male reproductive organs, female reproductive organs, salivary glands		Crotti et al., 2009
Saccharicoccus sacchari	Homoptera: Pseudococcidae	Acetobacter sp.	A ♀, J ♀	Whole insect		Ashbolt et al., 1990
			A ♀, J ♀	Whole insect		Franke et al., 1999
		Gluconobacter sp.	A ♀, J ♀	Whole insect		Ashbolt et al., 1990
			A ♀, J ♀	Whole insect		Franke et al., 1999
		Gluconacetobacter sp.	A ♀, J ♀	Whole insect		Ashbolt et al., 1990
			A ♀, J ♀	Whole insect		Franke et al., 1999
			A ♀, J ♀	Whole insect		Franke et al., 2000
			A ♀, J ♀	Whole insect		Franke-Whittle et al., 2005
Scaphoideus titanus	Hemiptera: Cicadellidae	Asaia sp.	A	Whole insect		Marzorati et al., 2006
Pieris rapae	Lepidoptera: Pieridae	Asaia sp.	A	Whole insect	ABR: 4.9%	Crotti et al., 2009
			L	Midgut	Clones in the libraries: 5%	Robinson et al., 2010

[a] A, adults; Lp, larvae close to pupation; L, larvae; J, juveniles; E, eggs.

[b] Where available.

[c] ABR, Asaia to bacteria 16S rRNA gene copy ratio.

in honeybees and ripening honey by Ruiz-Argueso and Rogriguez-Navarro dates to the early 1970s (Ruiz-Argueso and Rodriguez-Navarro, 1973, 1975). They mainly isolated two groups of bacteria, classified as *Gluconobacter* and *Lactobacillus*. *Gluconobacter* reached numbers ranging from 10^3 up to 10^6 cells per bee, being equally distributed between the intestine and the rest of the body, with the midgut as the main reservoir within the alimentary tract. Microbial counts of the nectar extracted from the bee stomach have been performed for *Gluconobacter*, revealing that bacteria isolated from the nectar extracted from bee stomach were more numerous than those isolated from honeycomb: an average of 10^5 cells of *Gluconobacter* in stomach were found vs. 3×10^3 in honeycomb samples with the highest moisture. In 1981, Lambert and coworkers investigated with cultivation-dependent methods the presence of *Gluconobacter* in bees (*Apis mellifera* subsp. *mellifera*), flowers visited by bees (*Solidago canadensis*, *Heleniurn* sp., *Campanula patula*), honey, and other materials from the hive, including in the study also some wasps (Lambert et al., 1981). The samples were collected at different locations in northern Belgium. Fifty-six strains of *Gluconobacter* spp. were isolated, and one isolate classified as *Acetobacter* was found in a wasp. On the basis of the different electrophoretic mobility of the soluble cell proteins, the *Gluconobacter* isolates belonged to two types, one of which was more frequently found in the sampled hives.

More recently, the characterization of the microbial community associated with bees has been analyzed with cultivation-independent techniques. In 2003, Jeyaprakash and colleagues analyzed the gut bacterial community from adult workers of the honeybee *Apis mellifera*, subspp. *capensis* and *scutellata*, collected in South Africa (Jeyaprakash et al., 2003). These authors found, shared between both *mellifera* subspecies, 10 unique 16S rRNA gene sequences that defined 6 bacterial genera. Among these, one sequence clustered with a sequence from the α-Proteobacterium *Gluconacetobacter sacchari*, previously identified as a gut symbiont located in the cecal bacteriocytes of the pink sugarcane mealybug (Franke et al., 2000). In 2006, Mohr and Tebbe studied the structural diversity and variability of bacteria in the gut of three bee species (the honeybee *Apis mellifera* subsp. *carnica*, the bumble bee *Bombus terrestris*, and the red mason bee *Osmia bicornis*) sampled at a flowering oilseed rape field in Germany. By genetic profiling with single-strand conformation polymorphism (SSCP) of polymerase chain reaction (PCR)-amplified partial (370 bp) 16S rRNA genes, among the quantitatively most important groups, a phylotype clustering with an uncultured *Gluconacetobacter* with a similarity higher than 97% represented 6% of all cloned 16S rRNA sequences (Mohr and Tebbe, 2006). Two other phylotypes had a similarity higher than 95% to 16S rRNA gene sequences of *Acetobacter* spp. and other *Gluconacetobacter* spp. These AAB were detected only in *Ap. mellifera* larvae and adults. In an additional study, the same authors isolated 96 different bacterial strains from the same three bee species reported above (Mohr and Tebbe, 2007). Six AAB detected with molecular ecology methods were all found in *Ap. mellifera*. Several of them were closely related to *Saccharibacter floricola*, an osmophilic bacterium previously isolated from pollen (Jojima et al., 2004), or to *Gluconobacter oxydans*. The bacterial community structure in the intestine of *Ap. mellifera* subsp. *mellifera* honeybees reared in the Zurich area or collected in the field at two locations in Switzerland has also been evaluated by Babendreier and collaborators (Babendreier et al., 2007). PCR amplification of bacterial 16S rRNA gene fragments and terminal restriction fragment length polymorphism analyses revealed a total of 17 distinct terminal restriction fragments (T-RFs), which were highly consistent between laboratory-reared and free-flying honeybees. One of these T-RFs contained clones that clustered with sequences of *Acetobacter* spp. and *Gluconobacter* spp., as well as an uncultured clone identified as *Gluconacetobacter*, related to a clone previously identified from the honeybee intestinal bacteria by Jeyaprakash et al. (2003). Moreover, recently the presence of *Acetobacter pasteurianus* and *Acidomonas methanolica* has been reported in the endoparasitoid wasp *Asobara tabida* (Zouache et al., 2009).

D. *melanogaster* is one of the most studied organisms, and AAB have been reported by several authors to inhabit the fruit fly body. Cox and Gilmore (2007) analyzed the bacterial community associated with laboratory-reared and wild-captured *D. melanogaster*. The analysis of 16S rRNA gene libraries, performed on pools of flies, showed that one of the most abundant genera was *Acetobacter*, with 29% of all the analyzed clones. The following species were identified: *Acetobacter aceti*, *Ac. cerevisiae*, *Ac. pasteurianus*, *Ac. pomorum*, and *Ac. peroxydans*, together with several species of *Gluconobacter* and *Gluconacetobacter*. By analyzing the degree of overlap of three 16S rRNA gene libraries established from three pools of individuals, one from wild animals and two from laboratory-reared individuals, three phylotypes were found in all three libraries. Of these, two were AAB, *Ac. aceti* and *Ac. pasteurianus* (Cox and Gilmore, 2007). Similar results were obtained, using a sequence-based approach, on the bacterial composition and diversity among 11 natural populations of *D. melanogaster* collected across a latitudinal cline on the East Coast of the United States. Also in this study, the most abundant group of bacteria fell into the α-Proteobacteria subclass (17.2%), accounting for 15 operational taxonomic units (OTUs). Twelve of these OTUs were sequences related to AAB, and many of them grouped closely within the *Gluconacetobacter* genus (Corby-Harris et al., 2007).

In another study AAB were identified not only from the interior, but also from the surface of *Drosophila* flies (Figure 3.2). The authors reported the presence of several *Acetobacter* species (*Ac. aceti*, *Ac. tropicalis*, and *Ac. pasteurianus*) by cultivation-dependent and -independent methods; moreover, by the use of a scanning electron microscope (SEM), they observed on the insect surface the presence of acetic acid bacterial-like cells (Ren et al., 2007).

Another dipteran shown to host an AAB symbiont is the olive fruit fly *Bactrocera oleae*, one of the major pests of the olive tree, which strongly affects olive production worldwide. The nature of the olive fly-associated microbial community is controversial and debated, due firstly to the few studies performed on its microbiota. Since the beginning of the last century *Pseudomonas savastanoi* has been suspected to be a mutualist of *B. oleae* (Petri, 1909), and in following studies by the use of both traditional microbiological approaches and cultivation-independent techniques, various species belonging to the different phyla Firmicutes, Actinobacteria, and Proteobacteria have been reported to be associated with the olive fruit fly. Their identification was not always consistent among the different studies. Recently, Kounatidis et al. (2009), by analyzing the microbiota associated with *B. oleae* using approaches that combined cultivation-dependent and -independent techniques, together with ultrastructure and microscopic analyses, recognized the acetic acid bacterium *Acetobacter tropicalis* as a dominant bacterium within the microbiota of *B. oleae*. Using an *Ac. tropicalis*-specific PCR assay, the symbiont has been detected in all laboratory and field-collected individuals, originating from different locations in Greece. This acetic acid bacterium has been successfully isolated in culture. Typing analyses, carried out on a collection of isolates, have revealed that multiple *Ac. tropicalis* strains are present per fly. Its capability to colonize and lodge in the gut system of both larvae and adults was demonstrated using an *Ac. tropicalis* strain labeled with a green fluorescent protein (Gfp). Interestingly, the Gfp-labeled *Ac. tropicalis* cells were found to be restricted to a dense matrix of brown color, probably the peritrophic membrane, within the gut.

The pink sugarcane mealybug *Saccharicoccus sacchari* (Homoptera: Pseudococcidae) is an arthropod found in all sugarcane-growing countries. It is almost the only species of mealybug found on commercially grown sugarcane in Australia. Ashbolt and Inkerman (1990), studying the acetic acid bacterial biota of this mealybug, found that the insect supports the growth of members of the Acetobacteraceae. Strains were isolated from all stages of the life cycle of *S. sacchari* on sugarcane, both above and below ground (eggs were not examined, as they hatch within a few minutes of oviposition). An adult female mealybug actively feeding on aerial storage tissue typically carried at least 10^6 acetic acid bacteria, whereas less active adults as well as individuals from the underground mealybug population generally maintained fewer than 10^4 bacteria per insect. Unidentified AAB were the predominant isolates from the pink sugarcane mealybug and surrounding tissues when cultured at pH 3. The presence of AAB has also been demonstrated for other mealybugs, such as *Planococcus* sp. and *Dysmicoccus brevis*. Franke et al. (1999) identified the mealybug AAB isolates as belonging to the species *Ac. aceti*, *Gluconacetobacter diazotrophicus*, and *Ga. liquefaciens*, and also identified a new species, *Gluconacetobacter sacchari*. The authors came to the conclusion that these AAB species represent only a relatively small proportion of the microbial community associated with the insect (Franke-Whittle et al., 2004, 2005; Franke et al., 2000).

Among Diptera, mosquitoes are one of the major studied insect groups due to their importance for public health, being vectors of viruses, parasites, and other microorganisms causing human diseases. Nevertheless, only recently the microbiota associated with these arthropods have become objects of investigation (Pumpuni et al., 1996, Khampang et al., 1999, Luxananil et al., 2001, Pidiyar et al., 2002, 2004; Gonzalez-Ceron et al., 2003; Lindh et al., 2005; Gusmão et al., 2007).

The microbiota associated with *An. stephensi* (Diptera: Culicidae) has been studied by Favia et al. (2007), who reported a stable association with acetic acid bacteria of the genus *Asaia* with both the juvenile stages and adults of this mosquito. The dominance of *Asaia* within the mosquito microbiota was proven by clone prevalence in 16S rRNA gene libraries obtained from the insect metagenome, quantitative PCR, transmission electron microscopy and *in situ* hybridization with specific probes targeting the 16S rRNA gene (Figure 3.3). 16S rRNA gene clones of *Asaia* sp. represented 90, 20, and 5% of the total clones, respectively, in laboratory-reared *An. stephensi*, field-collected *An. maculipennis* (central Italy), and *An. gambiae* (Burkina Faso). In *An. stephensi*, *Asaia*

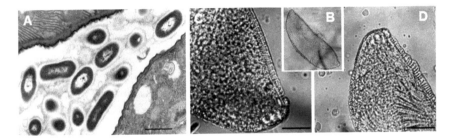

FIGURE 3.3 See color insert. (A) TEM micrograph of *Asaia* sp. able to colonize the gut epithelium of *An. stephensi*. Bar, 1.2 μm. (B–D) High magnification of an *An. gambiae* egg showing high concentration of Cy3-labeled *Asaia* (in yellow). Phase contrast (B) of an egg and contrast/fluorescence merge images (C,D) of the egg apical portions. Bars in (C) and (D), 25 μm.

sp. amounted to 9.8×10^5 colony forming units (CFUs) per female and 9.8×10^4 CFUs per male individuals. The capability of *Asaia* sp. to colonize the body of *An. stephensi* has been demonstrated, using an *Asaia* strain labeled with a green fluorescent protein. Colonization experiments of the mosquito body with Gfp-tagged *Asaia* sp. provided with a sugar-based diet indicated that this bacterial symbiont is able to efficiently lodge in the gut, salivary glands, and male and female reproductive organs (Favia et al., 2007). Similar results have been obtained for *An. gambiae*: *Asaia* has been detected in the gut, the salivary glands, and the reproductive organs of this mosquito species. Using fluorescent strains of *Asaia*, it was shown that the bacterium promptly colonizes organs that are of interest in connection with *Plasmodium* development, like the gut and salivary glands, or in connection with symbiont transmission, such as the reproductive organs (Damiani et al., 2010).

Asaia was also identified in *Aedes aegypti* and *Aedes albopictus* (Chouaia et al., 2010). It was shown that in both *Anopheles* and *Aedes* different strains can coexist in the same individual (Chouaia et al., 2010).

Asaia has been also identified in the hymenopteran *Marietta leopardina*, a hyperparasitoid (Matalon et al., 2007), and in the lepidopteron *Pieris rapae*, the cabbage white butterfly (Robinson et al., 2009), by culture-independent methods.

While investigating the bacterial community associated with the leafhopper *Scaphoideus titanus* (Hemiptera, Cicadellidae), the vector of a phytoplasma causing Flavescence Dorèe in grapes, Marzorati and colleagues (2006) noted the presence of the acetic acid bacterium *Asaia*. *Asaia* had been previously isolated from wine grapes from several Australian vineyards (Bae et al., 2006), suggesting that the bacterium can be acquired from the environment. Its presence in *S. titanus* has been confirmed by Crotti et al. (2009). Using quantitative real-time PCR with *Asaia*-specific and universal primers for bacteria, the average *Asaia*-to-bacteria ratio of the 16S rRNA genes has been determined, indicating that *Asaia* 16S rRNA gene copies constituted 4.9% of the total bacterial 16S rRNA gene copies per leafhopper. *In situ* hybridization (ISH) with *Asaia*-specific probes showed that *Asaia* is harbored in testicles, intermixed within the spermatic bundles and in the Malpighian tubules associated to brochosomes.

Fluorescent ISH (FISH) showed that *Asaia* massively colonizes the surface of ovarian eggs, forming thick biofilms.

Supporting evidence for acetic acid bacteria as symbionts of leafhoppers comes from the aforementioned work of Ashbolt and Inkerman (1990), in which AAB associated with mealybugs have been analyzed. This study also included leafhoppers such as *Perkinsiella saccharidica*, in which an acetic bacterial community of 5×10^3 cells per insect has been reported. The species or genera of the AAB were not reported.

INTERACTION OF AAB WITH THE INSECT IMMUNE SYSTEM

Recently AAB have been implicated in the regulation of the immune system homeostasis of several insect hosts, i.e., *D. melanogaster* and *An. gambiae*.

Ryu et al. (2008) studied, in wild type *Drosophila*, the relationship between antimicrobial peptide production by the insect host and the resident intestinal microbiota. Five commensal species dominated, among which *Ac. pomorum*, *Gluconobacter* sp. strain EW707, and Acetobacteraceae EW911 were found to be associated with the midgut of the fruit fly by culture-independent techniques. Other AAB considered as minor commensals, by quantitative PCR, were identified as *Gluconacetobacter rhaeticus*, *Ga. intermedius*, *Ga. europaeus*, and *Ac. malorum*. The relative abundance of the major AAB associated with the *D. melanogaster* was 5×10^4, 1.4×10^5, and approximately 800 CFUs per gut for *Acetobacter pomorum*, strain EW911, and *Gluconobacter* sp. strain EW707, respectively. Notably, a delicate equilibrium was found between the gut indigenous bacteria and the fly immune system. The normal commensal community structure maintains the pathogenic *Gluconobacter* sp. strain EW707 at a relatively low level, whereas an increase in the number of this microbe leads to gut apoptosis. Recently, strain EW911 and *Gluconobacter* sp. strain EW707 have been reclassified as *Commensalibacter intestini* and *Gluconobacter morbifer*, respectively (Roh et al., 2008).

In the *An. gambiae* Keele strain, silencing of a gene involved in the mosquito innate immune response (AgDscam) by RNA interference (iRNA) allows proliferation of bacteria of the species *Asaia bogorensis* in the hemolymph (Dong et al., 2006). Abundance measurements of this microorganism within the insect hemolymph showed up to 10^4 CFUs μl^{-1} of this AAB in the hemolymph of mosquitoes with silenced AgDscam, while in control mosquitoes low numbers of *Asaia* were counted (Dong et al., 2006). In an uncompromised host *Asaia* has been shown to have the mosquito gut as a primary niche, being able to line the midgut epithelium and being embedded in a gelatinous matrix (Favia et al., 2007). AgDscam immune reaction contains the bacterium in the midgut, and the depletion of this essential hypervariable receptor could allow *Asaia* to initiate a translocation process across the midgut epithelium and to invade the hemolymph where a massive proliferation occurs.

Concluding, within the insect body, not only the pathogens have to face the insect innate immune system, but also beneficial and commensal symbionts, like AAB. In the case of the aforementioned study about *Drosophila* and its acetic acid bacterial symbionts (Ryu et al., 2008), AAB have been shown to modulate the gut homeostasis of the fly, interacting with the innate immune system of the host. A lack of this homeostasis could lead to gut apoptosis in the host. This suggests that AAB could be tools for interfering with the immune response and hence the fitness of the host. Thus, an in-depth investigation of the relationships established between AAB and the innate immune system is required.

TRANSMISSION ROUTES OF AAB

Favia et al. (2007) provided the first description of the transmission routes followed by the acetic acid bacterium *Asaia* in *Anopheles* mosquitoes. The *Anopheles-Asaia* system has been seen as a borderline situation where acquisition from the environment was likely the most common source of infection for both preadult and adult stages, but where transmission from mother to offspring and from male to female mosquitoes can also occur. However, in this case a clear-cut distinction between environmental acquisition and vertical transmission of the symbiont cannot be established. The situation is similar to that of microorganisms living in the gut of wood-feeding cockroaches and termites where vertical and horizontal transmissions coexist, and their relative importance is not clear yet (Nalepa et al., 2001). Recently, it has been shown that a paternal transmission route of *Asaia* to the progeny occurs in *An. stephensi* (Damiani et al., 2008). This mode of transmission is uncommon in arthropods, and these results represent the first unambiguous demonstration of paternal transmission of a bacterial symbiont in mosquitoes. Previously, this mode of transmission had been only demonstrated for beneficial symbionts in aphids (Moran and Dunbar, 2006).

Evidence of a transstadial transfer of the symbiotic bacterium *Asaia* from larvae to pupae and from pupae to adults of *An. stephensi* has also been obtained. For a long time, such bacterial transfers in mosquitoes have been a controversial issue (Lindh et al., 2008; Moll et al., 2001; Pumpuni et al., 1996; Jadin et al., 1966). Moreover, experiments using molecular tools presented by Damiani et al. (2008) showed a lack of detection of *Asaia* symbionts in the larval breeding water, thus strengthening the explanation of a transfer of the symbionts from one mosquito stage to the next one. In contrast, a recent study by Lindh et al. (2008) demonstrated the transmission of the symbiont *Pantoea stewartii* from larvae to pupae, but not from pupae to adults.

The mechanism by which *Asaia* is proposed to be transmitted to the offspring is the egg-smearing one, according to which the symbiotic bacteria are smeared by the mother on the surface of the eggs. This transmission process has been recently described for *Asaia* symbionts of *An. gambiae* and *S. titanus* (Damiani et al., 2010; Crotti et al., 2009).

The horizontal transmission of *Asaia* symbionts among insects that belong to phylogenetically distant orders indicates another important ecological feature of AAB symbionts of insects (Crotti et al., 2009). *Asaia* has been shown to be present and capable of cross-colonizing other sugar-feeding insects of phylogenetically distant genera and orders besides *Anopheles* spp. PCR, real-time PCR, and *in situ* hybridization experiments showed *Asaia* in the body of the mosquito *Aedes aegypti* and the leafhopper *Scaphoideus titanus* (Marzorati et al., 2006), respectively vectors of human viruses and grapevine phytoplasma. Cross-colonization patterns of the body of *Ae. aegypti*, *An. stephensi*, and *S. titanus* have been documented with *Asaia* strains isolated from *An. stephensi* or *Ae. aegypti*, and labeled with plasmid- or chromosome-encoded fluorescent proteins.

ARE AAB VIRULENT IN HUMANS?

Bacteria in the Acetobacteraceae family are not common human pathogens, although in the last few years some AAB have been reported as potential causes of bacteremia. At present, five AAB species have been reported in the medical literature as possible responsible agents of infections: *Asaia bogorensis*, *Asaia lannaensis*, *Granulibacter bethesdensis*, *Acetobacter cibinogensis*, and *Ac. indonesiensis* (Abdel-Haq et al., 2009; Bittar et al., 2008; Rodríguez López et al., 2008; Gouby et al., 2007; Tuuminen et al., 2006, 2007; Greenberg et al., 2006a, 2006b; Snyder et al., 2004). Because AAB have never been isolated from the human microbiota, the source of human contamination by AAB remains unknown. Many unanswered questions still remain, like the origin of the infectious AAB and if they act as primary or secondary pathogens (Fredricks and Ramakrishnan, 2006). The administration of antibiotics or drug abuse may significantly contribute to the infection potential of AAB. These compounds might destabilize the human microbiota by eliminating susceptible commensal species, opening new niches for colonization by other species. A similar explanation has been suggested for the disturbance of the highly evolved consortium associated with human gastrointestinal tract (Gilmore and Ferretti, 2003). Not surprising in this context is the high resistance to different classes of antibiotic shown by AAB strains isolated from the mentioned patients.

The current literature suggests no strong relationship between AAB and human diseases, since all the cases of infections reported so far have been related to patients with a compromised immune system, making them susceptible to bacteria that would otherwise have low pathogenicity.

SEQUENCED GENOMES OF AAB, AND THE METAGENOME OF INSECT HOST SYMBIONTS

Since ancient times of human history, AAB have been exploited by man, and still nowadays they are used in many applications even on an industrial scale (e.g., production of vinegar and cocoa, or industrial production of polysaccharides). AAB have a unique metabolism, governed by specialized enzymes, which allow their use in various fields of biotechnology, such as food production, pharmaceutics, medicinal biotechnology, biotransformations, biosensorics, alternative energy source production, fine chemicals production, etc. (Raspor and Goranovič, 2008).

Much of the knowledge needed for a proper understanding of the processes catalyzed by AAB remains unavailable. However, gaining access to AAB genomes could really boost our understanding of AAB and their interactions with the insect host. For instance, genome sequencing has been successfully used in order to understand the interactions existing between some insects and their primary symbionts. As in the case of *Candidatus* Sulcia muelleri and *Candidatus* Baumania cicadellinicola, a metabolic interdependence has been recently deduced from the genome organization

(McCutcheon and Moran, 2007). Moreover, genome sequencing of primary symbionts made it possible to understand that the genome degradation and reduction reflect a strict adaptation to the constant environment supplied from the insect host. Examples include three genomes of *Buchnera aphidicola* strains from different aphid hosts, two of *Candidatus* Blochmannia species from ants, one of *Wigglesworthia glossinidia* from tsetse flies, and one each of *Candidatus* Baumannia cicadellinicola and *Candidatus* Sulcia muelleri from leafhoppers (McCutcheon and Moran, 2007; Dale and Moran, 2006). More recently, complete or partial genome sequences have been obtained from a number of facultative symbionts, which appear to have established associations with hosts in the more recent evolutionary past.

In the case of AAB, to date the complete genomes of *Gluconobacter oxydans* (Prust et al., 2005), *Granulibacter bethesdensis* (Greenberg et al., 2007), *Gluconacetobacter diazotrophicus* (Bertalan et al., 2009), and *Acetobacter pasteurianus* NBRC 3283 (Azuma et al., 2009) have been recently published. Moreover, noteworthy is that the symbiont role(s) in a host is generally investigated by the administration of antibiotics to the host, thus understanding which host biological features are affected by the removal of the symbiont. In the case of AAB, attempts to understand their role by the administration of antibiotics pose several problems like the difficulty to effectively remove environmental bacteria and a rapid reacquisition of AAB after the treatment. Genome sequencing of AAB could allow achieving clues about their role(s) as symbionts of sugar-feeding insects.

SYMBIOTIC CONTROL APPROACH EXPLOITING AAB

Knowing the microbiota associated with disease-carrying vectors, such as *Anopheles* mosquitoes, or insect pests for a crop, such as *B. oleae*, is a prerequisite for the development of strategies based on symbiotic control. The symbiotic control approach could utilize bacteria capable of colonizing the insect body to produce effector molecules (natural or transgenic in the paratransgenic mode) that kill or inhibit the causative agent of the disease or interfere with the survival of parasitic insects (Beard et al., 2001). Considering the localization in the insect body, the capability of colonizing very different hosts, the culturability, and the genetic transformability, AAB are potentially interesting agents for natural or paratransgenic symbiotic control. Nevertheless, it is worth mentioning that in the development of a symbiotic control strategy, multiple transmission routes and cross-colonizing capability are potential biosafety risks, but also are advantageous for a successful disease transmission control.

The ability of *Asaia* to efficiently cross-colonize insects that belong to distantly related orders suggests that it has evolved mechanisms for the colonization of the insect body independently from the evolutionary history of the host, and that it is able to survive different immune systems. Such a feature, besides having interesting implications for understanding how bacteria tackle and escape the response of the immune system during host invasion, represents a key point for the development of an approach of symbiotic control of vector-borne pathogens. The symbiotic bacterium that should be used for controlling a pathogen in the insect vector should be capable of quickly and competitively colonizing different insect hosts, thus increasing the chances to become fixed in natural populations of the vector.

OPEN QUESTIONS

Despite the fact that AAB are frequently detected in different insects, such as fruit flies, bees, and mosquitoes, in some of which they represent a significant part of the insect-associated bacterial community, and thus likely play important roles for the insect hosts, there is a lack of in-depth studies about the nature of the symbiosis established between AAB and their hosts. First, a better knowledge of the community structure and the AAB ecology in the insect hosts is required. AAB abundance and prevalence data are not available for all the investigated insects. Also, the precise localization of these symbionts in the insect body and tissues is unknown in most cases. Investigations of the biodiversity of these strains and the most likely different colonization capabilities are lacking. Another interesting and not well-understood feature is the transmission route followed by the AAB symbionts. Many different modes of transmission of insect symbionts have been described so far, including the intracellular vertical (maternal or paternal) and lateral ones, as well as other mechanisms, like the superficial bacterial contamination of eggs (egg smearing), the probing of the mother's excrement (coprophagy), the deposition of bacterium-containing capsules with the eggs (capsule transmission), or the environmental acquisition (Kikuchi et al., 2005).

Other unanswered questions concern the interactions between the AAB symbionts and the hosts, as well as the determination of the physiological role(s) exerted by these γ-Proteobacteria. Some of the important questions in this context are related to the interaction of AAB with the host immune system, the mode of adhesion to the host epithelia, and the balance that these bacteria establish with other microorganisms living in the insect host, including eventual primary or secondary symbionts. Considering the peculiar localization of AAB in mosquitoes, for instance, in male and female gonoducts (Favia et al., 2007, 2008), it would be important to understand which factors allow the colonization of these organs.

ACKNOWLEDGMENTS

We thank for financial support the European Union in the ambit of project BIODESERT (European Community's Seventh Framework Programme CSA-SA REGPOT-2008–2 under grant agreement 245746) and the Italian Ministry for Research (MIUR) in the ambit of project PRIN 2007 (Caratterizzazione del microbiota associato a *Scaphoideus titanus* e *Hyalesthes obsoletus*, cicaline vettrici di fitoplasmi nella vite ed isolamento e studio della localizzazione di batteri acetici simbionti). Kostas Bourtzis is grateful to the European Union (European Community's Seventh Framework Programme CSA-SA_REGPOT-2007–1 under grant agreement 203590), the International Atomic Energy Agency, and the University of Ioannina, which have supported the research from his laboratory. C.B., K.B., A.C., E.Cr., and D.D. benefited from travel grants from Cost Action FA0701: *Arthropod Symbiosis: From Fundamental Studies to Pest and Disease Management.*

REFERENCES

Abdel-Haq, N., S. Savaşan, M. Davis, B. I. Asmar, T. Painter, and H. Salimnia. 2009. *Asaia lannaensis* bloodstream infection in a child with cancer and bone marrow transplantation. *J. Med. Microbiol.* 58:974–976.

Ashbolt, N. J., and P. A. Inkerman. 1990. Acetic acid bacteria biota of the pink sugar cane mealybug, *Saccharococcus sacchari*, and its environs. *Appl. Environ. Microbiol.* 56:707–712.

Azuma, Y., A. Hosoyama, M. Matsutani, et al. 2009. Whole-genome analyses reveal genetic instability of *Acetobacter pasteurianus*. *Nucleic Acids Res.* 37:5768–5783.

Babendreier, D., D. Joller, J. Romeis, F. Bigler, and F. Widmer. 2007. Bacterial community structures in honeybee intestines and their response to two insecticidal proteins. *FEMS Microbiol. Ecol.* 59:600–610.

Bae, S., G. H. Fleet, and G. M. Heard. 2006. Lactic acid bacteria associated with wine grapes from several Australian vineyards. *J. Appl. Microbiol.* 100:712–727.

Barak, J. D., C. E. Jahn, D. L. Gibson, and A. O. Charkowsky. 2007. The role of cellulose and O-antigen capsule in the colonization of plants by *Salmonella enterica*. *Mol. Plant Microbe Interact.* 20:1083–1091.

Beard, C. B., E. M. Dotson, P. M. Pennington, S. Eichler, C. Cordon-Rosales, and R. V. Durvasula. 2001. Bacterial symbiosis and paratransgenic control of vector-borne Chagas disease. *Int. J. Parasitol.* 31:621–627.

Behar, A., B. Yuval, and E. Jurkevitch. 2005. Enterobacteria-mediated nitrogen fixation in natural population of the fruit fly *Ceratitis capitata*. *Mol. Ecol.* 14:2637–2643.

Bertalan, M., R. Albano, V. de Pádua, et al. 2009. Complete genome sequence of the sugarcane nitrogen-fixing endophyte *Gluconacetobacter diazotrophicus* Pal5. *BMC Genomics* 10:450.

Bittar, F., M. Reynaud-Gaubert, P. Thomas, S. Boniface, D. Raoult, and J.-M. Rolain. 2008. *Acetobacter indonesiensis* pneumonia after lung transplant. *Emerg. Infect. Dis.* 14:997–998.

Bodenstein, D., K. W. Cooper, G. F. Ferris, A. Miller, D. F. Poulson, and B. P. Sonnenblick. 1950. *Biology of Drosophila*. New York: John Wiley & Sons.

Cavalcante, V. A., and J. Döbereiner. 1988. A new acid-tolerant nitrogen-fixing bacterium associated with sugarcane. *Plant Soil* 108:23–31.

Chouaia, B., P. Rossi, M. Montagna, et al. 2010. Typing of *Asaia* spp. bacterial symbionts in four mosquito species: molecular evidence for multiple infections. *Appl. Environ. Microbiol.* 76:7444–7450.

Cleenwerck, I., and P. De Vos. 2008. Polyphasic taxonomy of acetic acid bacteria: an overview of the currently applied methodology. *Int. J. Food Microbiol.* 125:2–14.

Corby-Harris, V., A. C. Pontaroli, L. J. Shimkets, J. L. Bennetzen, K. E. Habel, and D. E. Promislow. 2007. Geographical distribution and diversity of bacteria associated with natural populations of *Drosophila melanogaster*. *Appl. Environ. Microbiol.* 73:3470–3479.

Cox, C., and M. Gilmore. 2007. Native microbial colonization of *Drosophila melanogaster* and its use as a model of *Enterococcus faecalis* pathogenesis. *Infect. Immun.* 75:1565–1576.

Crotti, E., C. Damiani, M. Pajoro, et al. 2009. *Asaia*, a versatile acetic acid bacterial symbiont, capable of cross-colonizing insects of phylogenetically-distant genera and orders. *Environ. Microbiol.* 11:3252–3264.

Dale, C., and N. Moran. 2006. Molecular interaction between bacterial symbionts and their hosts. *Cell* 126:453–465.

Damiani, C., I. Ricci, E. Crotti, et al. 2008. Paternal transmission of symbiotic bacteria in malaria vectors. *Curr. Biol.* 18:R1087–R1088.

Damiani, C., I. Ricci, E. Crotti, et al. 2010. Mosquito-bacteria symbiosis: the case of *Anopheles gambiae* and *Asaia*. *Microb. Ecol.* 60:644–654.

Dapples, C. G., and A. O. Lea. 1974. Inner surface morphology of the alimentary canal in *Aedes aegypti* (L.) (Diptera: Culicidae). *Int. J. Insect. Morphol. Embryol.* 3:433–442.

Dillon, R. J., and V. M. Dillon. 2004. The gut bacteria of insects: non-pathogenic interactions. *Annu. Rev. Entomol.* 49:71–92.

Dong, Y., H. E. Taylor, and G. Dimopoulos. 2006. AgDscam, a hypervariable immunoglobulin domain-containing receptor of the *Anopheles gambiae* innate immune system. *PLoS Biol.* 4:1137–1146.

Dutta, D., and R. Gachhui. 2006. Novel nitrogen-fixing *Acetobacter nitrogenifigens* sp. nov., isolated from Kombucha tea. *Int. J. Syst. Evol. Microbiol.* 56:1899–1903.

Dutta, D., and R. Gachhui. 2007. Nitrogen-fixing and cellulose-producing *Gluconacetobacter kombuchae* sp. nov., isolated from Kombucha tea. *Int. J. Syst. Evol. Microbiol.* 57:353–357.

Favia, G., I. Ricci, C. Damiani, et al. 2007. Bacteria of the genus *Asaia* stably associate with *Anopheles stephensi*, an Asian malarial mosquito vector. *Proc. Natl. Acad. Sci. USA* 104:9047–9051.

Favia, G., I. Ricci, M. Marzorati, et al. 2008. Bacteria of the genus *Asaia*: a potential paratransgenic weapon against malaria. *Adv. Exp. Med. Biol.* 627:49–59.

Feldhaar, H., and R. Gross. 2009. Insects as hosts for mutualistic bacteria. *Int. J. Med. Microbiol* 299:1–8.

Franke, I. H., M. Fegan, C. Hayward, G. Leonard, and L. I. Sly. 2000. Molecular detection of *Gluconacetobacter sacchari* associated with the pink sugarcane mealybug *Saccharicoccus sacchari* (Cockerell) and the sugarcane leaf sheath microenvironment by FISH and PCR. *FEMS Microbiol. Ecol.* 31:61–71.

Franke, I. H., M. Fegan, C. Hayward, G. Leonard, E. Stackebrandt, and L. I. Sly. 1999. Description of *Gluconacetobacter sacchari* sp. nov., a new species of acetic acid bacterium isolated from the leaf sheath of sugar cane and from the pink sugar-cane mealy bug. *Int. J. Syst. Bacteriol.* 49:1681–1693.

Franke-Whittle, I. H., M. G. O'Shea, G. J. Leonard, and L. I. Sly. 2004. Molecular investigation of the microbial populations of the pink sugarcane mealybug, *Saccharicoccus sacchari. Ann. Microbiol.* 54:455–470.

Franke-Whittle, I. H., M. G. O'Shea, G. J. Leonard, and L. I. Sly. 2005. Design, development, and use of molecular primers and probes for the detection of *Gluconacetobacter* species in the pink sugarcane mealybug. *Microb. Ecol.* 50:128–139.

Fredricks, D., and L. Ramakrishnan. 2006. The Acetobacteraceae: extending the spectrum of human pathogens. *PLoS Pathog.* 2:249–250.

Fuentes-Ramírez, L. E., R. Bustillos-Cristales, A. Tapia-Hernandez, et al. 2001. Novel nitrogen-fixing acetic acid bacteria, *Gluconacetobacter johannae* sp. nov. and *Gluconacetobacter azotocaptans* sp. nov., associated with coffee plants. *Int. J. Syst. Evol. Microbiol.* 51:1305–1314.

Gilliam, M. 1997. Identification and roles of non-pathogenic microflora associated with honey bees. *FEMS Microbiol. Lett.* 155:1–10.

Gillis, M., K. Kersters, B. Hoste, et al. 1989. *Acetobacter diazotrophicus* sp. nov., a nitrogen fixing acetic acid bacterium associated with sugarcane. *Int. J. Syst. Bacteriol.* 39:361–364.

Gilmore, M. S., and J. J. Ferretti. 2003. The thin line between gut commensal and pathogen. *Science* 299:1999–2001.

Gonzalez-Ceron, L., F. Santillan, M. H. Rodriguez, D. Mendez, and J. E. Hernandez-Avila. 2003. Bacteria in midguts of field-collected *Anopheles albimanus* block *Plasmodium vivax* sporogonic development. *J. Med. Entomol.* 40:371–374.

Gouby, A., C. Teyssier, F. Vecina, H. Marchandin, C. Granolleras, I. Zorgniotti, and E. Jumas-Bilak. 2007. *Acetobacter cibinongensis* bacteremia in human. *Emerg. Infect. Dis.* 13:784–785.

Greenberg, D. E., L. Ding, A. M. Zelazny, et al. 2006a. A novel bacterium associated with lymphadenitis in a patient with chronic granulomatous disease. *PLoS Pathog.* 2:e28.

Greenberg, D. E., S. F. Porcella, F. Stock, et al. 2006b. *Granulibacter bethesdensis* gen. nov., sp. nov., a distinctive pathogenic acetic acid bacterium in the family Acetobacteraceae. *Int. J. Syst. Evol. Microbiol.* 56:2609–2616.

Greenberg, D. E., S. F. Porcella, A. M. Zelazny, et al. 2007. Genome sequence analysis of the emerging human pathogenic acetic acid bacterium *Granulibacter bethesdensis. J. Bacteriol.* 189:8727–8736.

Gusmão, D. S., A. D. Santos, D. C. Marini, et al. 2007. First isolation of microorganisms from the gut diverticulum of *Aedes aegypti* (Diptera: Culicidae): new perspectives for an insect-bacteria association. *Mem. Inst. Oswaldo Cruz* 102:919–924.

Indiragandhi, P., R. Anandham, M. Madhaiyan, et al. 2007. Cultivable bacteria associated with larval gut of prothiofos-resistant, prothiofos-susceptible, and field-caught populations of diamondback moth, *Plutella xylostella*, and their potential for antagonism towards entomopathogenic fungi and host insect nutrition. *J. Appl. Microbiol.* 103:2664–2675.

Jadin, J., I. H. Vincke, A. Dunjic, et al. 1966. Role of *Pseudomonas* in the sporogenesis of the hematozoon of malaria in the mosquito. *Bull. Soc. Pathol. Exot. Filiales.* 59:514–525.

Jeyaprakash, A, M. A. Hoy, and M. H. Allsopp. 2003. Bacterial diversity in worker adults of *Apis mellifera capensis* and *Apis mellifera scutellata* (Insecta: Hymenoptera) assessed using 16S rRNA sequences. *J. Invertebr. Pathol.* 84:96–103.

Jojima, Y., Y. Mihara, S. Suzuki, K. Yokozeki, S. Yamanaka, and R. Fudou. 2004. *Saccharibacter floricola* gen. nov., sp. nov., a novel osmophilic acetic acid bacterium isolated from pollen. *Int. J. Syst. Evol. Microbiol.* 54:2263–2267.

Katsura, K., H. Kawasaki, W. Potacharoen, et al. 2001. *Asaia siamensis* sp. nov., an acetic acid bacterium in the α-Proteobacteria. *Int. J. Syst. Evol. Microbiol.* 51:559–563.

Kersters, K., P. Lisdiyanti, K. Komagata, and J. Swings. 2006. The family Acetobacteraceae: the genera *Acetobacter, Acidomonas, Asaia, Gluconacetobacter, Gluconobacter*, and *Kozakia*. In *The prokaryotes*, ed. M. Dworkin, S. Falkow, E. Rosenberg, K.-H. Schleifer, and E. Stackebrandt, 163–200. 3rd ed., vol. 5. New York: Springer.

Khampang, P., W. Chungjatupornchai, P. Luxananil, and S. Panyim. 1999. Efficient expression of mosquito-larvicidal proteins in a Gram negative bacterium capable of recolonization in the guts of *Anopheles dirus* larva. *Appl. Microbiol. Biotechnol.* 51:79–84.

Kikuchi, Y., X.-Y. Meng, and T. Fukatsu. 2005. Gut symbiotic bacteria of the genus *Burkholderia* in the broad-headed bugs *Riptortus clavatus* and *Leptocorisa chinensis* (Heteroptera: Alydidae). *Appl. Environ. Microbiol.* 71:4035–4043.

Kounatidis, I., E. Crotti, P. Sapountzis, et al. 2009. *Acetobacter tropicalis* is a major symbiont in the olive fruit fly (*Bactrocera oleae*). *Appl. Environ. Microbiol.* 75:3281–3288.

Lambert, B., K. Kersters, F. Goosele, J. Swings, and J. De Ley. 1981. Gluconobacters from honey bees. *Antonie van Leeuwenhoek J. Microbiol. Serol.* 47:147–157.

Lindh, J. M., A.-K. Borg-Karlsonb, and I. Faye. 2008. Transstadial and horizontal transfer of bacteria within a colony of *Anopheles gambiae* (Diptera: Culicidae) and oviposition response to bacteria-containing water. *Acta Tropica* 107:242–250.

Lindh, J. M., O. Terenius, and I. Faye. 2005. 16S rRNA gene-based identification of midgut bacteria from field-caught *Anopheles gambiae* sensu lato and *A. funestus* mosquitoes reveals new species related to known insect symbionts. *Appl. Environ. Microbiol.* 71:7217–7223.

Lisdiyanti, P., H. Kawasaki, Y. Widyastuti, et al. 2002. *Kozakia baliensis* gen. nov., sp. nov., a novel acetic acid bacterium in the α-Proteobacteria. *Int. J. Syst. Evol. Microbiol.* 52:813–818.

Lisdiyanti, P., R. R. Navarro, T. Uchimura, and K. Komagata. 2006. Reclassification of *Gluconacetobacter hansenii* strains and proposals of *Gluconacetobacter sacchariv-orans* sp. nov. and *Gluconacetobacter nataicola* sp. nov. *Int. J. Syst. Evol. Microbiol.* 56:2101–2111.

Loganathan, P., and S. Nair. 2004. *Swaminathania salitolerans* gen. nov., sp. nov., a salt-tolerant, nitrogen-fixing and phosphate-solubilizing bacterium from wild rice (*Porteresia coarctata* Tateoka). *Int. J. Syst. Evol. Microbiol.* 54:1185–1190.

Luxananil, P., H. Atomi, S. Panyim, and T. Imanaka. 2001. Isolation of bacterial strains coloni-sable in mosquito larval guts as novel host cells for mosquito control. *J. Biosci. Bioeng.* 92:342–345.

Malimas, T., P. Yukphan, M. Takahashi, et al. 2008. *Asaia lannaensis* sp. nov., a new acetic acid bacterium in the Alphaproteobacteria. *Biosci. Biotechnol. Biochem.* 72:666–667.

Marzorati, M., A. Alma, L. Sacchi, et al. 2006. A novel Bacteroidetes symbiont is localized in *Scaphoideus titanus*, the insect vector of Flavescence Dorée in *Vitis vinifera*. *Appl. Environ. Microbiol.* 72:1467–1475.

Matalon, Y., N. Katzir, Y. Gottlieb, V. Portnoy, and E. Zchori-Fein. 2007. *Cardinium* in *Plagiomerus diaspidis* (Hymenoptera: Encyrtidae). *J. Invertebr. Pathol.* 96:106–108.

McCutcheon, J. P., and N. A. Moran. 2007. Parallel genomic evolution and metabolic interde-pendence in an ancient symbiosis. *Proc. Natl. Acad. Sci. USA* 104:19392–19397.

Mohr, K. I., and C. C. Tebbe. 2006. Diversity and phylotype consistency of bacteria in the guts of three bee species (Apoidea) at an oilseed rape field. *Environ. Microbiol.* 8:258–272.

Mohr, K. I., and C. C. Tebbe. 2007. Field study results on the probability and risk of a horizon-tal gene transfer from transgenic herbicide-resistant oilseed rape pollen to gut bacteria of bees. *Appl. Microbiol. Biotechnol.* 75:573–582.

Moll, R. M., W. S. Romoser, M. C. Modrzakowski, A. C. Moncayo, and K. Lerdthusnee. 2001. Meconial peritrophic membranes and the fate of midgut bacteria during mosquito (Diptera: Culicidae) metamorphosis. *J. Med. Entomol.* 38:29–32.

Moran, N. A., and H. E. Dunbar. 2006. Sexual acquisition of beneficial symbionts in aphids. *Proc. Natl. Acad. Sci. USA* 103:12803–12806.

Muthukumarasamy, R., I. Cleenwerck, G. Revathi, et al. 2005. Natural association of *Gluconacetobacter diazotrophicus* and diazotrophic *Acetobacter peroxydans* with wet-land rice. *Syst. Appl. Microbiol.* 28:277–286.

Nalepa, C. A., D. E. Bignell, and C. Bandi. 2001. Detritivory, coprophagy and the evolution of digestive mutualisms in *Dictyoptera*. *Insectes Sociaux* 48:194–201.

Nardi, J. B., R. I. Mackie, and J. O. Dawson. 2002. Could microbial symbionts of arthro-pod guts contribute significantly to nitrogen fixation in terrestrial ecosystems? *J. Insect Physiol.* 48:751–763.

Oliver, K. M., J. A. Russell, N. A. Moran, and M. S. Hunter. 2003. Facultative bacterial symbionts in aphids confer resistance to parasitic wasps. *Proc. Natl. Acad. Sci. USA* 100:1803–1807.

Petri, L. 1909. Ricerche Sopra i Batteri Intestinali della Mosca Olearia. Roma: Memorie della Regia Stazione di Patologia Vegetale di Roma (in Italian).

Pidiyar, V. J., K. Jangid, M. S. Patole, and Y. Shouche. 2004. Studies on cultured and uncul-tured microbiota of wild *Culex quinquefasciatus* mosquito midgut based on 16S ribo-somal RNA gene analysis. *Am. J. Trop. Med. Hyg.* 70:597–603.

Pidiyar, V. J., A. Kaznowski, N. N. Badri, M. S. Patole, and Y. S. Shouche. 2002. *Aeromonas culicicola* sp. nov., from the midgut of *Culex quinquefasciatus*. *Int. J. Syst. Evol. Microbiol.* 52:1723–1728.

Prust, C., M. Hoffmeister, H. Liesegang, et al. 2005. Complete genome sequence of the acetic acid bacterium *Gluconobacter oxydans*. *Nature Biotechnol.* 23:195–200.

Pumpuni, C. B., J. Demaio, M. Kent, J. R. Davis, and J. C. Beier. 1996. Bacterial population dynamics in three anopheline species: the impact on *Plasmodium* sporogonic development. *Am. J. Trop. Med. Hyg.* 54:214–218.

Raspor, P., and D. Goranovič. 2008. Biotechnological application of acetic acid bacteria. *Crit. Rev. Biotechnol.* 28:101–124.

Ren, C., P. Webster, S. E. Finkel, and J. Tower. 2007. Increased internal and external bacterial load during *Drosophila* aging without life-span trade-off. *Cell Metab.* 6:144–152.

Robinson, C. J., P. Schloss, Y. Ramon, K. Raffa, and J. Handelsman. 2009. Robustness of the bacterial community in the cabbage white butterfly larval midgut. *Microb. Ecol.* 59:199–211.

Rodríguez López, F. C., F. Franco-Álvarez de Luna, M. C. Gamero Delgado, I. Ibarra de la Rosa, S. Valdezate, J. A. Saez Nieto, and M. Casal. 2008. *Granulibacter bethesdensis* isolated in a child patient with chronic granulomatous disease. *J. Infect.* 57:275–277.

Roh, S. W., Y.-D. Nam, H.-W. Chang, et al. 2008. Characterization of two novel commensal bacteria involved with innate immune homeostasis in *Drosophila melanogaster*. *Appl. Environ. Microbiol.* 74:6171–6177.

Ruiz-Argueso, T., and A. Rodriguez-Navarro. 1973. Gluconic acid producing bacteria from honey bees and ripening honey. *J. Gen. Microbiol.* 76:211–216.

Ruiz-Argueso, T., and A. Rodriguez-Navarro. 1975. Microbiology of ripening honey. *Appl. Microbiol.* 30:893–896.

Ryu, J.-H., S.-H. Kim, H.-Y. Lee, et al. 2008. Innate immune homeostasis by the homeobox gene caudal and commensal-gut mutualism in *Drosophila*. *Science* 319:777–782.

Snyder, R. W., J. Ruhe, S. Kobrin, et al. 2004. Case report: *Asaia bogorensis* peritonitis identified by 16S ribosomal RNA sequence analysis in a patient receiving peritoneal dialysis. *Am. J. Kidney Dis.* 44:E15–E17.

Tuuminen, T., T. Heinäsmäki, and T. Kerttula. 2006. First report of bacteremia by *Asaia bogorensis*, in a patient with a history of intravenous drug abuse. *J. Clin. Microbiol.* 44:3048–3050.

Tuuminen, T., A. Roggenkamp, and J. Vuopio-Varkila. 2007. Comparison of two bacteremic *Asaia bogorensis* isolates from Europe. *Eur. J. Clin. Microbiol. Infect. Dis.* 26:523–524.

White, P. B. 1921. The normal bacterial flora of the bee. *J. Pathol. Bacteriol.* 24:64–78.

Yamada, Y., K.-I. Hoshino, and T. Ishikawa. 1997. The phylogeny of acetic acid bacteria based on the partial sequences of 16S ribosomal RNA: the elevation of the subgenus *Gluconoacetobacter* to the generic level. *Biosci. Biotechnol. Biochem.* 61:1244–1251.

Yamada, Y., K.-I. Hoshino, and T. Ishikawa. 1998. Validation of publication of new names and new combinations previously effectively published outside the IJSB. List No. 64: *Gluconoacetobacter* nom. corrig. (*Gluconoacetobacter* [sic]). *Int. J. Syst. Bacteriol.* 48:327–328.

Yamada, Y., K. Katsura, H. Kawasaki, et al. 2000. *Asaia bogorensis* gen. nov., sp. nov., an unusual acetic acid bacterium in the α-Proteobacteria. *Int. J. Syst. Evol. Microbiol.* 50:823–829.

Yukphan, P., T. Malimas, Y. Muramatsu, et al. 2008. *Tanticharoenia sakaeratensis* gen. nov., sp. nov., a new osmotolerant acetic acid bacterium in the alpha-Proteobacteria. *Biosci. Biotechnol. Biochem.* 72:672–676.

Yukphan, P., T. Malimas, Y. Muramatsu, et al. 2009. *Ameyamaea chiangmaiensis* gen. nov., sp. nov., an acetic acid bacterium in the alpha-Proteobacteria. *Biosci. Biotechnol. Biochem.* 73:2156–2162.

Yukphan, P., T. Malimas, W. Potacharoen, S. Tanasupawat, M. Tanticharoen, and Y. Yamada. 2005. *Neoasaia chiangmaiensis* gen. nov., sp. nov., a novel osmotolerant acetic acid bacterium in the a-Proteobacteria. *J. Gen. Appl. Microbiol.* 51:301–311.

Yukphan P., W. Potacharoen, S. Tanasupawat, M. Tanticharoen, and Y. Yamada. 2004. *Asaia krungthepensis* sp. nov., an acetic acid bacterium in the α-Proteobacteria. *Int. J. Syst. Evol. Microbiol.* 54:313–316.

Zouache, K., D. Voronin, V. Tran-Van, and P. Mavingui. 2009. Composition of bacterial communities associated with natural and laboratory populations of *Asobara tabida* infected with *Wolbachia*. *Appl. Environ. Microbiol.* 75:3755–3764.

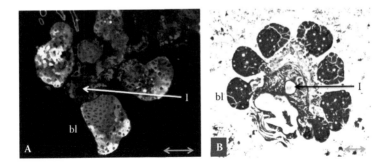

FIGURE 1.4 Bacteriome morphology of *Hylobius abietis* (A) and *Metamasius hemipterus* (B). Shown is the structure of isolated larval bacteriome lobelets from the foregut-midgut junction. Slide A was hybridized with a specific probe designed on *Hylobius* 16S rDNA and stained with DAPI. (Modified from Conord, C., et al., *Mol. Biol. Evol.*, 25, 859–868, 2008.) Slide B was stained with Tolouid blue. bl, bacteriome lobelets; I, intestine. Scale bar: 50 μm.

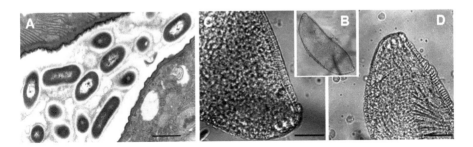

FIGURE 3.3 (A) TEM micrograph of *Asaia* sp. able to colonize the gut epithelium of *An. stephensi*. Bar, 1.2 μm. (B–D) High magnification of an *An. gambiae* egg showing high concentration of Cy3-labeled *Asaia* (in yellow). Phase contrast (B) of an egg and contrast/ fluorescence merge images (C,D) of the egg apical portions. Bars in (C) and (D), 25 μm.

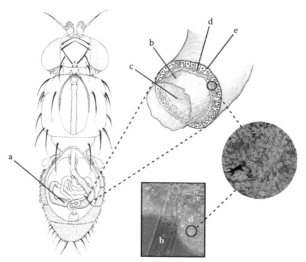

FIGURE 5.2 Anatomical location of symbiotic bacteria in *Tephritis matricariae*. The drawing, portraying an insect observed under a dissecting microscope, shows the position of the midgut tract in which symbiotic bacteria reside (yellow portion). (a) Close-up detail (right) shows the coaxial presence of an inner thin vessel: the peritrophic membrane (b), in whose lumen (c) regular alimentary bolus transit occurs and many nonspecific easily culturable bacteria can be found. Instead, in the interstitial gap space that runs all along (d), between the peritrophic tube and the outer midgut epithelium (e), resident bacteria are observed, which constitute the target of the present analysis. BacLight assays for bacterial cell viability under fluorescent light microscopy; live cells fluoresce green; dead cells fluoresce red. (From Mazzon, L., Piscedda, A., Simonato, M., Martinez-Sañudo, I., Squartini, A., and Girolami V., *Int. J. Syst. Evol. Microbiol.*, 55, 1641–1647, 2008. With permission.)

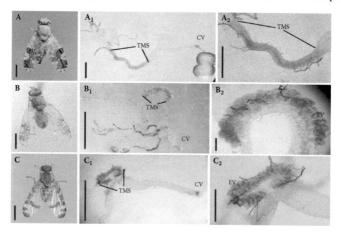

FIGURE 5.4 Three representative species of Tephritinae flies with their respective dissected alimentary tract and detail of midgut structures adapted to harbor symbiotic bacteria: (A) *Tephritis bardanae*, (B) *Acanthiophilus helianthi*, and (C) *Sphenella marginata*. Bars = 1.5 mm (A, A$_1$, A$_2$, B, B$_1$, C, and C$_1$), 0.1 mm (B$_2$), and 0.5 mm (C$_2$). Abbreviations: EV, evaginations of the midgut full of symbiotic bacteria; TMS, tract of the midgut adapted to harbor symbiotic bacteria; CV, cardial valve.

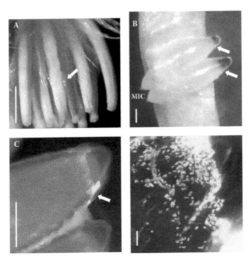

FIGURE 5.6 (A,B) Eggs of *Tephritis matricariae* inserted in a flowerhead of *Crepis vesicaria*. (C,D) BacLight assays for bacterial cell viability under fluorescent light microscopy. Bacteria can be observed in the chorion of the eggs mainly in the opposite pole of the micropyle, partially embedded in a mucilaginous layer. Arrows indicate areas where the majority of bacteria are commonly found. Abbreviations: MIC, micropyle. Bars = 2.3 mm (A), 100 μm (B and C), and 10 μm (D).

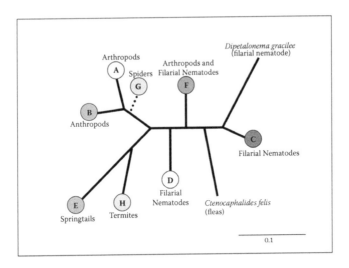

FIGURE 8.2 Schematic diagram of *Wolbachia pipientis* phylogeny based on various phylogenetic studies of the genes *ftsZ*, *groEL*, *gltA*, and *dnaA*. Letters represent supergroups that have been confirmed on the basis of these four genes. The position of supergroup G is tentative since it was estimated using the *wsp* and 16S rRNA genes. Host species are indicated next to each clade or lineage. Two lineages, from *Dipetalonema gracile* and *Ctenocaphalides felis*, have not yet been classified into supergroups. The status of supergroup G is currently being debated. Bar indicates 0.1 substitution per site and is an approximation based on a concatenated gene analysis of these four genes. (Modified from Lo et al. 2007.)

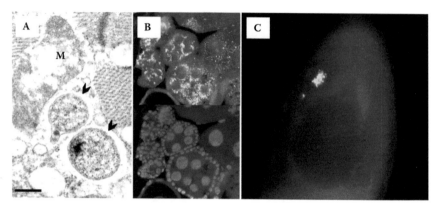

FIGURE 8.3 Localization of *Wolbachia* within various host tissues. (A) Electron micrograph showing *Wolbachia* in ultra-thin sections of *Rhagoletis cerasi* thorax; scale bar = 0.5 μm, M = Mitochondria; *Wolbachia* are indicated by black arrowheads. (B) Fluorescent *in situ* hybridization performed on 5 μm paraffin sections of *R. cerasi* ovaries. *Wolbachia* (green) localize with nuclei of nurse cells (blue) in egg chambers. (C) Immunostainings performed on blastodermal (stage 9) embryos of *Drosophila willistoni*. *Wolbachia* (*w*Wil), indicated in green, localize in the primordial germ cells (PGCs).

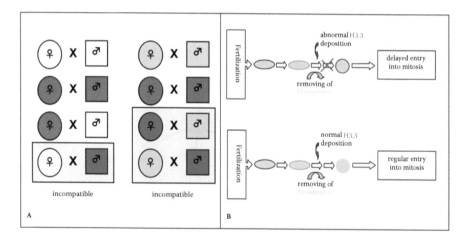

FIGURE 8.4 (A) Schematic presentation of cytoplasmic incompatibility. Depending on *Wolbachia* strain and host species, unidirectional CI may result when *Wolbachia*-infected males (red) mate with uninfected females (white). These matings result in few viable offspring; all other crosses are compatible. Bidirectional CI may result when *Wolbachia*-infected males (red) mate with females harboring a different *Wolbachia* strain (green). (B) Proposed model of key events in the transformation of sperm to male pronucleus in embryos from normal and CI crosses: in normal crosses proper H3.3 deposition is not inhibited; in CI crosses abnormal H3.3 deposition leads to formation of a ring of histone H3.3 encompassing the paternal pronucleus (indicated in red). (Modified from Landmann, F., Orsi, G. A., Loppin, B., Sullivan, W., *PLoS Pathog.*, 5, e1000343, 2009.)

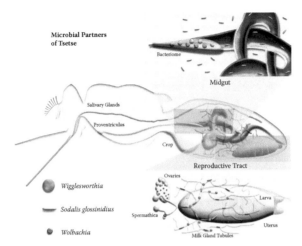

Microbial Partners of Tsetse

Bacteriome

Midgut

Salivary Glands

Proventriculus

Crop

Reproductive Tract

Ovaries

Larva

Spermathica

Uterus

Milk Gland Tubules

● Wigglesworthia

— Sodalis glossinidius

● Wolbachia

FIGURE 9.1 Localization of symbiotic bacteria in tsetse flies. Tsetse harbors three distinct vertically transmitted endosymbionts, *Wigglesworthia glossinidia*, *Sodalis glossinidius*, and members of the genus *Wolbachia*. Two distinct populations of *Wigglesworthia* are found in tsetse. The first is located within specialized bacteriocyte cells that together comprise an organ called the bacteriome. Tsetse's bacteriome is localized in the anterior midgut. The second *Wigglesworthia* population is extracellular in the milk gland lumen of female flies. *Sodalis* has a broad tissue distribution and can be found both intra- and extracellularly in tsetse's midgut, fat body, muscle, hemolymph, milk gland, and salivary glands of certain species. *Wolbachia*, a parasitic bacterium, is also intracellular and can be found within the fly's reproductive tract.

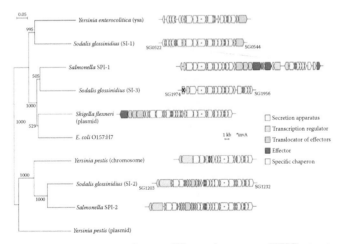

FIGURE 9.3 *Sodalis*'s genome encodes type III secretion system (TTSS) structures that are phylogenetically similar to those from pathogenic microbes. The tree was based on amino acid sequence alignments of the *invA* gene product. Scale bars representing branch lengths and bootstrap values are displayed at each internal node. Genes associated with each cluster are depicted as arrows that indicate the direction of transcription. The arrows with a cross denote pseudogenes. The light blue bars between loci indicate the regions of sequence similarity and gene order conservation. The different colors depict the functional roles of their putative products. (Permission to reprint this figure was granted by CSH Press.)

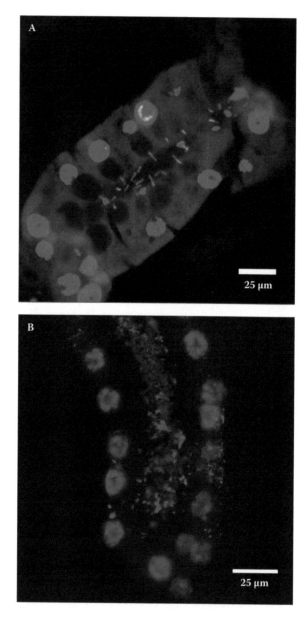

FIGURE 9.4 *In situ* staining of *Wigglesworthia* and *Sodalis* in pregnant female milk gland. Tissue sections were stained with DAPI and *Wigglesworthia*- and *Sodalis*-specific DIG-labeled 16S ribosomal RNA probe. (A) The staining pattern shows *Wigglesworthia* extracellularly in secretory cell ducts and in the milk gland lumen. (B) Concentrated *Sodalis* are visible in the extracellular space of the lumen as well as intracellularly within the cytoplasm of the secretory cells.

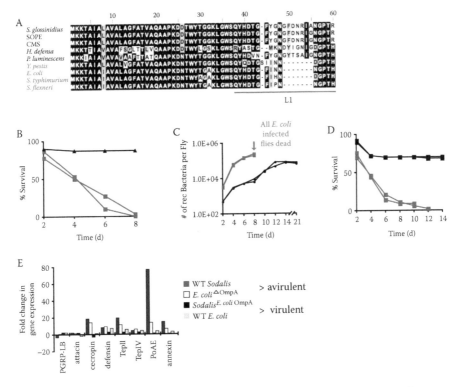

FIGURE 9.5 Mechanism of tsetse's tolerance of *Sodalis*. (A) Multiple sequence alignment of the first 60 N-terminal putative amino acids of OmpA from *Sodalis*, SOPE (*Sitophilus oryzae* principal endosymbiont), CMS (*C. melbae* symbiont), *H. defensa*, *P. luminescens*, *Y. pestis*, *E. coli* 536 (UPEC), *S. typhimurium*, and *S. flexneri*. External loop 1 is underlined. Conserved residues are outlined in black, and substitutions in grey. (B) Systemic infection of tsetse with 1×10^3 CFU of *Sodalis* (red squares) genetically modified to express *E. coli* OmpA. Controls (black triangles) received 1×10^3 CFU of WT *Sodalis*. (C) Number of cells per fly over time after injecting 1×10^3 CFU of luciferase-expressing *Sodalis* (black circles) and *E. coli* (red squares) into tsetse's hemocoel. All rec*E. coli*$_{pII}$-infected flies perished by 10 dpi. (D) Survival of tsetse infected with mutant *E. coli* K12 that do not express OmpA (black squares) and mutant *E. coli* K12 transcomplemented to express their native OmpA (red squares). In B–D, d = days postinfection. (E) Fold change in the expression of tsetse immunity-related genes following infection with avirulent and virulent bacterial strains. All genes were normalized against the constitutively expressed tsetse β-tubulin gene.

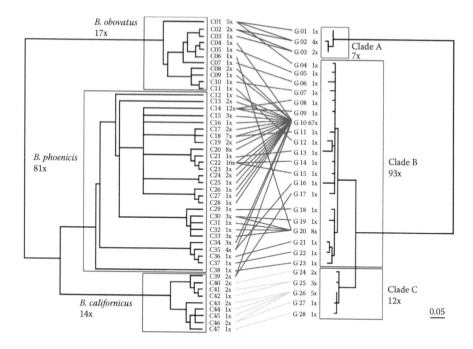

FIGURE 11.3 Mite *COI* tree (cladogram) facing symbiont *gyr*B tree (phylogram). The names of the mite species and symbiont clades are followed by the number of samples that comprise that species or clade. Haplotypes are numbered; haplotype numbers are preceded by a C for *COI*, and a G for *gyr*B. Each haplotype name is followed by the number of times the haplotype was encountered. Both trees are midpoint rooted. The colored lines connect individual mites with their symbionts. When several lines connect a single mite haplotype to various symbionts, this means that several mites with the same haplotypes were found to contain different symbionts. The bar with the *gyr*B phylogram indicates the branch length that represents 5% maximum likelihood distance.

4 Facultative Tenants from the Enterobacteriaceae within Phloem-Feeding Insects

Tom L. Wilkinson

CONTENTS

Genus	*Serratia*
Species	*symbiotica*
Family	Enterobacteriaceae
Order	Enterobacteriales
Description year	2005 (*Candidatus*)
Origin of name	The specific epithet refers to the symbiotic association of these bacteria with insects—they are the only members of the genus that have an intracellular location and are transmitted vertically from mother to offspring.
Description	
Reference	Moran et al. (2005)

Genus	*Hamiltonella*
Species	*defensa*
Family	Enterobacteriaceae
Order	Enterobacteriales
Description year	2005 (*Candidatus*)
Origin of name	Generic name in honor of the evolutionary biologist William D. Hamilton; specific name refers to the role of these symbionts in defending hosts against natural enemies
Description	
Reference	Moran et al. (2005)

Genus	*Regiella*
Species	*insecticola*
Family	Enterobacteriaceae
Order	Enterobacteriales
Description year	2005 (*Candidatus*)
Origin of name	Generic name in honor of the entomologist Reginald ("Reg") F. Chapman; specific name in reference to the insect hosts
Description	
Reference	Moran et al. (2005)

INTRODUCTION

The occurrence of facultative microorganisms in phloem-feeding insects, particularly aphids (Hemiptera, Aphidoidea), has been recognized for many years (see Buchner, 1965), but it is only recently that their phylogenetic position and biological function have been explored in detail. Undoubtedly, their newfound fame is largely a result of advances in molecular biology, particularly the ability to characterize unculturable microorganisms, but whole insect studies to manipulate the bacterial tenants have also played a role in assessing the contribution of the bacteria to insect biology.

The symbiotic microorganisms fall into two categories. The vast majority of phloem-feeding insects harbor so-called primary symbionts (in aphids these belong to the genus *Buchnera*) in specialized insect cells called mycetocytes, or bacterio-cytes in the abdominal hemocoel. These bacteria belong to the γ-Proteobacteria and supplement the nutritionally poor diet of phloem sap through the provision of essential amino acids (Douglas 2006). The metabolic interplay between aphids and *Buchnera* is relatively well understood (e.g., Wilson et al. 2010; Ramsey et al. 2010). In addition, other bacterial symbionts have been detected in some, but not all, phloem-feeding insects, and these fall into the loose assemblage referred to as sec-ondary or accessory symbionts, reflecting the facultative nature of the relationship for the insect host. This chapter concerns three such secondary symbionts, "*Candidatus* Regiella insecticola," "*Candidatus* Hamiltonella defensa," and "*Candidatus* Serratia symbiotica" (see Moran et al. 2005), that are united by their phylogenetic posi-tion and distribution; they are all members of the Enterobacteriaceae (within the γ-Proteobacteria) and, to date, have only been detected in phloem-feeding hemipter-ans, including aphids, whiteflies, and psyllids (Table 4.1; Darby et al. 2001; Thao and Baumann 2004; Russell et al. 2003; Moran et al. 2005; Hansen et al. 2007). Other facultative bacteria have been described from aphids, including a *Spiroplasma* sp. (Fukatsu et al. 2001), an α-Proteobacterium informally designated PAR or S-type (Chen et al. 2000), *Wolbachia* (Gómez-Valero et al. 2004), *Arsenophonus* (Wille and Hartman 2009), and the recently discovered pea aphid X-type symbiont (PAXS; Gauy et al. 2009), but the relationship between these bacteria and aphids is poorly characterized, and they are not considered in detail here.

TABLE 4.1
Facultative Symbiotic Bacteria from the Enterobacteriaceae that Reside in Phloem-Feeding Insects (see Moran et al. 2005)

Candidate Name	Alternative Designation	Insect Host and Tissue Location	Phenotypic Impact on Insect Host
"*Candidatus* Serratia symbiotica"	R-type PASS (pea aphid secondary symbiont) S-sym	Aphids Bacteriocytes, sheath cells, hemolymph	Resistance to high temperature and parasitoids
"*Candidatus* Hamiltonella defensa"	T-type PABS (pea aphid *Bemisia*-like symbiont) "*Candidatus* Consessoris aphidicola"	Aphids, whiteflies, psyllids Bacteriocytes, sheath cells, hemolymph	Resistance to parasitoids and high temperature
"*Candidatus* Regiella insecticola"	U-type PAUS (pea aphid U-type symbiont) "*Candidatus* Adiaceo aphidicola"	Aphids Bacteriocytes, sheath cells, hemolymph	Host plant specialization and resistance to fungal pathogens

HISTORICAL PERSPECTIVE

Before the 1990s the secondary symbionts of aphids that are recognized today as belonging to at least five different species were considered together as accessory bacteria residing in various locations adjacent to the bacteriocytes (Buchner 1965). They were distinguished from *Buchnera* on the basis of morphology; *Buchnera* are coccoid bacteria with a diameter of 2–5 μm, whereas the secondary bacteria, if present, were described as either smaller cocci or rods and filaments with variable lengths of between 8 and 30 μm. The presence of two or sometimes three readily distinguishable bacterial morphologies within aphids (described as disymbiotic and trisymbiotic aphids, respectively) was recognized in the nineteenth century (e.g., Krassilstschik 1889, cited in Buchner, 1965; see Figure 153, p. 304), and detailed descriptions of tissue distributions and transmission pathways in various species were published throughout the twentieth century. Examination using the electron microscope revealed ultrastructural details not possible with light microscopy (e.g., Griffiths and Beck 1973) and confirmed the presence of multiple bacterial morphotypes, and in the last two decades of the twentieth century molecular techniques were used to visualize and identify the secondary bacteria. Current areas of research are now more focused on understanding the biological interactions between the secondary symbionts and aphid host.

INTEGRATION INTO APHID BIOLOGY

An individual aphid may contain none to multiple secondary symbionts within its tissues (e.g., Haynes et al. 2003; Sandstrom et al. 2001; Russell et al. 2003), but irrespective of the number of bacterial species present, the natural microbiota within aphids is dominated by *Buchnera*, and the absolute number of secondary bacteria is significantly smaller (although there is an important exception to this generalization in the cedar aphid *Cinara cedri*; see the "Biological Effects" section). Prior to their identification through sequence similarity, examinations using the electron microscope demonstrated that the secondary symbionts in the pea aphid (*Acyrthosiphon pisum*) were localized to flattened syncytial sheath cells surrounding the bacteriocytes, and that each individual rod-shaped bacterium was enclosed in a membrane of host origin (Griffiths and Beck 1973; McLean and Houk 1973). However, the secondary bacteria in other aphid species were also observed in bacteriocytes similar to those containing *Buchnera* by both electron microscopy (Hinde 1971) and immunohistochemistry (Fukatsu and Ishikawa 1998), and it is now apparent that the secondary symbionts can be found in multiple locations: sheath cells, so-called secondary bacteriocytes, and free in the hemocoel (Buchner 1965; Fukatsu and Ishikawa 1998; Chen et al. 1996; Sandstrom et al. 2001; Fukatsu et al. 2000; Fukatsu 2001). A representative electron micrograph depicting "*Candidatus* Regiella insecticola" in sheath cells from the black bean aphid *Aphis fabae* is shown in Figure 4.1.

The bacteria are maintained in aphid populations largely by vertical transmission, although occasional horizontal transfer must occur to explain the current distribution patterns between divergent hosts (Sandstrom et al. 2001; Russell et al. 2003). Possible routes for horizontal transmission include phloem sap and parasitoid wasps

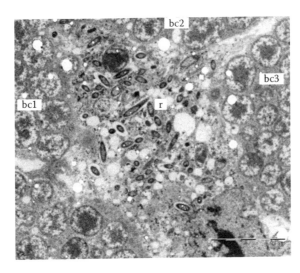

FIGURE 4.1 Transmission electron micrograph of embryos from the black bean aphid *Aphis fabae* showing rod-shaped "*Candidatus* Regiella insecticola" (r) residing between four bacteriocyte cells (bc1-bc4) containing coccoid *Buchnera* cells (b). Scale bar = 5 μm. (Image by J. Bermingham and T.L. Wilkinson.)

(Fukatsu et al. 2001; Sandstrom et al. 2001; Darby and Douglas 2003; Tsuchida et al. 2005). As an example, "*Ca.* Hamiltonella defensa" is vertically transmitted to asexual and sexual morphs and sexually produced eggs of *A. pisum*, but aphids naturally lacking "*Ca.* Hamiltonella defensa" can acquire the bacterium orally (Darby and Douglas 2003), and relatively low levels of horizontal transmission are predicted to maintain the natural prevalence of "*Ca.* Hamiltonella defensa" in the population. However, the occurrence of distinct secondary bacterial menageries within three aphid species that naturally colonize the same host plant (Tsuchida et al. 2006) indicates the rarity of horizontal transmission events. Vertical transmission of the secondary symbionts was described by the early microscopists (Buchner 1965), and follows the route of the primary symbionts *Buchnera* into early embryos or developing eggs (see also Moran and Dunbar 2006).

DEFINING CHARACTERISTICS

Table 4.1 lists the alternate names by which the various secondary symbionts in aphids have been designated, highlighting the different nomenclatures that were found in the literature prior to 2005. Information from 16S rRNA genes and partial sequences of *recA* and *gyrB* was used to clarify the phylogenetic position and assign candidate names to "*Ca.* Serratia symbiotica," "*Ca.* Hamiltonella defensa," and "*Ca.* Regiella insecticola" (Moran et al., 2005). All three bacteria were confirmed to belong to the Enterobacteriaceae, with "*Ca.* Hamiltonella defensa" and "*Ca.* Regiella insecticola" as sister species (Degnan and Moran 2008).

CANDIDATUS SERRATIA SYMBIOTICA

This was the first secondary symbiont of aphids for which DNA sequences were published (Unterman et al. 1989). The bacterium belongs to the genus *Serratia* that includes pathogens of humans (e.g., *S. macrescens*) and insects (e.g., *S. entomophila*) (Grimont and Grimont 1978), although "*Ca.* Serratia symbiotica" is the only member of the genus to date that can persist in an intracellular location and is maintained through vertical transmission from mother to offspring. "*Ca.* Serratia symbiotica" has been found in many aphid species from Europe, Japan, and North America (Chen and Purcell 1997; Darby and Douglas 2003; Haynes et al. 2003; Russell et al. 2003; Sandstrom et al. 2001; Tsuchida et al. 2002, 2006; Lamelas et al. 2008; Burke et al. 2009). In the pea aphid (*A. pisum*) the bacteria are rod shaped (with a variable length of 2–9 µm and a width of approximately 0.6 µm) and are found within bacteriocytes, sheath cells, and extracellularly in the hemocoel (Fukatsu et al. 2000; Moran et al. 2005), whereas in the aphid *Cinara cedri* (Lachnini) "*Ca.* Serratia symbiotica" cells are large, round, and pleomorphic (Gomez-Valero et al. 2004; Pérez-Brocal et al. 2006), not unlike the morphology of the primary symbiont *Buchnera*. This variation in morphology coincides with phylogenetic analyses that place "*Ca.* Serratia symbiotica" into two distinct sister clades (Lamelas et al. 2008). The hypothesis that the relationship between "*Ca.* Serratia symbiotica" and *C. cedri* is more obligate than other secondary symbionts (Gosalbes et al. 2008) is discussed in the "Biological Effects" section.

CANDIDATUS HAMILTONELLA DEFENSA

This bacterial species is the only facultative symbiont to date that has been identified in other sap-feeding insects (notably psyllids and whitefly; Clark et al. 1992; Russell et al. 2003; Sandstrom et al. 2001) in addition to aphids. It is divergent from other members of the Enterobacteriaceae with "*Ca.* Regiella insecticola" its closest relative (Moran et al. 2005). The morphology of the cells and their tissue distribution within hosts is similar to that of "*Ca.* Serratia symbiotica," and likewise its prevalence within populations is variable. In *A. pisum*, the most widely surveyed host, "*Ca.* Hamiltonella defensa" is present to varying degrees in populations from North America and Europe (Darby and Douglas, 2003; Haynes et al. 2003; Russell et al. 2003; Sandstrom et al. 2001), but it is apparently absent from *A. pisum* in Japan (Tsuchida et al. 2002). This species, together with "*Ca.* Regiella insecticola," has been successfully cultured and maintained in insect cell lines (Darby et al. 2005).

CANDIDATUS REGIELLA INSECTICOLA

"*Ca.* Regiella insecticola" shares a close sister relationship with "*Ca.* Hamiltonella defensa" within the Enterobacteriaceae, and the two are phylogenetically distinct from "*Ca.* Serratia symbiotica" but share a close relationship with *Photorhabdus* and *Xenorhabdus* species that are associated with entomopathogenic nematodes (Moran et al. 2005). "*Ca.* Regiella insecticola" has only been detected from aphids, but has been found in a wide range of subfamilies from Europe, North America, Japan, and Australia (Haynes et al. 2003; Russell et al. 2003; Sandstrom et al. 2001; Tsuchida et al. 2005; von Burg et al. 2008). The morphology and tissue distribution of "*Ca. Regiella insecticola*" are similar to those of the previous species, with the bacteria inhabiting bacteriocytes (i.e., the same location as the primary symbiont *Buchnera*), syncitial sheath cells surrounding the bacteriocytes (see Figure 4.1), and the hemolymph (e.g., Tsuchida et al. 2005).

COMPARATIVE GENOMICS

The molecular characteristics of individual genes from all three facultative bacterial species have been explored in detail over the last decade, and further insights have recently emerged following the publications of the full genome sequence of both "*Ca.* Hamiltonella defensa" and "*Ca.* Regiella insecticola" (Degnan et al. 2009a, 2009b). The genome of "*Ca.* Serratia symbiotica" from the cedar aphid *Cinara cedri* is currently being sequenced (Gosalbes et al. 2008). In common with other endosymbiotic bacteria (Moran 1996), "*Ca.* Serratia symbiotica," "*Ca.* Hamiltonella defensa," and "*Ca.* Regiella insecticola" have a low G and C content in coding sequences, confirmed by the relatively low G and C content of the complete genomes of "*Ca.* Hamiltonella defensa" and "*Ca.* Regiella insecticola," and comparisons of gene sequences with related free-living bacteria indicate that "*Ca.* Hamiltonella defensa" and "*Ca.* Regiella insecticola" exhibit accelerated sequence evolution (Russell et al. 2003).

The genomes of "*Ca.* Hamiltonella defensa" and "*Ca.* Regiella insecticola," at 2.11 and 2.07 Mbp, respectively, are both small compared to their closest free-living relatives, and despite the close phylogenetic relationship of the bacteria, the genomes have only 55% of genes in common. The overall picture is one of gene rearrangement, inactivation, and insertion, with a high propensity of mobile DNA and horizontal gene transfer maintaining the dynamic nature of these facultative genomes. More specifically, "*Ca.* Hamiltonella defensa" and "*Ca.* Regiella insecticola" have both lost the ability to synthesize 8 of the 10 essential amino acids, although they retain active uptake mechanisms to import the missing amino acids, which are presumably derived from the primary symbiont *Buchnera*. In addition, these bacteria (unlike *Buchnera*) can be transmitted horizontally, and therefore they retain abundant so-called pathogenicity loci (even though these bacteria are not pathogenic), including type 3 secretion systems and toxin homologs that may facilitate invasion of novel hosts (Degnan et al. 2009a, 2009b).

BIOLOGICAL EFFECTS

Facultative associates represent a cost to the host because inevitably they will sequester resources away from growth and reproduction. Consequently, one would predict significant benefits associated with housing secondary symbionts, which in theory should lead to their eventual fixation in natural populations. However, the patchy distribution of the facultative symbionts demonstrates that this is not the case, and much current research in aphid secondary associates is concerned with revealing the costs and benefits the host harboring the bacteria. The phenotypic response of the aphid host in the presence of facultative bacteria to various environmental insults has been well characterized, but the underlying mechanistic basis behind all but a few of these interactions remains obscure.

Not surprisingly, given the patchy distribution and loss of genes for the synthesis of essential amino acids (see the "Comparative genomics" section), the secondary bacteria do not appear to play a significant role in the nutrition of most aphids (Douglas et al. 2006), although the striking case of tryptophan provision in cedar aphids (*Cinara cedri*) may be the exception (Gosalbes et al. 2008). In this species, "*Ca.* Serratia symbiotica" always occurs within bacteriocytes adjacent to those housing the primary symbiont *Buchnera*, and the two bacterial populations occur at similar densities, casting doubt on the facultative nature of this strain of "*Ca.* Serratia symbiotica" (Gómez-Valero et al. 2003; Lamelas et al. 2008). Furthermore, the genome of *Buchnera* in *C. cedri* is the most degenerate of the *Buchnera* genomes sequenced to date, and specifically has lost the ability to synthesize two essential nutrients, the amino acid tryptophan and the vitamin riboflavin (Pérez-Brocal et al. 2006). A subset of the genes involved in tryptophan synthesis are retained on a plasmid in *Buchnera* from *C. cedri*, and the remaining genes are present on the chromosome of "*Ca.* Serratia symbiotica," suggesting that both bacteria (a symbiotic consortium) are required to provide tryptophan in this aphid (Gosalbes et al. 2008). A detailed analysis of secondary symbionts within the subfamily Lachninae, however, indicates that *C. cedri* from Chile does not harbor any secondary bacteria, and the *Cinara* subgenus as a whole does not exhibit parallel coevolution with "*Ca.* Serratia symbiotica," as would be expected in an obligate nutritional mutualist (Burke et al. 2009), suggesting that there is some variability in the facultative nature of the relationship, perhaps driven by the degenerative genome of *Buchnera* in this aphid. In all other aphids studied to date, the selective advantage of maintaining facultative bacterial associates is linked to ecologically important traits such as expanded host-plant range, thermal tolerance, and resistance to parasitoid wasps or entomopathogenic fungi (so-called symbiont-mediated protection; Haine 2008). These phenotypic effects are discussed in turn below.

FACULTATIVE BACTERIA AND THE HOST PLANT RANGE OF APHIDS

Surveys to detect the presence of symbiotic bacteria in populations of aphids restricted to specific host plants have demonstrated nonrandom distributions of secondary symbionts, implying facultative bacterial associates provide fitness benefits to aphids feeding on certain plants within the host plant range. Once again, pea aphids provide the most comprehensive data set (from Europe, North America, and Japan), and also the most compelling; pea aphids infected with "*Ca.* Regiella insecticola" are more abundant on *Trifolium* (clover) than the numerous alternative legumes within the host plant range (Tsuchida et al. 2002; Leonardo and Miura 2003; Simon et al. 2003; Ferrari et al. 2004). Experimental manipulation of the symbiosis (Tsuchida et al. 2004) demonstrated that pea aphids cured of "*Ca.* Regiella insecticola" by antibiotic treatment performed poorly on *T. repens* (red clover), but the performance of the same aphids once "*Ca.* Regiella insecticola" had been reinstated by microinjection was equivalent to the original naturally infected insects. Importantly, aphid performance was not affected by any of these manipulations when the insects were reared on a different host plant (Tsuchida et al. 2004). However, similar experiments to cure aphids of either "*Ca.* Regiella insecticola" or "*Ca.* Serratia symbiotica" (Leonardo 2004) could find no fitness benefit to pea aphids of harboring secondary symbionts when feeding on *T. repens* and *Medicago sativa*, respectively. The impact of secondary symbionts on host plant specialization and other life history traits of pea aphids (such as dispersal and mating) is therefore complex and governed by a combination of both aphid and bacterial genotypes (Leonardo and Mondor 2006; Ferrari et al. 2007).

The only candidate factor linking facultative secondary bacteria and aphid host plant use has been identified in the black bean aphid *Aphis fabae*, a polyphagous species that feeds on a wide variety of plant taxa. Field populations contain either no facultative bacteria or one of "*Ca.* Serratia symbiotica," "*Ca.* Hamiltonella defensa," or "*Ca.* Regiella insecticola," in addition to the primary symbiont *Buchnera* (Chandler et al. 2008). One particular host plant, *Lamium purpureum*, supports relatively poor performance of the aphid, and infection with either "*Ca.* Hamiltonella defensa" or "*Ca.* Regiella insecticola" exacerbates the impact of the plant on aphid performance, irrespective of aphid genotype. In addition, there is an increase in the density of secondary bacteria in aphids reared on *L. purpureum* (Wilkinson et al. 2001; Chandler et al. 2008) that mirrors an increase in the density of secondary bacteria in *A. fabae* (Chandler et al. 2008) and *A. pisum* (Wilkinson et al. 2007) reared on low-nitrogen diets. It is suggested that low nitrogen in the phloem sap of *L. purpureum* is responsible for the deleterious impact of the secondary bacteria on aphid performance (Chandler et al. 2008).

FACULTATIVE BACTERIA AND APHID TOLERANCE OF HIGH TEMPERATURE

The influence of secondary symbionts on the ability of aphids to withstand elevated temperatures was one of the first positive phenotypic impacts of the facultative bacteria to be described (Chen et al. 2000). In this study, one of three clonal lineages of *A. pisum* exhibited increased fitness when reared at 25°C, but only when the aphids were infected with a secondary symbiont (either "*Ca.* Serratia symbiotica" or a species of *Rickettsia* designated PAR (pea aphid *Rickettsia*)). Further analysis indicated that "*Ca.* Serratia symbiotica" had the greatest ameliorative effect on pea aphids exposed to heat stress and occurred at higher frequencies in natural populations during warm summer months (Montllor et al. 2002). Recent metabolic analysis suggests that the protective effects of "*Ca.* Serratia symbiotica" arise through the delivery of protective metabolites to insect cells following heat exposure (Burke et al. 2010). "*Ca.* Hamiltonella defensa" has also been shown to confer tolerance to high temperature (Russell and Moran 2006), but a positive effect of either bacteria depends on when and for how long the heat stress occurs during nymphal development and the aphid and bacterial genotypes involved in the interaction. However, it is clear that the presence of facultative bacteria has the potential to shape the seasonal and geographical distribution of aphid hosts, and how the aphids may respond to environmental change (Harmon et al. 2009).

FACULTATIVE BACTERIA AND APHID RESISTANCE TO PATHOGENS

The specific epithet of "*Ca.* Hamiltonella defensa" provides a clue to the phenotypic trait that infection with this bacterium bestows on aphid hosts, namely, an improved resistance against hymenopteran parasitoid wasps that are ecologically important natural enemies. Female wasps deposit an egg in the abdomen of living aphids, and the larvae develop to pupation before emerging as adults from the mummified remains of the aphid host. One lineage of pea aphid containing "*Ca.* Hamiltonella defensa" was equally likely to be parasitized, but less likely to support development of the parasitoid *Aphidius ervi* (Oliver et al. 2003). Similar, although not as pronounced, effects were observed for aphids infected with "*Ca.* Serratia symbiotica." In population cage experiments, the frequency of infection with "*Ca.* Hamiltonella defensa" increased under sustained parasitoid pressure, and decreased when parasitoids were absent (Oliver et al. 2008), demonstrating that natural enemies can play an important role in shaping the dynamics of facultative symbiont infections. As an example, a survey of pea aphids from the UK demonstrated that increased resistance to *Aph. ervi* and the closely related *Aph. eadyi* was correlated with the presence of "*Ca.* Hamiltonella defensa" across 47 lineages of pea aphid (Ferrari et al. 2004). Variation in the susceptibility of aphid genotypes to parasitoids has been described frequently (e.g., Ferrari et al. 2001), but it appears that in pea aphids it is the symbiont that determines the level of resistance (Oliver et al. 2005). Interestingly, resistance to parasitoids increases to even higher levels when both "*Ca.* Hamiltonella defensa" and "*Ca.* Serratia symbiotica" are present in an artificially constructed superinfection in the laboratory (Oliver et al. 2006), but a severe reduction in aphid fitness and the overproliferation of "*Ca.* Serratia symbiotica" may explain why such multiple infections are uncommon in wild populations of pea aphid (Oliver et al. 2006).

The protective phenotype of "*Ca.* Hamiltonella defensa" is linked to an active bacteriophage, first identified in a European strain of pea aphid and referred to as APSE (van der Wilk et al. 1999). Several variants of APSE have been identified (designated APSE-1 to APSE-7) that encode toxin homologs that are known to target eukaryotic tissue, such as cytolethal distending toxin (*cdtB*) and a Shiga-like toxin (*stxB*; Moran et al. 2005; Degnan and Moran 2008). Manipulation of the bacteria-phage complex in genetically identical aphid backgrounds has demonstrated that APSE is required for protection, and that the toxin homologs are highly expressed, directly implicating the phage-derived toxins in the protective mechanism. In addition, the phage is readily lost from "*Ca.* Hamiltonella defensa"-infected lines, resulting in increased susceptibility to parasitism (Oliver et al. 2009). The significance of these interactions is best summarized by quoting Oliver et al. (2009, p. 994): "The evolutionary interests of phages, bacterial symbionts, and aphids are all aligned against the parasitoid wasp that threatens them all."

The defensive role of facultative symbionts in other aphid species has only recently begun to be explored. In the black bean aphid, *Aphis fabae*, "*Ca.* Hamiltonella defensa" also appears to provide protection against the parasitoid *Lysiphlebus fabarum* (Vorburger et al. 2010), but the interactions are more complex than in pea aphids, with substantial genetic variation for innate defenses in the aphid coupled with acquired defenses of the symbionts, presenting two lines of defense against the parasitoid. Significant variation in susceptibility to parasitoids was also demonstrated in a set of 17 clones of the peach-potato aphid *Myzus persicae* (von Burg et al. 2008), but the only aphid clone that was completely resistant was also the only one that harbored a facultative symbiont, in this case "*Ca.* Regiella insecticola." This specific isolate, unlike other isolates of "*Ca.* Regiella insecticola," has also since been shown to increase resistance to parasitoids in different aphid species (Castañeda et al. 2010). Although the mechanisms are not clear, this study highlights the potential for multiple facultative bacteria to evolve the ability to protect their host from natural enemies. Of particular relevance is the finding that "*Ca.* Regiella insecticola" can also protect aphids from another pathogen, the Entomophthorales fungus *Pandora neoaphidis* (Scarborough et al. 2005).

CONCLUDING REMARKS

The facultative symbionts in aphids were at one time treated as something of a nuisance, adding at best another level of unexplored complexity to the interactions between the primary symbiont and host. However, it is increasingly apparent that these bacteria play very important roles in shaping aphid responses to a variety of ecological scenarios, including host plant specialization, adaptation to heat stress, and defense against natural enemies, and future studies investigating phenotypic variability in the response of phloem-feeding insects to these and other environmental stimuli should bear in mind the potential mediating role of secondary symbionts. The majority of conclusions that have been drawn to date are inferred from studies on the pea aphid *Acyrthosiphon pisum*, and it would be premature to expect similar phenotypic effects in other aphids until the range of species examined is much greater. Recent studies in *Aphis fabae* and *Myzus persicae* are encouraging, and a positive relationship between the presence of an unidentified secondary symbiont and parasitism pressure in a psyllid (Hansen et al. 2007) suggests that the defensive role of facultative symbionts may extend beyond aphids.

    A. pisum is seen as an emerging model insect (as reflected in the publication and ongoing annotation of the full genome sequence; International Aphid Genomics Consortium 2010), and the availability of the genome sequence of *Buchnera* (Shigenobu et al. 2000) and "*Ca.* Hamiltonella defensa" (Degnan et al. 2009a), and the almost complete sequence of "*Ca.* Regiella insecticola" (Degnan et al. 2009b), all obtained from *A. pisum*, will provide unparalleled insights into the molecular interactions between these symbiotic partners. The mechanistic basis of the phenotypic effects observed in *A. pisum* remain largely unknown, although bacteriophages are undoubtedly important in defensive interactions, and a thorough understanding of the evolutionary significance of the facultative symbionts across aphid lineages will require a theoretical approach that is currently entirely lacking. These are exciting times to be exploring the enigmatic world of facultative endosymbionts.

REFERENCES

Buchner, P. (1965). *Endosymbiosis of animals with plant micro-organisms.* John-Wiley & Sons, Chichester, UK.

Burke, G.R., Fiehn, O., and Moran, N.A. (2010). Effects of facultative symbionts and heat stress on the metabolome of pea aphids. *ISME Journal* 4, 242–252.

Burke, G.R., Normarck, B.B., Favret, C., and Moran, N.A. (2009). Evolution and diversity of facultative symbionts from the aphid subfamily Lachninae. *Applied and Environmental Microbiology* 75, 5328–5335.

Castañeda, L.E., Sandrock, C., and Vorburger, C. (2010). Variation and covariation of life history traits in aphids are related to infection with the facultative bacterial endosymbiont *Hamiltonella defensa. Biological Journal of the Linnean Society* 100, 237–247.

Chandler, S.M., Wilkinson, T.L., and Douglas, A.E. (2008). Impact of plant nutrients on the relationship between a herbivorous insect and its symbiotic bacteria. *Proceedings of the Royal Society of London Series B* 275, 565–570.

Chen, D.Q., Campbell, B.C., and Purcell, A.H. (1996). A new *Rickettsia* from a herbivorous insect, the pea aphid *Acyrthosiphon pisum* (Harris). *Current Microbiology* 33, 123–128.

Chen, D.Q., Montllor, C.B., and Purcell, A.H. (2000). Fitness effects of two facultative endosymbiotic bacteria on the pea aphid *Acyrthosiphon pisum*, and the blue alfalfa aphid, *A. kondoi. Entomologia Experimentalis et Applicata* 95, 315–323.

Chen, D.Q., and Purcell, A.H. (1997). Occurrence and transmission of facultative endosymbionts in aphids. *Current Microbiology* 34, 220–225.

Clark, M.A., Baumann, L., Munson, M.A., Baumann, P., Campbell, C., Duffus, J.E., Osborne, L.S., and Moran, N.A. (1992). The eubacterial endosymbionts of whiteflies (Homoptera: Aleyrodoidea) constitute a lineage distinct from the endosymbionts of aphids and mealybugs. *Current Microbiology* 25, 119–123.

Darby, A.C., Birkle, L.M., Turner, S.L., and Douglas, A.E. (2001). An aphid-borne bacterium allied to the secondary symbionts of whitefly. *FEMS Microbiology Ecology* 36, 43–50.

Darby, A.C., Chandler, S.M., Welburn, S.C., and Douglas, A.E. (2005). Aphid-symbiotic bacteria cultured in insect cell lines. *Applied and Environmental Microbiology* 71, 4833–4839.

Darby, A.C., and Douglas, A.E. (2003). Elucidation of the transmission patterns of an insect-borne bacterium. *Applied and Environmental Microbiology* 69, 4403–4407.

Degnan, P.H., Leonardo, T.E., Cass, B.N., Hurwitz, B., Stern, D., Gibbs, R., Richards, S., and Moran, N.A. (2009b). Dynamics of genome evolution in facultative symbionts of aphids. *Environmental Microbiology*, 12, 2060–2069.

Degnan, P.H., and Moran, N.A. (2008). Diverse phage-encoded toxins in a protective insect endosymbiont. *Applied and Environmental Microbiology* 74, 6782–6791.

Degnan, P.H., Yu, Y., Sisneros, N., Wing, R.A., and Moran, N.A. (2009a). *Hamiltonella defensa*, genome evolution of protective bacterial endosymbiont from pathogenic ancestors. *Proceedings of the National Academy of Sciences of the United States of America* 106, 9063–9068.

Douglas, A.E. (2006). Phloem sap feeding by animals: problems and solutions. *Journal of Experimental Botany* 57, 747–754.

Douglas, A.E., François, C.L.M.J., and Minto, L.B. (2006). Facultative 'secondary' bacterial symbionts and the nutrition of the pea aphid *Acyrthosiphon pisum. Physiological Entomology* 31, 262–269.

Ferrari, J., Darby, A.C., Daniell, T.J., Godfray, H.C.J., and Douglas, A.E. (2004). Linking the bacterial community in pea aphids with host-plant use and natural enemy resistance. *Ecological Entomology* 29, 60–65.

Ferrari, J., Müller, C.B., Kraaijeveld, A.R., and Godfray, H.C.J. (2001). Clonal variation and covariation in aphid resistance to parasitoids and a pathogen. *Evolution* 55, 1805–1814.

Ferrari, J., Scarborough, C.L., and Godfray, H.C.J. (2007). Genetic variation in the effect of a facultative symbiont on host-plant use by pea aphids. *Oecologia* 153, 323–329.

Fukatsu, T. (2001). Secondary intracellular symbiotic bacteria in aphids of the genus *Yamatocallis* (Homoptera: Aphididae: Drepanosiphinae). *Applied and Environmental Microbiology* 67, 5315–5320.

Fukatsu, T., and Ishikawa, H. (1998). Differential immunohistochemical visualisation of the primary and secondary intracellular symbiotic bacteria of aphids. *Applied Entomology and Zoology* 33, 321–326.

Fukatsu, T., Nikoh, N., Kawai, R., and Koga, R. (2000). The secondary endosymbiotic bacterium of the pea aphid *Acyrthosipon pisum* (Insecta: Homoptera). *Applied and Environmental Microbiology* 66, 2748–2758.

Fukatsu, T., Tsuchida, T., Nicoh, N., and Koga, R. (2001). *Spiroplasma* symbiont of the pea aphid *Acyrthosiphon pisum* (Insecta: Homoptera). *Applied and Environmental Microbiology* 67, 1284–1291.

Gauy, J.F., Boudreault, S., Michaud, D., and Cloutier, C. (2009). Impact of environmental stress on aphid clonal resistance to parasitoids: role of *Hamiltonella defensa* bacterial symbiosis in association with a new facultative symbiont of the pea aphid. *Journal of Insect Physiology* 55, 919–926.

Gómez-Valero, L., Soriano-Navarro, M., Pérez-Brocal, V., Heddi, A., Moya, A., Garcia-Verdugo, J.M., and Latorre, A. (2004). Coexistence of *Wolbachia* with *Buchnera aphidicola* and a secondary symbiont in the aphid *Cedri cinara*. *Journal of Bacteriology* 186, 6626–6633.

Gosalbes, M.J., Lamelas, A., Moya, A., and Latorre, A. (2008). The striking case of tryptophan provision in the cedar aphid *Cinara cedri*. *Journal of Bacteriology* 190, 6026–6029.

Griffiths, G. W., and Beck, S. D. (1973). Intracellular symbiotes of the pea aphid, *Acyrthosiphon pisum*. *Journal of Insect Physiology* 19, 75–84.

Grimont, P.A., and Grimont, F. (1978). The genus *Serratia*. *Annual Review of Microbiology* 32, 221–248.

Haine, E.R. (2008). Symbiont-mediated protection. *Proceedings of the Royal Society Series B* 275, 353–361.

Hansen, A.K., Jeong, G., Paine, T.D., and Stouthamer, R. (2007). Frequency of secondary symbiont infection in an invasive psyllid relates to parasitism pressure on a geographic scale in California. *Applied and Environmental Microbiology* 73, 7531–7535.

Harmon, J.P., Moran, N.A., and Ives, A.R. (2009). Species response to environmental change: impacts of food web interactions and evolution. *Science* 323, 1347–1350.

Haynes, S., Darby, A.C., Daniell, T.C., Webster, G., van Veen, F.J., Godfray, H.C., Prosser, J.L., and Douglas, A.E. (2003). Diversity of bacteria associated with natural aphid populations. *Applied and Environmental Microbiology* 69, 7216–7223.

Hinde, R. (1971). The control of the mycetome symbiosis of the aphids *Brevicoryne brassicae*, *Myzus persicae*, and *Macrosiphum rosae*. *Journal of Insect Physiology* 17, 1791–1800.

International Aphid Genomics Consortium. (2010). Genome sequence of the pea aphid *Acyrthosiphon pisum*. *PLoS Biology* 8, e1000313.

Lamelas, A., Pérez-Brocal, V., Gómez-Valero, L., Gosalbes, M.J., Moya, A., and Latorre, A. (2008). Evolution of the secondary symbiont 'Candidatus Serratia symbiotica' in aphid species of the subfamily Lachninae. *Applied and Environmental Microbiology* 74, 4236–4240.

Leonardo, T.E. (2004). Removal of a specialisation-associated symbiont does not affect aphid fitness. *Ecology Letters* 7, 461–468.

Leonardo, T.E., and Miura, G.T. (2003). Facultative symbionts are associated with host plant specialization in pea aphid populations. *Proceedings of the Royal Society Series B* 270, S209–S212.

Leonardo, T.E., and Mondor, E.B. (2006). Symbiont modifies host life-history traits that affect gene flow. *Proceedings of the Royal Society Series B* 273, 1079–1084.

McLean, D.L., and Houk, E.J. (1973). Phase contrast and electron microscopy of the myceto-cytes and symbiotes of the pea aphid *Acyrthosiphon pisum*. *Journal of Insect Physiology* 19, 625–633.

Montllor, C.B., Maxmen, A., and Purcell, A.H. (2002). Facultative bacterial endosymbionts benefit pea aphids *Acyrthosiphon pisum* under heat stress. *Ecological Entomology* 27, 189–195.

Moran, N.A. (1996). Accelerated evolution and Muller's ratchet in endosymbiotic bacteria. *Proceedings of the National Academy of Sciences of the United States of America* 93, 2873–2878.

Moran, N.A., and Dunbar, H.E. (2006). Sexual acquisition of beneficial symbionts in aphids. *Proceedings of the National Academy of Sciences of the United States of America* 103, 12803–12806.

Moran, N.A., Russell, J.A., Koga, R., and Fukatsu, T. (2005). Evolutionary relationships of three new species of Enterobacteriaceae living as symbionts of aphids and other insects. *Applied and Environmental Microbiology* 71, 3302–3310.

Oliver, K.M., Campos, J., Moran, N.A., and Hunter, M.S. (2008). Population dynamics of defensive symbionts in aphids. *Proceedings of the Royal Society Series B* 275, 293–299.

Oliver, K.M., Degnan, P.H., Hunter, M.S., and Moran, N.A. (2009). Bacteriophages encode factors required for protection in a symbiotic mutualism. *Science* 325, 992–994.

Oliver, K.M., Moran, N.A., and Hunter, M.S. (2005). Variation in resistance to parasitism in aphids is due to symbionts not host genotype. *Proceedings of the National Academy of Sciences of the United States of America* 102, 12795–12800.

Oliver, K.M., Moran, N.A., and Hunter, M.S. (2006). Costs and benefits of a superinfection of facultative symbionts in aphids. *Proceedings of the Royal Society Series B* 273, 1273–1280.

Oliver, K.M., Russell, J.A., Moran, N.A., and Hunter, M.S. (2003). Facultative bacterial symbionts in aphids confer resistance to parasitic wasps. *Proceedings of the National Academy of Sciences of the United States of America* 100, 1803–1807.

Pérez-Brocal, V., Gil, R., Ramos, S., Lamelas, A., Postigo, M., Michelana, J.M., Silva, F.J., Moya, M., and Latorre, A. (2006). A small microbial genome: the end of a long symbi-otic relationship? *Science* 314, 312–313.

Ramsey, J.S., MacDonald, S.J., Jander, G., Nakabachi, A., Thomas, G.H., and Douglas, A.E. (2010). Genomic evidence for complementary purine metabolism in the pea aphid *Acyrthosiphon pisum* and its symbiotic bacterium *Buchnera aphidicola*. *Insect Molecular Biology* 19 (S2), 241–248.

Russell, J.A., Latorre, A., Sabater-Muñoz, B., Moya, A., and Moran, N.A. (2003). Side-stepping secondary symbionts: widespread horizontal transfer across and beyond the Aphidoidea. *Molecular Ecology* 12, 1061–1075.

Russell, J.A., and Moran, N.A. (2006). Costs and benefits of symbiont infection in aphids: variation among symbionts and across temperatures. *Proceedings of the Royal Society Series B* 273, 603–610.

Sandstrom, J.P., Russell, J.A., White, J.P., and Moran, N.A. (2001). Independent origins and horizontal transfer of bacterial symbionts of aphids. *Molecular Ecology* 10, 217–228.

Scarborough, C.L., Ferrari, J., and Godfray, H.C. (2005). Aphid protected from pathogen by endosymbiont. *Science* 310, 1781.

Simon, J.C., Carre, S., Boutin, N., Prunier-Leterme, N., Sabater-Munoz, B., Latorre, A., and Bournoville, R. (2003). Host-based divergence in populations of the pea aphid: insights from nuclear markers and the prevalence of facultative symbionts. *Proceedings of the Royal Society Series B* 270, 1703–1712.

Thao, M.L., and Baumann, P. (2004). Evidence for multiple acquisition of *Arsenophonus* by whitefly species (Sternorrhyncha: Aleyrodidae). *Current Microbiology* 48, 140–144.

Tsuchida, T., Koga, R., and Fukatsu, T. (2004). Host plant specialization governed by facultative symbiont. *Science* 303, 1989.

Tsuchida, T., Koga, R., Meng, X.Y., Matsumoto, T., and Fukatsu, T. (2005). Characterization of a facultative endosymbiotic bacterium of the pea aphid *Acyrthosiphon pisum*. *Microbial Ecology* 49, 126–133.

Tsuchida, T., Koga, R., Sakurai, M., and Fukatsu, T. (2006). Facultative bacterial endosymbionts of three aphid species, *Aphis craccivora*, *Megoura crassicauda* and *Acyrthosiphon pisum* symbpatrically found on the same host plants. *Applied and Environmental Zoology* 41, 129–137.

Tsuchida, T., Koga, R., Shibao, H., Matsumoto, T., and Fukatsu, T. (2002). Diversity and geographical distribution of secondary endosymbiotic bacteria in natural populations of the pea aphid, *Acyrthosiphon pisum*. *Molecular Ecology* 11, 2123–2135.

Unterman, B., Baumann, P., and McLean, D.L. (1989). Pea aphid symbiont relationships established by analysis of 16S rRNAs. *Journal of Bacteriology* 171, 2970–2974.

van der Wilk, F., Dullemans, A.M., Verbeek, M., and van den Heuvel, J.F. (1999). Isolation and characterization of APSE-1, a bacteriophage infecting the secondary endosymbiont of *Acyrthosiphon pisum*. *Virology* 262, 104–113.

von Burg, S., Ferrari, J., Müller, C.B., and Vorburger, C. (2008). Genetic variation and covariation of susceptibility to parasitoids in the aphid *Myzus persicae*: no evidence for trade-offs. *Proceedings of the Royal Society Series B* 275, 1089–1094.

Vorburger, C., Gehrer, L., and Rodriguez, P. (2010). A strain of bacterial symbiont *Regiella insecticola* protects aphids against parasitoids. *Biology Letters* 6, 109–111.

Vorburger, C., Sandrock, C., Gouskov, A., Castañeda, L.E., and Ferrari, J. (2009). Genotypic variation and the role of defensive endosymbionts in an all-parthenogenetic host-parasitoid interaction. *Evolution* 63, 1439–1450.

Wilkinson, T.L., Adams, D., Minto, L.B., and Douglas, A.E. (2001). The impact of host plant on the abundance and function of symbiotic bacteria in an aphid. *Journal of Experimental Biology* 204, 3027–3038.

Wilkinson, T.L., Koga, R., and Fukatsu, T. (2007). Role of host nutrition in symbiont regulation: impact of dietary nitrogen on proliferation of obligate and facultative bacterial endosymbionts of the pea aphid *Acythosiphon pisum*. *Applied and Environmental Microbiology* 73, 1326–1366.

Wille, B.D., and Hartman, G.L. (2009). Two species of symbiotic bacteria present in the soybean aphid (Hemiptera: Aphididae). *Environmental Entomology* 38, 110–115.

Wilson, A.C.C., Ashton, P.D., Calevro, F., Charles, H., Colella, S., Febvay, G., Jander, G., Kushlan, P., Macdonald, S.A., Schwartz, J., Thomas, G.H., and Douglas, A.E. (2010). Genomic insight into the amino acid relations of the pea aphid *Acyrthosiphon pisum* with its symbiotic bacterium *Buchnera aphidicola*. *Insect Molecular Biology* 19 (S2), 249–258.

5 *Stammerula* and Other Symbiotic Bacteria within the Fruit Flies Inhabiting Asteraceae Flowerheads

Luca Mazzon, Isabel Martinez-Sañudo,
Claudia Savio, Mauro Simonato,
and Andrea Squartini

CONTENTS

Genus	*Stammerula*
Species	*tephritidis*
Family	Enterobacteriaceae
Order	Enterobacteriales
Description year	2008
Origin of name	In honor of the German zoologist Hans-Jürgen Stammer (1899–1968), who first discovered bacteria associated with Tephritinae flies
Description	
Reference	Mazzon et al. (2008)

Hans-Jürgen Stammer. (Photo from Postenr 1970.)

INTRODUCTION

The many instances of associations between insects and microorganisms observed in nature have long since stimulated the curiosity of scientists. In the past, such studies mostly relied upon microscopy to define the morphohistological features of symbiotic organs and describe the bacteria hosted within. Indeed, many insect-microbe associations were covered in Buchner's renowned treatise (Buchner 1965). However, as most of the prokaryotic microsymbionts are not culturable *ex situ*, their characterization and taxonomical placement has had to wait until the advent of biomolecular techniques that have enabled 16S rRNA-based taxonomy via gene amplification and sequencing.

In the case of flies belonging to the family Tephritidae, the olive fly *Bactrocera oleae* (Rossi) (subfamily Dacinae) was the first species for which a bacterial symbiosis was described, by the Italian plant pathologist Lionello Petri (Petri 1909). As regards the subfamily Tephritinae (Diptera, Tephritidae), the subject of this chapter, the presence of symbiotic bacteria was described for the first time by the German zoologist Hans-Jürgen Stammer (Stammer 1929).

In this chapter we review our recent work focused on the presence of hereditary and nonculturable symbiotic bacteria in the subfamily Tephritinae. This study concerns different species of the subfamily, exploring the examples described by Stammer (1929) and extending the analysis to uncover the identities of the symbionts via 16S rRNA sequencing. This approach enables us to verify the existence of a strict host-symbiont specificity and to trace the possible phylogenetic relationships occurring among hosted bacteria.

This review gives us the opportunity to briefly trace a historical overview to the steps leading up to the present knowledge of bacterial symbiosis in the family Tephritidae, and in the Tephritinae subfamily. This chapter will also focus on fly-rearing methods under microbiologically controlled conditions in order to obtain monoxenic adults. To conclude, we discuss the biological meaning of the symbiosis and suggest critical aspects that could be explored in the future. For further details, the following original literature can be referred to: Stammer (1929), Girolami (1973, 1983), Mazzon et al. (2008).

THE SUBFAMILY TEPHRITINAE

The subfamily Tephritinae includes about 200 genera with over 1,800 species from all zoogeographical regions (Norrbom et al. 1999). The Tephritinae are almost ubiquitous and have penetrated even to subarctic and mountain tundras, as well as alpine and arid deserts (Korneyev 1999). The larvae predominantly infest flowerheads of the Asteraceae, and for this reason Tephritinae are considered a nonfrugivorous group of tephritids (Zwölfer 1983; Straw 1989; Headrick and Goeden 1998). Tephritinae is considered to be the most specialized subfamily of Tephritidae (Korneyev 1999; Merz 1999). This host specialization is a key reason for which this subfamily has been regarded as a monophyletic group in the past (Han et al. 2006).

With some rare exceptions, species of this group of tephritids are not considered pests of economic importance. As a consequence, most of the scientific interest has been focused on other subfamilies of tephritid flies grouping important pests (e.g., *Ceratitis*, *Anastrepha*, *Rhagoletis*). Nevertheless, some species of subfamily Tephritinae have been tested as possible biological control agents for weeds (Zwölfer 1983).

THE DISCOVERY OF SYMBIOTIC BACTERIA IN THE SUBFAMILY TEPHRITINAE AND IN OTHER TEPHRITID FLIES

The first example of hereditary bacterial symbiosis in the fruit flies dates back to the beginning of the twentieth century. Petri (1909) showed that a cephalic organ (esophageal bulb), connected to the pharynx of the adult of olive fly *Bactrocera* (= *Dacus*) *oleae* (Rossi) (subfamily Dacinae), harbors symbiotic bacteria. These bacteria continuously multiply within this organ, forming masses that are discharged into the midgut. The female, endowed with contractile perianal glands that become filled with bacteria, discharges symbionts to the chorion egg's surface during oviposition. Subsequently, the bacteria multiply inside intestinal ceca of all larval stages. The same author suggested that the symbiont might be "*Bacterium*" (*Pseudomonas*) *savastanoi* Smith, the causal agent of the olive knot disease.

Thanks to the molecular approach, Capuzzo et al. (2005) proved that the symbiont of the olive fly is not *P. savastanoi* and proposed a novel inheritable, nonculturable bacterial species, designated as "*Candidatus* Erwinia dacicola" (hereafter *Erwinia dacicola*). Subsequent recent studies regarding symbiosis in olive fly have confirmed the presence of *Erwinia dacicola* in specimens coming from different geographical areas (Sacchetti et al. 2008; Estes et al. 2009; Kounatidis et al. 2009).

In the case of flies belonging to the subfamily Tephritinae, the presence of bacterial symbiosis has been described by Stammer (1929) in several genera (Table 5.1), whereas in other genera it has been excluded. Stammer analyzed a total of 37 species belonging to 3 subfamilies (Dacinae, Trypetinae, and Tephritinae). In particular, based on microscopy techniques, he postulated the presence of symbiotic bacteria in eight genera: *Paroxyna* (= *Campiglossa*), *Tephritis*, *Oxyna*, *Acanthiophilus*, *Sphenella*, *Trypanea* (= *Trupanea*), *Schistopterum*, and *Euarestella*. For the last two, only the larvae could be studied (Table 5.1).

Stammer further reported that in adult flies, symbiotic bacteria are harbored in a particular tract of the midgut that features specialized structures for this aim. Moreover, he observed the presence of anal glands in the ovipositor of these species, similar to those of *B. oleae* described by Petri. As in *B. oleae*, these anal glands contribute to the transmission of bacteria to the egg, ensuring the vertical transmission of symbiotic bacteria. Larval stages maintain bacteria in their intestinal ceca similarly to the larvae of *B. oleae*.

Up to present no hereditary and unculturable symbiotic bacteria have been described in other tephritid flies. Several authors report the presence of communities of "associated bacteria" in the alimentary tract of tephritids flies belonging mostly to the genera *Klebsiella*, *Enterobacter*, and *Pseudomonas* (e.g., Drew and Lloyd 1987; Daser and Brandl 1992; Lauzon 2003; Behar et al. 2009).

THE ESOPHAGEAL BULB IN TEPHRITINAE AND IN THE OTHER TEPHRITID FLIES

The esophageal bulb is a cephalic diverticulum of adult fruit flies (Girolami 1973) connected by a short tract with the foregut. The presence of this organ in "Trypetidae" (= Tephritidae) was long discussed. For a long time it was considered unique to *B. oleae*. Petri (1909) in his treatise of bacterial symbiosis in the olive fly wrote: "I've examined the adults of *Rhagoletis cerasi* and *Ceratitis capitata* without finding any dorsal diverticulum of the pharynx or the esophagus equivalent to that of the olive fly."* Later Stammer (1929) similarly affirmed: "We sought without success a cephalic organ in *Tephritis conura* or in any species studied."† Therefore, the presence of an esophageal bulb different from that of *B. oleae* in Tephritidae was denied by both Petri and Stammer. Also, Buchner (1965) reported that with the exception of *B. oleae* and the Tephritinae in which Stammer had pointed out the presence of symbiotic bacteria in a specialized tract of the midgut, the remaining

---

* "Ho esaminato gli adulti della Rhagoletis cerasi e della Ceratitis capitata senza trovare alcun diverticolo dorsale della faringe o dell'esofago omologo a quello della mosca olearia." [As originally written.]
† "Vergeblich suchen wir bei T. conura oder auch bei irgendeiner der anderen in der Folge beschriebenen Arten nach einem Kopforgan." [As originally written.]

TABLE 5.1

Species of Tephritinae and Relative Host Plants in Which the Presence of Symbiotic Bacteria Has Been Pointed Out Using Traditional Microscopy (Stammer, 1929) or Molecular Techniques (16S rRNA Gene)

	Stammer 1929		Mazzon et al. 2008	
Tephritinae Species	**Presence of Symbiosis**	**Host Plant**	**Presence of Symbiosis**	**Host Plant**
Acanthiophilus helianthi Rossi	x	*Centaurea* sp.	x	*Centaurea jacea* *Cirsium arvense*
Campiglossa (= Paroxyna) absinthii Fabr.[a]	x	*Bidens* sp.	—	—
Campiglossa doronici Loew	—	—	x	*Doronicum austriacum*
Campiglossa guttella Rond.	—	—	x	*Hieracium murorum* *H. glaucum*
Campiglossa tessellata Loew (= *Paroxyna tessalata* Loew)	x	*Taraxacum* *Leontodon* *Crepis* *Sonchus*[b]	—	—
Capitites ramulosa Loew	—	—	x	*Phagnalon saxatile*
Dioxyna bidentis Rob.-Des.	—	—	x	*Bidens tripartita*
Euarestella iphionae Effl.[c]	x	?	—	—
Noeeta bisetosa Merz	—	—	x	*Hieracium piloselloides*
Noeeta pupillata Fall. (= *Noëta pupillata* Rob.-Des.)	N.D.	*Hieracium umbellatum*	x	*Hieracium porrifolium* *H. murorum* *H. umbellatum*
Oxyna flavipennis Loew	—	—	x	*Achillea millefolium*
Oxyna nebulosa Wied.	x	*Chrysanthemum* sp.	—	—
Schistopterum moebiusi Becker[c]	x	?	—	—
Sphenella marginata Fall.	x	*Senecio vulgaris* *S. jacobaea*	x	*Senecio inaequidens* *S. vulgaris* *S. alpinus*
Tephritis arnicae Linn.	—	—	x	*Arnica montana*
Tephritis bardanae Schrank	x	*Lappa tomentosa* (= *Arctium tomentosum*)	x	*Arctium minus*
Tephritis cometa Loew	x	*Cirsium arvense*	x	*Cirsium arvense*
Tephritis divisa Rondani	—	—	x	*Picris echioides*
Tephritis conura Loew	x	*Cirsium oleraceum*	x	*Cirsium oleraceum* *C. erisithales*

(continued)

TABLE 5.1 (continued)
Species of Tephritinae and Relative Host Plants in Which the Presence of Symbiotic Bacteria Has Been Pointed Out Using Traditional Microscopy (Stammer, 1929) or Molecular Techniques (16S rRNA Gene)

	Stammer 1929		Mazzon et al. 2008	
Tephritinae Species	**Presence of Symbiosis**	**Host Plant**	**Presence of Symbiosis**	**Host Plant**
Tephritis formosa Loew	—	—	x	*Sonchus oleraceus*
Tephritis heiseri Frau.	x	*Carduus crispus*	—	—
Tephritis hendeliana Hering	—	—	x	*Carduus chrysacanthus C. nutans*
Tephritis hyoscyami Linn.	—	—	x	*Carduus personata*
Tephritis leontodontis De Geer	x	*Leontodon hispidum (= hispidus) L. autumnalis*	x	*Leontodon autumnalis L. hispidus*
Tephritis matricariae Loew	—	—	x	*Crepis vesicaria*
Hyalotephritis (= Tephritis) planiscutellata B.	x	*Pluchea (Coryza) dioscoridis*	—	—
Tephritis postica Loew	—		x	*Onopordum acanthium*
Tephritis ruralis Loew	x	*Hieracium pilosella*	—	
Tephritis separata Rondani	—	—	x	*Picris hieracioides*
Tephritis sp.	—	—	x	*Crepis chondrilloides*
Tephritis fallax Loew	x	*Leontodon hispidum (= hispidus)*	x	*Leontodon hispidus*
Trupanea (= Trypanea) amoena Frau.	x	*Several Asteraceae*	x	*Lactuca serriola Reichardia picroides*
Trupanea (= Trypanea) stellata Fuessl.	x	*Matricaria inodora Senecio jacobea S. vulgaris*	x	*Erigeron annuus Crepis foetida*

Note: N.D., not detected; —, not tested; ?, unknown.

[a] *Paroxyna absinthii* from *Bidens* sp. (Stammer 1929) probably corresponds with the current species *Dioxyna bidentis*.

[b] Most of these records probably refer to *Campiglossa producta*.

[c] Observations carried out only in larval stages.

tephritids (*Chaetostomella, Xyphosia, Noeeta, Ditricha, Ceratitis,* and *Rhagoletis*) feature "larvae and imagines that both lack specific symbiotic organs."

In a following part (p. 117), the same author, considering the description of the alimentary tract of *Rhagoletis pomonella* (Dean 1933, 1935), indicated for this species the presence of a "pharyngeal bulb like that of *Dacus*," hypothesizing its implication with bacterial symbioses.

Subsequently, specific morphohistological studies showed that the esophageal bulb is always present in tephritid flies, even with shapes and sizes different from those of *B. oleae* and not always related to the presence of symbiosis. Hence, the esophageal bulb is always present in the subfamily Tephritinae, but in this case it appears small and devoid of bacteria (Girolami 1973, 1983). Girolami (1973) identified three principal types of "pharyngeal diverticulum" (= esophageal bulb) in the fruit flies, differing from a morphohistological point of view and for the presence or absence of bacteria within (Figure 5.1):

1. "*Dacus* type," distinctive of *B. oleae* according to the description of Petri (1909). This type is large and spherical, with a base provided of columnar epithelial cells and full of symbiotic bacteria.
2. "*Ceratitis* type," distinctive of *Ceratitis capitata* and other species of the subfamilies Trypetinae (e.g., genus *Rhagoletis*) and Dacinae (with the exception of *B. oleae*). It is also spherical but smaller than the first type, often containing bacteria and with a columnar epithelium in the apex.
3. "*Ensina* type," distinctive of some genera of the subfamily Tephritinae (*Acanthiophilus, Trupanea, Ensina, Noeeta, Tephritis, Urophora, Xyphosia, Campiglossa, Oxyna, Sphenella*). It is small with an ovoid shape, a wide muscle tunic, and without columnar epithelium in the apex. Bacteria have been never found in this kind of esophageal bulb. Interestingly, all the Tephritinae in which symbiotic bacteria have been detected in a specialized tract of the midgut present this type of esophageal bulb.

Girolami (1973) also described another type of esophageal bulb characteristic of other Tephritinae (*Chaetorellia, Chaetostomella, Orellia, Terellia*) and defined it "*Chaetorellia* type." It is an intermediate esophageal bulb, resembling the "*Ceratitis* type" in the apex spherical part and the "*Ensina* type" in the basal part (Figure 5.1).

Besides the inheritable and nonculturable symbionts of the esophageal bulb of *B. oleae* ("*Ca.* Erwinia dacicola"), several studies regarding the identity of bacteria present in the esophageal bulb of *Ceratitis capitata* and some species of the genus *Rhagoletis* have been carried out using culture-dependent methods. All studies confirm that this organ harbors a discrete community of free-living bacteria belonging predominantly to the genera *Klebsiella, Enterobacter,* and *Pseudomonas* (Rossiter et al. 1983; Marchini et al. 2002; Howard et al. 1985; Lauzon et al. 1998). In the majority of cases these bacterial communities have been commonly found in the intestinal lumen of the same insects species using both cultivation-dependent and molecular techniques (Lauzon 2003; Behar et al. 2009). The same bacterial species have been detected in some Tephritinae, which present an esophageal bulb "type *Chaetorellia*" (Figure 5.1). Stammer (1929) did not report the presence of a mesointestinal symbiosis in that species.

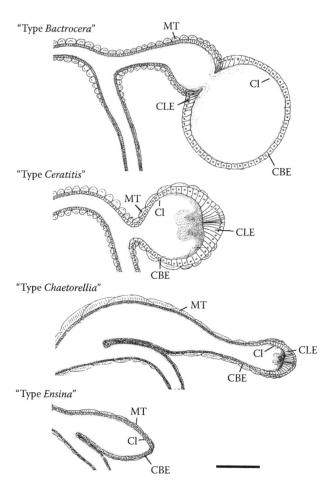

FIGURE 5.1 Different types of esophageal bulbs of tephritid flies. Abbreviations: CBE, cuboidal epithelium; CLE, columnar epithelium; CI, cuticular intima; MT, muscle tunic. Bar = 0.1 mm. (From Girolami, V., *Redia*, 54, 269–294, 1973. With permission.)

The characteristics of the esophageal bulb (e.g., spherical apex) seems to be related to the aptitude to host bacteria inside, regardless of whether bacteria are truly inheritable and unculturable symbionts ("type *Dacus*") or if they are free-living bacteria ("type *Ceratitis*" and "type *Chaetorellia*") (Girolami 1973; 1983; Capuzzo et al. 2005). An exhaustive study of the esophageal bulb ultrastructure has been carried out by Ratner and Stoffolano (1984) in *Rhagoletis pomonella*. A thorough study comparing the morphohistology of the esophageal bulb and the identity of the bacteria harbored could provide additional interesting information useful for phylogenetic studies of fruit flies and also for attempts to estimate the time of establishment of the symbiosis.

MIDGUT BACTERIAL SYMBIOSIS IN THE TEPHRITINAE SUBFAMILY

In the adults of species in which the symbiosis is present, bacteria occupy an extracellular location and are visible in the first part of the abdominal midgut. For these species Girolami (1973, 1983) reports that their esophageal bulb appears small and devoid of bacteria (Figure 5.1). The bacteria, according to Stammer (1929), adhere to specific tracts of the intestinal epithelium but, as later reported by Girolami (1983), external to the peritrophic membrane, and therefore not inside the intestinal lumen in direct contact with the food bolus (Figure 5.2). The peritrophic membrane, which is present in different insect species, is a membranous film that forms a thin lining layer that surrounds the food bolus and separates it from the delicate midgut epithelium.

Morphologically the tract of the midgut where the symbiosis occurs features a larger size and, in most species (e.g., genera *Tephritis* and *Trupanea*), is uniformly covered by outgrowths called crypts that harbor the symbiotic bacteria (Figures 5.3A, B and 5.4A).

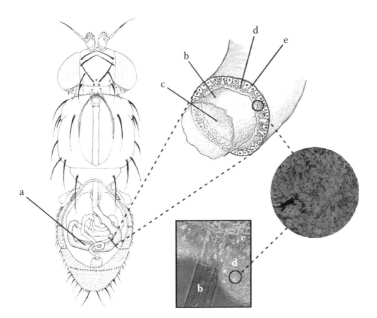

FIGURE 5.2 See color insert. Anatomical location of symbiotic bacteria in *Tephritis matricariae*. The drawing, portraying an insect observed under a dissecting microscope, shows the position of the midgut tract in which symbiotic bacteria reside (yellow portion). (a) Close-up detail (right) shows the coaxial presence of an inner thin vessel: the peritrophic membrane (b), in whose lumen (c) regular alimentary bolus transit occurs and many nonspecific easily culturable bacteria can be found. Instead, in the interstitial gap space that runs all along (d), between the peritrophic tube and the outer midgut epithelium (e), resident bacteria are observed, which constitute the target of the present analysis. BacLight assays for bacterial cell viability under fluorescent light microscopy; live cells fluoresce green; dead cells fluoresce red. (From Mazzon, L., Piscedda, A., Simonato, M., Martinez-Sañudo, I., Squartini, A., and Girolami V., *Int. J. Syst. Evol. Microbiol.*, 55, 1641–1647, 2008. With permission.)

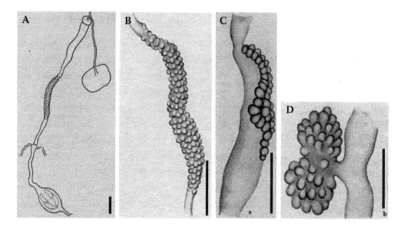

FIGURE 5.3 Early descriptions of the midgut structures that harbor symbiotic bacteria in adults of Tephritinae flies. (A) Alimentary tract of *Tephritis conura*. (B) Detail of the midgut region of *T. conura* showing the "crypts" where symbiotic bacteria are located. (C) Detail of the midgut of an adult of *Acanthiophilus helianthi* showing epithelial evaginations located only in one side of the midgut. (D) Detail of the midgut of *Sphenella marginata* showing bilobate evaginations with slightly separated position with respect to the midgut. Bars = 500 μm. (Modified from Stammer, H.J., *Zoomorphology*, 15, 481–523, 1929. With permission.)

In *Acanthiophilus helianthi* and *Sphenella marginata* these crypts are more evident and peculiar. In particular, *A. helianthi* has 20–30 evaginations along one side of the midgut only (Figures 5.3C and 5.4B), whereas *S. marginata* houses its symbiont in a two-lobed villiferous evagination (Figures 5.3D and 5.4C) (Stammer, 1929).

Anyhow, bacteria are consistently found in extracellular placements in the enclosed niche between the epithelium and the peritrophic membrane even when they are hosted inside a voluminous specialized evagination of the midgut (Girolami 1983; Mazzon et al. 2008). In the larval stages, symbiotic bacteria are hosted inside the intestinal ceca located at the beginning of the midgut, similarly to *B. oleae*. Their structure was thoroughly studied by Stammer, mainly in the species *Tephritis heiseri*. During the first larval instars there are four intestinal ceca, but in the subsequent larval instars they become more complex, and novel constrictions appear, causing further subdivisions. Large masses of bacteria are easily visible inside the intestinal ceca (Figure 5.5).

MECHANISM OF VERTICAL TRANSMISSION OF THE SYMBIONT

Stammer (1929), studying the ovipositor of the Tephritinae with symbiotic bacteria, detected revealing structures related to the vertical transmission mechanisms of symbionts.

As described in *B. oleae* by Petri (1909), this ovipositor presents "anal glands" full of symbiotic bacteria. During the oviposition the egg comes into contact with these glands, which release symbiotic bacteria, smearing the eggs, in order to ensure

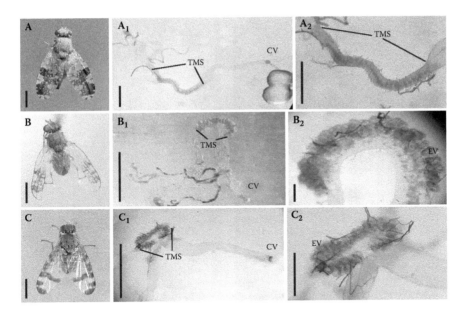

FIGURE 5.4   See color insert. Three representative species of Tephritinae flies with their respective dissected alimentary tract and detail of midgut structures adapted to harbor symbiotic bacteria: (A) *Tephritis bardanae*, (B) *Acanthiophilus helianthi*, and (C) *Sphenella marginata*. Bars = 1.5 mm (A, A_1, A_2, B, B_1, C, and C_1), 0.1 mm (B_2), and 0.5 mm (C_2). Abbreviations: EV, evaginations of the midgut full of symbiotic bacteria; TMS, tract of the midgut adapted to harbor symbiotic bacteria; CV, cardial valve.

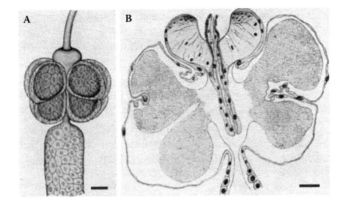

FIGURE 5.5   Larval intestinal ceca. (A) Intestinal ceca of a young larva of *Tephritis conura*. (B) Longitudinal section of the proventricle and intestinal ceca containing symbiotic bacteria in a larva of *Tephritis bardanae*. Bars = 100 µm (A) and 500 µm (B). (From Stammer, H.J., *Zoomorphology*, 15, 481–523, 1929. With permission.)

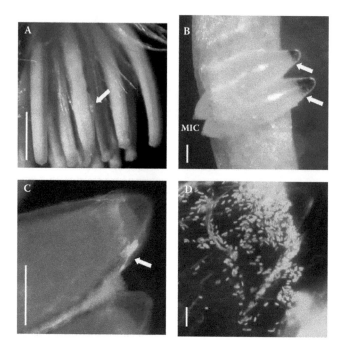

FIGURE 5.6   See color insert. (A,B) Eggs of *Tephritis matricariae* inserted in a flowerhead of *Crepis vesicaria*. (C,D) BacLight assays for bacterial cell viability under fluorescent light microscopy. Bacteria can be observed in the chorion of the eggs mainly in the opposite pole of the micropyle, partially embedded in a mucilaginous layer. Arrows indicate areas where the majority of bacteria are commonly found. Abbreviations: MIC, micropyle. Bars = 2.3 mm (A), 100 µm (B and C), and 10 µm (D).

their vertical transmission to the offspring. However, in some genera (e.g., *Noeeta*) the presence of these glands is not yet clear, and additional histological studies need to be carried out in order to clarify these aspects.

In the egg's chorion, bacteria can be detected mostly opposite of the micropyle, partially embedded within a mucilaginous layer (Figure 5.6). These bacteria are more visible in freshly laid eggs. It is worth noticing that although the newly formed larva has its cephalic extremity positioned toward the micropylar side, just before hatching the larva turns inside the egg, and erupts through the opposite pole of the micropyle (Stammer 1929), thereby coming into contact with the bacteria.

In order to study the possible coincidence of these bacteria with *Stammerula*, the application of fluorescence *in situ* hybridization (FISH) is in progress.

DETECTION OF SYMBIONTS IN TEPHRITINAE FLIES

LABORATORY HANDLING OF THE FLIES

In order to carry out biological and molecular experiments, the different species of Tephritinae need to be kept and reared in the laboratory. Thus, we have obtained

FIGURE 5.7 Rearing cages for adults of Tephritinae flies.

adults, rearing them from mature larvae or pupae present in field-collected infested flowerheads. The adults obtained were routinely maintained in tulle-lined cages (Figure 5.7) and fed with a 50% sucrose solution (w/v) containing 0.2% benzoic acid and 0.05% sorbic acid as antimicrobial agents. From our experience, the majority of the species reared as described above can easily live in the laboratory at room conditions, for at least 2 or 3 months after their emergence. Interestingly, species belonging to the genus *Tephritis* are extremely long-lived. In particular, *Tephritis matricariae* can be kept alive for over a year.

REARING UNDER MICROBIOLOGICALLY CONTROLLED CONDITIONS

Obtaining adults from surface-sterilized pupae and keeping them in axenic conditions has allowed us to eliminate the contamination of free-living bacteria present in the environment, and therefore unrelated to the hereditary symbiosis. This rearing technique, devised during the studies of the symbiosis in the olive fly (Capuzzo et al. 2005), has been improved and streamlined over time. The optimized protocol described below is in our experience the most efficient. The first and most delicate step consists in extracting the pupa from the flowerhead. This needs to be carried out carefully using the stereomicroscope, in order to avoid damage to the pupa. Once taken out pupae were surface sterilized by a 5 min immersion in 1% sodium hypochlorite. They were subsequently rinsed in sterile water at least twice, air-dried in sterile conditions, and kept in sterile small vials until the emergence of the adults (Figure 5.8A). The resulting flies were kept under microbiologically controlled conditions to avoid contamination with foreign microorganisms. The flies were then aseptically transferred under a laminar flow hood into larger glass vials (30 mm in diameter by 150 mm in height) containing a layer of plate count agar

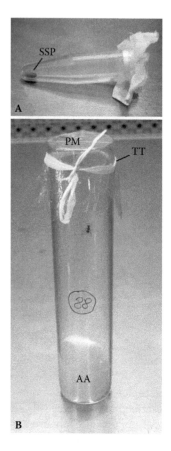

FIGURE 5.8 Microbiological setup for single fly rearing. (A) Sterile vial for adult emerging. (B) Sterile glass vial for adult rearing (30 mm in diameter by 150 mm in height). Abbreviations: AA, agar media; PM, permeable membrane; SSP, surface-sterilized pupa; TT, Teflon® tape.

(PCA) on the bottom, as a sterility check, and sealed with a sterilized transparent gas-permeable cellulose membrane for dialysis (MEMBRA-CEL dialysis tubing MD77, Viskase, Inc., Chicago, IL. U.S.) (Figure 5.8B). A drop of sterile glucose solution was placed on the internal side of the membrane to allow insect feeding. The drop of solution was rewetted, whenever necessary, by spraying water on the external surface of the cellulose membrane at the top of the vials. These flies can survive without problems for weeks. Adult flies were used for the analysis after at least a week of such rearing.

In retrospect, a methodological insight was achieved. Identical bacterial 16S rRNA sequences, within each species tested, were obtained from all the samples analyzed irrespective of the rearing history of the individuals. In fact, adults from surface-sterilized pupae hatched in sterile vials or those that emerged in nonsterile conditions, or even wild ones that were occasionally captured, yielded the same bacterial 16S rRNA gene sequence.

As a consequence, rearing in a microbiologically controlled situation is not always necessary, and a further simplified procedure could be useful for future studies of this kind. In such case, however, if the adults have been obtained without microbiologically controlled conditions, it is important to avoid newly hatched ones and to work with insects that are at least 7 days old. In the former, there are usually too few bacteria and a polymerase chain reaction (PCR) may be template-limited. Stammer (1929) noticed that bacteria become abundant and are easily detected in insects about a week after emergence.

INSECT DISSECTION

The following protocol was adopted in our studies. After being reared for a week under microbiologically controlled conditions, the flies were dissected. Flies were handled under a stereomicroscope in a laminar flow hood using sterile equipment and sterile water.

The abdomen of the insects was opened and the whole intestine was extracted. The midgut tract was selected from the intestine by sectioning between the cardial valve and the malpighian tubes. The resulting segment was transferred into a sterile Eppendorf tube and used in part for bacterial culturability tests and in part for bacterial DNA extraction and amplification.

As reported above, for bacterial DNA extraction it is possible to also use adults that have not been reared under microbiologically controlled conditions (either captured in the field or emerged from the flowerhead in the laboratory). In these cases, the peritrophic membrane was pulled off the midgut and discarded, and the epithelium was gently rinsed in sterile water. These operations were intended to minimize the contamination of bacteria from the alimentary bolus in these specimens, which had not been fed with sterile solutions. These additional procedures of removing the peritrophic membrane and rinsing the midgut could ensure a clean PCR outcome even in a nongnotobiotic situation, supposedly due to a prevailing amount of the specific putative endosymbiont DNA.

ATTEMPTS TO CULTURE *EX SITU* SYMBIONTS OF TEPHRITINAE HAVE FAILED SO FAR

When we performed the LIVE/DEAD BacLight bacterial viability test on the bacteria of the midgut in order to verify the actual presence and viability of bacterial cells in the specimens, the majority of cells stained green, indicating a substantially viable population.

However, attempts to culture the bacteria by plating on different standard microbiological media did not yield colonies from any of the host species tested. In the same way, in the majority of the cases, microbial colonies did not develop on the PCA culture medium on the bottom of the vials in which the adults from surface-sterilized pupae were introduced and reared for a week. Besides being a control for the aseptic conditions of the rearing technique, this also indicated the absence of culturable bacteria that were released in the feces or in other excreta. Similar results were observed in many other bacterial-insect associations (e.g., *Erwinia dacicola*), suggesting that the prokaryotic partner is not yet culturable *ex situ* on standard microbiological media (Mazzon et al. 2008). In fact, several studies have concluded that a loss of the ability for multiplication outside the host correlates with symbiotic coevolution between insects and bacteria (Baumann and Moran 1997).

SPECIFICITY OF THE BACTERIAL NUCLEOTIDE SEQUENCE

A bacterial small subunit ribosomal 16S rRNA gene fragment was amplified by PCR and sequenced (about 1,300 bp). The starting material for the molecular taxonomical analysis of the insect-associated prokaryotes was the midgut tract in which bacteria reside.

It is remarkable to report that almost all the fly species investigated are associated with a specific bacterial sequence. In addition, 90% of the clones obtained from the midgut of *T. formosa* and *T. matricariae* corresponded to their specific bacterial sequence found with direct amplification of midgut bacteria. This fact supports the presence of a prevailing symbiotic bacterium in the above-mentioned intestinal tract. However we cannot exclude the presence of further minor unculturable bacteria. In three cases an identical bacterial sequence was shared by two or three species belonging to the same genus or at least systematically close: (1) *Tephritis postica, T. hendeliana, T. hyoscyami*; (2) *T. leontodontis, T. fallax*; and (3) *Trupanea stellata, Capitites ramulosa*. It is important to underline that there was a nearly complete reproducibility of the results for sequences of the bacteria isolated from the same insect species; i.e., the full-length 16S rRNA gene sequences of the bacteria inhabiting a given insect species always turned out to be identical, irrespective of the geographical origin (from the Alps to the central Apennine range, Italy), year of collection, or the host plant species for nonstrictly monophagous flies (e.g., *Acanthiophilus helianthi* and *Sphenella marginata*).

CAN BACTERIAL SPECIFICITY BECOME A TOOL TO DETECT GROUPS OF SIBLING SPECIES?

In our studies we have considered two of the three palaeartic species known for the small genus *Noeeta* (White 1988; Merz 1994): *N. bisetosa* and *N. pupillata*. While the former is monophagous, feeding only on *Hieracium piloselloides*, the latter is oligophagous, feeding on several species of *Hieracium*. For *N. pupillata*, slightly different sequences were observed from specimens obtained from flower-heads of three different plant species of the genus *Hieracium* (*H. porrifolium, H. umbellatum*, and *H. murorum*). As such sequences were repeatable in every host plant independently of the locality or the collection year, we think that this could also suggest the existence of a sympatric complex of sibling species. Traditional morphological studies of adults along with molecular analysis are necessary to address this issue.

PHYLOGENETIC POSITION OF THE SYMBIOTIC BACTERIA

The comparison of bacterial sequences in pairwise alignments indicated a range of similarity, from a maximum of 99% (within the genus *Tephritis* group) to a minimum of 92% (across the symbiont of *Acanthiophilus helianthi* and the symbiont group of the genus *Noeeta*).

A basic local alignment search tool (BLAST) analysis of the sequences revealed the degree of 16S rRNA gene sequence similarity with recognized taxa.

Results indicated that all of the samples belonged to the family Enterobacteriaceae in the class Gammaproteobacteria. Sequences of the bacterial symbionts from the genera *Tephritis*, *Acanthiophilus*, *Sphenella*, *Trupanea*, and *Capitites* shared no more than 96% similarity with database taxa. Instead, there were cases for which a similarity level of about 99% with known culturable species indicated a different relationship; this is the case of the *Campiglossa guttella* symbiont with *Erwinia persicina* (GenBank accession no. AM184098), of the *Dioxyna bidentis* symbiont with *Erwinia persicina* (Z96086), and of the symbionts from the *Noeeta* genus group with *Ewingella americana* (DQ383802).

Representative members of the closest known relatives indicated by the BLAST analysis were also included. It was noteworthy that the tree topology of the bacterial symbionts mostly corresponded to the phylogeny of the subfamily Tephritinae based on morphological features.

Bacterial 16S rRNA gene sequences were subjected to molecular phylogenetic analyses. Three main clades were clearly distinguished and supported by each of the clustering methods (maximum likelihood (ML), maximum parsimony (MP), and Bayesian inference (BI)) used (Figure 5.9).

Clade A included symbionts from all the analyzed species within the genera *Tephritis*, *Acanthiophilus*, *Sphenella*, *Trupanea*, *Capitites*, and *Oxyna*, as well as the symbiont of *Campiglossa doronici*. None of the other sequences present in GenBank grouped in this clade, suggesting that the symbionts in this group are probably monophyletic. The inference was well supported by the bootstrap values of all three methods used (77% for ML, 73% for MP, and 56% for BI). Inside this clade, the taxon assignment of host insects for the genus *Tephritis* was strongly aligned with the grouping of their symbiotic bacteria (ML 89%, MP 88%, and BI 96%). The same agreement (100% ML, 100% MP, and 100% BI) was seen for the two species of the genus *Trupanea*, although there was symbiont sequence similarity between *Trupanea stellata* and *Capitites ramulosa*. For this clade, the most similar free-living bacteria belonged to the genus *Erwinia*, sharing 95–96% sequence similarity.

Clade B, which grouped together with free-living members of the genus *Erwinia*, stemmed out as an apparent sister group of clade A. However, the bootstrap value (66%) given by only one method (ML) suggested a weak affinity. *Dioxyna bidentis* and *Campiglossa guttella* symbionts belonged to this clade, as well as our previously described "*Ca.* Erwinia dacicola," the olive fly symbiont (Capuzzo et al. 2005), with similarity values ranging from 97 to 99% with other species of the genus *Erwinia*.

It is noteworthy that symbionts of the two analyzed species of the genus *Campiglossa* belonged to different clades.

Clade C was statistically well supported (100% for all three methods) and included symbionts found in all species of the genus *Noeeta* analyzed. The most similar (99%) free-living culturable species to this group was the human pathogen *Ewingella americana*. The symbionts of the genus *Noeeta* appear to have been acquired

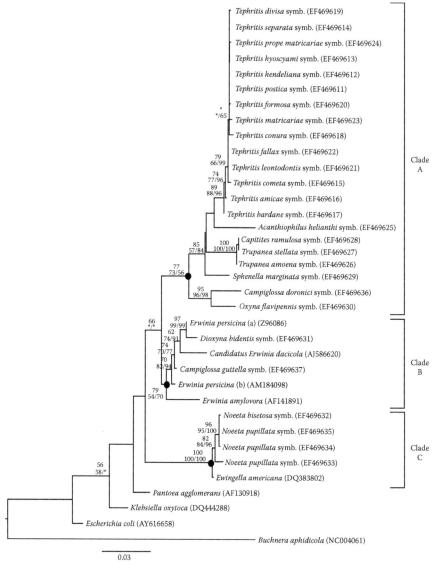

FIGURE 5.9 Phylogenetic tree of bacterial symbionts of the subfamily Tephritinae and close relatives within the class Gammaproteobacteria, based on the 16S rRNA gene sequences, constructed upon the alignment of a minimum common portion of 1,324 nucleotides (GenBank accession numbers given in parentheses). The ML tree is shown, with the bootstrap probabilities (ML, MP) and posterior probabilities (BI) reported on the nodes scoring a support higher than 50% in at least one of the three methods (asterisks indicate bootstrap values lower than 50%). Position of the values at nodes: ML, top; MP/BI, bottom. For the three cases in which identical symbiont sequences were found (*Tephritis postica–Tephritis hendeliana–Tephritis hyoscyami*; *Tephritis leontodontis–Tephritis fallax*; and *Trupanea stellata–Capitites ramulosa*), only one member is listed in the branch. Bar = number of substitutions per site for a unit branch length. (From Mazzon et al. 2008. With permission.)

independently from those of the other members of the subfamily Tephritinae considered in this study. The presence of bacteria hosted by this group of Tephritinae deserves to be studied in depth.

The bacterial symbiosis of the subfamily Tephritinae may therefore be considered not to be monophyletic (Mazzon et al. 2008). In this respect, it is important to recall that the genus *Noeeta* belongs to the tribe Noeetini, which is phylogenetically well separated (Han et al. 2006) from the tribe Tephritini, which includes all of the other species of clades A and B (with the exception of *E. dacicola*, which was found from one species belonging to the subfamily Dacinae).

The 16S rRNA gene sequences of *Klebsiella oxytoca* and *Pantoea agglomerans*, two of the most common fruit fly-associated intestinal bacterial species for the families Trypetinae and Dacinae (Rossiter et al. 1983; Lloyd et al. 1986), did not show particular similarity with any of these three clades.

PROPOSAL OF "*CANDIDATUS* STAMMERULA TEPHRITIDIS"

The phylogenetic analyses delineated above describe three main phenons. For one of these, namely, clade A, a deep and robust separation from the other branches exists, and the levels of similarity with known taxa are below 96%.

The degree of 16S rRNA gene sequence divergence that currently separates different genera within the family Enterobacteriaceae ranges between 2 and 8% (Moran et al. 2005). On the basis of these premises, we proposed the designation of a candidate genus, "*Candidatus* Stammerula," in honor of Prof. Hans-Jürgen Stammer, who first described bacteria associated with Tephritinae flies, to include symbionts of Tephritinae flies occurring in clade A (Mazzon et al. 2008).

Among these, an extremely coherent subset is represented by the sequences originating from all insects belonging to the genus *Tephritis*, sharing 99% gene sequence similarity (Figure 5.9). We proposed that this group is representative of a candidate species under the designation "*Candidatus* Stammerula tephritidis" (hereafter *Stammerula tephritidis*). The specific name indicates the insect genus (*Tephritis*) with which the bacteria are associated.

Regarding the more variable remaining sequences that also cluster in clade A, encompassing the symbionts from members of the genera *Acanthiophilus*, *Trupanea*, *Sphenella*, *Campiglossa*, and *Oxyna*, we propose (Mazzon et al. 2008) that these be gathered under the same designation, and we envisage a possible future assignment either to the candidate genus described above or to new ones once a larger number of insect hosts is investigated, in order to refine the rank attribution of their symbionts.

RATES OF EVOLUTION

When bacteria live confined within isolated contexts, a particularly fast rate of DNA sequence evolution is shown to take place. This is presumably due to the absence of recombination with external populations and because of a more pronounced effect of genetic drift (Moran 1996; Brynnel et al. 1998). Comparison of the mean of the observed number of nucleotide substitutions per site of symbionts of Tephritinae with their related free-living neighbors *Stammerula* exhibited a significantly high rate of substitution, ranging from 3.2- to 4.7-fold with respect to their free-living sister lineage (*Erwinia amylovora*). Relative substitution rates from *Campiglossa guttella* (clade B) and all *Noeeta* symbionts (clade C) were not significantly different from those of their nearest culturable bacteria (Mazzon et al. 2008).

CONSIDERATIONS AROUND THE SYMBIOSIS

Regarding hypotheses on the origin of this symbiosis, it appears that at least three distinct events could have taken place. The earliest concerns clade A and possibly involves the putative ancestor of the present representatives of the genus *Erwinia*. More recent independent events of lateral genetic transfer would instead be supported by the situation observed in the two other clades, in both of which Tephritinae symbionts appear intermingled with free-living members of the family Enterobacteriaceae (*Erwinia* in clade B and *Ewingella* in clade C) or with symbionts of more distantly related insects (e.g., *B. oleae*, belonging to the Dacinae subfamily). For the latter case, we envisage the possibility of an acquisition in relatively recent time, involving the descent of free-living species commonly occurring on vegetation. Other cases of insect-bacteria symbioses of apparently polyphyletic origin, invoking an interpretation based on lateral gene transfer events, are reported in the literature (e.g., Lefèvre et al. 2004).

In contrast to many symbiotic bacteria, whose location is intracellular, *Stammerula* and other Tephritinae symbionts are extracellular and are located between the peritrophic tube and the midgut epithelium. Thus, the chance of contact and possibly competition with the outer environment is higher than in endocellular symbionts.

CONCLUSIONS AND PERSPECTIVES

A thorough evaluation of the evolutionary relationships between hosts and symbionts will require the expansion of the research to several yet untouched but fascinating aspects. It will be interesting to extend the research to symbionts of species of the subfamily Tephritinae inhabiting other zoogeographic regions, and carry out a parallel analysis on the insect hosts by sequencing their 16S rRNA and other genes to verify the phylogenetic congruence with their symbiotic bacteria. Interestingly, Stammer pointed out that the presence of symbiotic bacteria is prerogative only of some and not all Tephritinae. If modern biomolecular techniques will confirm Stammer's work, the study of bacterial distribution in Tephritinae could help to further understand the above-discussed phylogenetic relationships among the tribes of the subfamily. This research, along with attempts to estimate the time of establishment of the symbionts and their evolutionary rate, will allow us to reconstruct the coevolutionary history between symbionts and insect hosts.

Furthermore, some aspects of the vertical transmission of symbiotic bacteria are still unclear. In particular, the fine mechanisms that ensure specific and efficient transmission of bacteria from the larval stages to the adult insect need to be clarified. In order to resolve this aspect, it is necessary to elucidate which is the location of symbiotic bacteria once larvae have rearranged their tissues during the metamorphosis. For such purposes, specific oligonucleotide probes have been designed and fluorescent *in situ* hybridization (FISH) tests are in progress for the different stages of the flies' life cycle (egg, larva, pupa, adult).

Moreover, it would be interesting to find answers to some intriguing questions like: Why is symbiosis present only in some species of the Tephritinae? What is the biological advantage of the symbiosis for the host during its life cycle? Is it functional for the larval development or for adult life? In this respect, it should be stressed that immature seeds and receptacles of Asteraceae, where larvae of Tephritinae develop, are considered nutritionally complete (Girolami 1983). The widely accepted tenet that symbiotic bacteria are commonly found in insects whose diet is not well balanced (Buchner 1965) allows us to consider that the Tephritinae symbiosis could be important not for the larval stage, but for the adult one. This view is supported by the fact that most of the Tephritinae that feature a symbiosis are reported to overwinter as adults (e.g., the *Tephritis* genus) (Merz 1999), while for others that overwinter, like larva or pupa in the flowerheads, no inheritable symbiont bacteria have yet been found.

ACKNOWLEDGMENTS

Special thanks to Prof. Vincenzo Girolami for critical revision of the manuscript and precious suggestions.

REFERENCES

Baumann, P., and Moran, N.A. (1997). Non-cultivable microorganisms from symbiotic associations of insects and other hosts. *Anton. Leeuw. Int. J. G.* 72: 39–48.

Behar, A., Ben-Yosef, M., Lauzon, C.R., Yuval, B., and Jurkevich, E. (2009). Structure and function of the bacterial community associated with the Mediterranean fruit fly. In *Insect symbiosis* (K. Bourtzis and T.A. Miller, Eds.), 251–271. CRC Press, Boca Raton, FL.

Brynnel, E.U., Kurland, C.G., Moran, N.A., and Andersson, S.G. (1998). Evolutionary rates for *tuf* genes in endosymbionts of aphids. *Mol. Biol. Evol.* 15: 574–582.

Buchner, P. (1965). *Endosymbiosis of animals with plant microorganisms*. Interscience Publishers, New York.

Capuzzo, C., Firrao, G., Mazzon, L., Squartini, A., and Girolami, V. (2005). 'Candidatus Erwinia dacicola', a coevolved symbiotic bacterium of the olive fly *Bactrocera oleae* (Gmelin). *Int. J. Syst. Evol. Microbiol.* 55: 1641–1647.

Daser, U., and Brandl, R. (1992). Microbial gut floras of eight species of tephritids. *Biol. J. Linn. Soc.* 45: 155–165.

Dean, R.W. (1933). Morphology of the digestive tract of the apple maggot fly, *Rhagoletis pomonella* Walsh. *Agr. Exp. Sta. N.Y. Bull.* 215.

Dean, R.W. (1935). Anatomy and postpupal development of the female reproductive system in the apple maggot fly, *Rhagoletis pomonella* Walsh. *Agr. Exp. Sta. N.Y. Bull.* 229.

Drew, R.A.I., and Lloyd, A.C. (1987). Relationship of fruit flies (Diptera: Tephritidae) and their bacteria to host plants. *Ann. Entomol. Soc. Am.* 80: 629–636.

Estes, A.M., Hearn, D.J., Bronstein, J.L., and Pierson, E.A. (2009). The olive fly endosymbiont, "Candidatus Erwinia dacicola," switches from an intracellular to an extracellular existence during host insect development. *Appl. Environ. Microbiol.* 75: 7097–7106.

Girolami, V. (1973). Reperti morfo-istologici sulle batteriosimbiosi del *Dacus oleae* Gmelin e di altri ditteri tripetidi, in natura e negli allevamenti su substrati artificiali. *Redia* 54: 269–294.

Girolami, V. (1983). Fruit fly symbiosis and adult survival: general aspects. In *Fruit flies of economic importance* (R. Cavalloro, Ed.), 74–76. Proceedings of the CEC/IOBC International Symposium, Athens, November 1982.

Han, H.Y., Ro, K.E., and McPheron, B.A. (2006). Molecular phylogeny of the subfamily Tephritinae (Diptera: Tephritidae) based on mitochondrial 16S rDNA sequences. *Mol. Cells* 22: 78–88.

Headrick, D.H., and Goeden, R.D. (1998). The biology of nonfrugivorous tephritid fruit flies. *Annu. Rev. Entomol.* 43: 217–241.

Howard, D.J., Bush, G.L., and Breznak, J.A. (1985). The evolutionary significance of bacteria associated with *Rhagoletis*. *Evolution*. 39: 405–417.

Korneyev, V.A. (1999). Phylogeny of the subfamily Tephritinae: relationships of the tribes and subtribes. In *Fruit flies (Tephritidae): phylogeny and evolution of behavior* (M. Aluja and A.L. Norrbom, Eds.), 549–580. CRC Press, Boca Raton, FL.

Kounatidis, I., Crotti, E., Sapountzis, P., Sacchi, L., Rizzi, A., Chouaia, B., Bandi, C., Alma, A., Daffonchio, D., Mavragani-Tsipidou, P., and Bourtzis, K. (2009). *Acetobacter tropicalis* is a major symbiont in the olive fruit fly *Bactrocera oleae*. *Appl. Environ. Microbiol.* 75: 3281–3288.

Lauzon, C.R. (2003). Symbiotic relationships of Tephritids. In *Insect symbiosis* (K. Bourtzis, and T.A. Miller, Eds.), 115–129. CRC Press, Boca Raton, FL.

Lauzon, C.R., Sjogren, R.E., Wright, S.E., and Prokopy, R.J. (1998). Attraction of *Rhagoletis pomonella* (Diptera: Tephritidae) flies to odor of bacteria: apparent confinement to specialized members of Enterobacteriaceae. *Environ. Entomol.* 27: 853–857.

Lefèvre, C., Charles, H., Vallier, A., Delobel, B., Farrell, B., and Heddi, A. (2004). Endosymbiont phylogenesis in the Dryophthoridae weevils: evidence for bacterial replacement. *Mol. Biol. Evol.* 21: 965–973.

Lloyd, A.C., Drew, R.A.I., Teakle, D.S., and Hayward, A.C. (1986). Bacteria associated with some *Dacus* species (Diptera: Tephritidae) and their host fruit in Queensland. *Aust. J. Biol. Sci.* 39: 361–368.

Marchini, D., Rosetto, M., Dallai, R., and Marri, L. (2002). Bacteria associated with the oesophageal bulb of the Medfly *Ceratitis capitata* (Diptera:Tephritidae). *Curr. Microbiol.* 44: 120–124.

Mazzon, L., Piscedda, A., Simonato, M., Martinez-Sañudo, I., Squartini, A., and Girolami, V. (2008). Presence of specific symbiotic bacteria in flies of the subfamily Tephritinae (Diptera Tephritidae) and their phylogenetic relationships: proposal of 'Candidatus Stammerula tephritidis.' *Int. J. Syst. Evol. Microbiol.* 58: 1277–1287.

Merz, B. (1994). Diptera Tephritidae. Schweizerischen Entomologischen Gesellschaft. *Insecta Helvetica Fauna*, 10: 1–198.

Merz, B. (1999). Phylogeny of the Palaearctic and Afrotropical genera of the *Tephritis* group (Tephritinae: Tephritini). In *Fruit flies (Tephritidae): phylogeny and evolution of behavior* (M. Aluja and A.L. Norrbom, Eds.), 629–669. CRC Press, Boca Raton, FL.

Moran, N.A. (1996). Accelerated evolution and Muller's ratchet in endosymbiotic bacteria. *Proc. Natl. Acad. Sci. USA* 93: 2873–2878.

Moran, N.A., Russell, J.A., Koga, R., and Fukatsu, T. (2005). Evolutionary relationships of three new species of enterobacteriaceae living as symbionts of aphids and other insects. *Appl. Environ. Microbiol.* 71: 3302–3310.

Norrbom, A.L., Carroll, L.E., Thompson, F.C., White, I.M., and Freidberg, A. (1999). Status of knowledge. In *Fruit fly expert identification system and systematic information database* (F.C. Thompson, Ed.), pp. 9–47. *Myia* (1998) 9: 524 pp. Backhuys Publisher, Leiden, Netherlands.

Petri, L. (1909). *Ricerche Sopra i Batteri Intestinali della Mosca Olearia*. Roma: Memorie della Regia Stazione di Patologia Vegetale di Roma.

Postenr, M. (1970). Professor Stammer zum Gedenken. *Anz. Schad.* 7: 107–108.

Ratner, S.S., and Stoffolano, J.G. (1984). Ultrastructural changes of the esophageal bulb of the adult female apple maggot, *Rhagoletis pomonella* (Walsh) (Diptera: Tephritidae). *Int. J. Insect. Morphol. Embryol.* 13: 191–208.

Rossiter, M.C., Howard, D.J., and Bush, G.L. (1983). Symbiotic bacteria of *Rhagoletis pomonella*. In *Fruit flies of economic importance* (R. Cavalloro, Ed.), 77–82. Proceedings of the CEC/IOBC International Symposium, Athens, November 1982.

Sacchetti, P., Granchietti, A., Landini, S., Viti, C., Giovannetti, L., and Belcari, A. (2008). Relationships between the olive fly and bacteria. *J. Appl. Entomol.* 132: 682–689.

Savio, C., Mazzon, L., Martinez-Sañudo, I., Simonato, M., Squartini, A., and Girolami, V. (2011). Evidence of two lineages of the symbiont "*Candidatus* Erwinia diaciola" in Italian populations of *Bactrocera oleae* (Rossi) basedon 16S rRNA gene sequence. *Int. J. Syst. Evol. Microbiol.* DOI 10.1099/ijs.0.030668-0. In press.

Stammer, H.J. (1929). Die bakteriensymbiose der trypetiden (Diptera). *Zoomorphology.* 15, 481–523.

Straw, N.A. (1989). Taxonomy, attack strategies and host relations in flowerhead Tephritidae: a review. *Ecol. Entomol.* 14: 455 – 462.

White, I.M. (1988). *Handbooks for the identification of british insects. Tephritid flies, Diptera: Tephritidae*, 1–134. Vol. X, Part 5(a). Royal Entomological Society, London.

Zwölfer, H. (1983). Life systems and strategies of resource exploitation in tephritids. In *Fruit flies of economic importance* (R. Cavalloro, Ed.), 16–30. Proceedings of the CEC/IOBC International Symposium, Athens, November 1982.

6 *Candidatus* Midichloria mitochondrii

Symbiont or Parasite of Tick Mitochondria?

*Dario Pistone, Luciano Sacchi,
Nathan Lo, Sara Epis, Massimo Pajoro,
Guido Favia, Mauro Mandrioli,
Claudio Bandi, and Davide Sassera*

CONTENTS

Genus	*Midichloria*
Species	*mitochondrii*
Family	Rickettsiaceae (or member of a novel family)
Order	Rickettsiales
Description year	2006
Origin of name	After George Lucas's *Star Wars* movie series, where the concept of symbiosis is presented as a close association between intracellular microorganisms (named Midichlorians) and living creatures
Description	
Reference	Sassera et al. (2006)
Anecdote	*M. mitochondrii* is the first intramitochondrial bacterium described. George Lucas allowed Sassera et al. (2006) to derive the name of these bacteria from the Midichlorians.

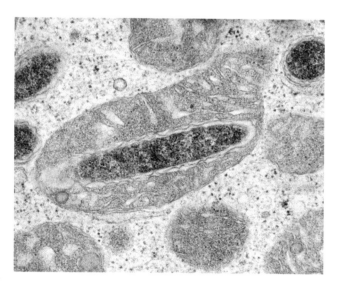

Mitochondrion of an oocyte of the tick Ixodes ricinus "infected" by Midichloria mitochondrii.

BASIC INFORMATION ON *CANDIDATUS* MIDICHLORIA MITOCHONDRII

Candidatus Midichloria mitochondrii is described as a Gram-negative, non-spore-forming intracellular bacterium, possessing a rippled outer envelope. It presents a bacillar shape and an average size of 0.5×1.2 μm (Lewis 1979, Zhu et al. 1992, Sacchi et al. 2004).

Candidatus Midichloria mitochondrii shows the unusual capacity of residing inside mitochondria. So far, the animal hosts in which *Candidatus* Midichloria mitochondrii has been observed inside mitochondria are hard ticks of the family Ixodidae

(Sassera et al. 2006, Epis et al. 2008). These microorganisms are localized in the intermembrane space of these organelles, where they replicate, leading to a reduction of the mitochondrial matrix, until the organelle assumes the shape of an empty sac full of bacteria. The fact that bacteria are able to invade mitochondria, organelles derived from bacteria over 1.5 billion years ago, raises interesting evolutionary questions on the nature of this relationship and on the origin of the invasion process of this organelle. Were bacteria phylogenetically related to *Candidatus* Midichloria mitochondrii able to invade the free-living ancestor of mitochondria, or has this capacity evolved only recently? (Davidov and Jurkevitch 2009).

From a taxonomic point of view *Candidatus* Midichloria mitochondrii represents a divergent lineage of the order Rickettsiales, bacteria belonging to the class α-Proteobacteria (Sassera et al. 2006). This order currently encompasses three families (Lee et al. 2005): the Anaplasmataceae (comprising the genera *Anaplasma*, *Ehrlichia*, *Wolbachia*, and *Neorickettsia*), the Rickettsiaceae (genera *Rickettsia* and *Orientia*), and the Holosporaceae (genera *Holospora* and *Caedibacter*). *Candidatus* Midichloria mitochondrii and other related bacteria are placed by recent phylogenetic analyses as the sister group of the Rickettsiaceae (Figure 6.1; Weinert et al. 2009). The genus *Midichloria* could thus be regarded either as a member of this family or as a representative of a novel family.

We will from here on out refer to *Candidatus* Midichloria mitochondrii of *Ixodes ricinus* and other phylogenetically related tick symbionts as *M. mitochondrii*. Other bacteria, associated with a variety of hosts and environments, that cluster as sister groups of *M. mitochondrii* will be referred to as *Midichloria*-like organisms (MLOs).

IXODES RICINUS AND *MIDICHLORIA MITOCHONDRII*, A SPECIAL RELATIONSHIP

Most of the observations thus far published on *M. mitochondrii* have been obtained in studies on the hard tick *I. ricinus*. This tick is the main vector of Lyme disease in Europe, and is also vector of a variety of other pathogenic microorganisms of medical and veterinary importance (Cotté et al. 2009). In several cell types (luminal cells, funicular cells, and oocytes) of the reproductive tissues of females of *I. ricinus*, *M. mitochondrii* is observed free in the cytoplasm or included in a host-derived membrane. In addition, in luminal cells and oocytes, a high proportion of these bacteria are observed within the mitochondria, in the space between the two membranes of these organelles (Sacchi et al. 2004). Transmission electron microscopy (TEM) images allow us to infer that the bacteria reduce the mitochondrial matrix while they multiply therein. There is thus evidence that these bacteria 'consume' the mitochondrial matrix. Molecular screenings thus far published indicate that *M. mitochondrii* is ubiquitous in females of *I. ricinus* across the geographical distribution of this species (100% prevalence), while lower prevalence is observed in males (44%; Lo et al. 2006). Since *M. mitochondrii* is vertically transmitted from the females to the eggs, with an efficiency of 100% (i.e., all eggs laid by infected females are infected), we might interpret the lower prevalence in males as the result of a decrease in the amount of bacteria during male development. The bacterial population could thus

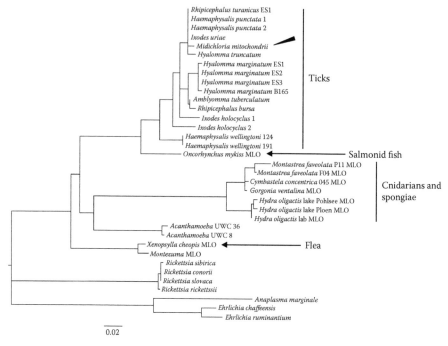

FIGURE 6.1   Representative 16S rDNA-based tree, showing the relationships between the *Midichloria mitochondrii* symbionts of ticks and other *Midichloria*-like organisms (MLOs). Arrowhead indicates *M. mitochondrii*. Bacteria of the genera *Anaplasma*, *Ehrlichia*, and *Rickettsia* were included as outgroups. The names at the terminal nodes of the tree are those of the eukaryotic host species, with the exception of bacterial species of the genera *Anaplasma*, *Ehrlichia*, and *Rickettsia*, and of Montezuma, which is the unofficial name of a MLO sequence retrieved from ticks and humans. The tree was generated using the neighbor-joining method after Kimura correction. The bar indicates the number of substitutions per site.

decline to an undetectable level. Variable numbers of *M. mitochondrii* have been observed within the mitochondria; in fact, from a single bacterium to over 20 bacteria can be present in a single ultra-thin section (Figure 6.2). Despite the high proportion of mitochondria parasitized by the bacterium, the eggs develop normally (Lo et al. 2006).

In order to discuss the possible role of *M. mitochondrii* in the biology of ticks, in the following paragraphs we will summarize the information available on this bacterium.

1. The prevalence of *M. mitochondrii* is 100% in wild-collected females of *I. ricinus*. In males and females from colonies that had been maintained in the laboratory for a few generations, prevalence is lower, even below 20% (Lo et al. 2006 and unpublished observations).

2. The distribution of *M. mitochondrii* in tick species and genera is not uniform. A recently published screening showed the presence of *M. mitochondrii* in 7 species of hard ticks of a total of 21 examined (Epis et al. 2008). Further tick species positive for *M. mitochondrii* have been

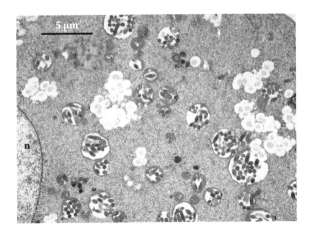

FIGURE 6.2 Developing oocyte of *Ixodes ricinus* with several *Midichloria* bacteria harbored in mitochondria that appear extremely swollen. Only remnants of the matrix can be observed. n = nucleus.

detected in other studies (Venzal et al. 2008, Mediannikov et al. 2004). There is evidence that prevalence of *M. mitochondrii* in tick species different from *I. ricinus* could be lower than 100%, even in females collected in the wild. It is also noteworthy that based on comparisons between the symbiont and host phylogenies, we can exclude coevolution between the two.

3. *M. mitochondrii* sensu stricto is the type species, the one described in *I. ricinus*. The information so far acquired indicates a certain degree of molecular variation of these bacteria in hard ticks (i.e., *M. mitochondrii* from different tick species can show up to about 3% nucleotide differences at the level of the 16S rDNA gene). However, within the same tick species, there is generally limited or no variation in *M. mitochondrii* 16S rDNA genes, with the exception of *I. holocyclus*, which harbors two strains with a 2.5% 16S rDNA nucleotide difference (Lo et al. 2006, Beninati et al. 2009). Throughout this chapter, we have chosen the option of referring to all of the tick bacteria that cluster within the same group of *M. mitochondrii* sensu stricto with less than 3% nucleotide substitution at the 16S rDNA level simply as *M. mitochondrii*; the acronym MLO is used in this chapter to refer to those bacteria that are phylogenetically related with *M. mitochondrii* and show a nucleotide difference on 16S rDNA up to 14%.

4. In general, congruence between host and symbiont phylogenies indicates that the two partners have coevolved (Lo et al. 2003), while lack of congruence might indicate some degree of symbiont autonomy and capacity of host exchange. The phylogeny of *M. mitochondrii* does not match the phylogeny of the host ticks (Epis et al. 2008). This observation emphasizes an evolutionary pattern that is different from the one observed in primary beneficial symbionts of arthropods and filarial nematodes, where phylogenies of the host and its symbiont are congruent (Funk et al. 2000, Dale and Moran

2006). This lack of consistency in the host and symbiont phylogenies might indicate that *M. mitochondrii* can be transmitted not only vertically from the female to the progeny, but also horizontally, possibly between ticks belonging to different species.

5. Even though over 10 tick species have so far been shown to harbor *M. mitochondrii*, ultra-structural information about the location of the bacteria is available only for three species. In the European ticks *I. ricinus* and *Rhipicephalus bursa*, bacteria have been observed in mitochondria (Epis et al. 2008). In the Australian tick *Ixodes holocyclus*, the bacteria have been observed only in the cytosol (Beninati et al. 2009).

6. There is definitive evidence that *M. mitochondrii* is vertically transmitted in *I. ricinus*, from adult females to the offspring. The bacteria then follow the development of the host in females, with an efficient transstadial transmission (i.e., bacteria are maintained through the molts from larva to nymph to adult). As already emphasized, maintenance and transstadial transmission in males is clearly less efficient, in that only less than 50% of the adult males are infected. The population dynamics of *M. mitochondrii* has been followed by quantitative real-time polymerase chain reaction (qPCR): bacterial numbers increase after the blood meal and decrease after the molts (Sassera et al. 2008).

The high prevalence of *M. mitochondrii* in wild-collected adult females of *I. ricinus* seems to suggest a beneficial role of this bacterium toward the host tick. However, the decrease in the number of bacteria in laboratory-reared ticks, without visible effects on the biology of the hosts, suggests that the presence of the symbionts, if beneficial, is facultative, at least in laboratory conditions.

M. mitochondrii could play an important role in natural conditions, where ticks might not be able to obtain a blood meal for extended periods. Under controlled laboratory conditions, bacteria may be lost when ticks are nourished regularly, or when antibiotics are placed in the supplied blood. The bacteria might synthesize metabolites that become necessary to the host in stressful situations, or under particular environmental conditions. For example, a complementation of the metabolic capacity of the host might be needed in situations of long starvation. Indeed, *I. ricinus* is a three-host tick (triphasic cycle); thus, larvae, nymphs, and adults feed on different hosts and, after each blood meal, detach and molt or lay eggs on the ground (adult females). This can lead to long time windows between two events of nutrition, up to 2 years. The amount of *M. mitochondrii* blooms after the blood meal in any transstadial stage, with the exception of the last molt in adult males, which usually do not take a blood meal. This fact might indicate a possible role for these bacteria in the host molting processes (Sassera et al. 2008) or in the process of heme detoxification (Graça-Souza et al. 2006). A further hypothesis regarding the role of *M. mitochondrii* in ticks could be that this bacterium confers resistance toward tick pathogens. There are currently no data supporting this role for *M. mitochondrii*. However, there is growing evidence that secondary symbionts of insects (i.e., beneficial but not obligatory symbionts) play a role in conferring protection toward different types of pathogens/parasites (Moran et al. 2005). Finally, we should remember that *M. mitochondrii*, due to its massive presence, is unquestionably an important member of the tick microbial community, likely interacting with other pathogenic and nonpathogenic bacteria and protozoa present in the host and transmitted to vertebrates, possibly even interfering with the transmission of these microorganisms (Baldridge et al. 2004).

At the moment, there is no evidence to suggest that *M. mitochondrii* is responsible for any form of manipulation of reproduction in ticks (e.g., male killing, parthenogenesis, feminization of male embryos, or cytoplasmic incompatibility). *M. mitochondrii* does not seem to cause sex ratio distortion in *I. ricinus* (Lo et al. 2006). The lower prevalence in adult males could be interpreted as a specialization of *M. mitochondrii* for females that ensures vertical transmission, while males represent a dead end for the bacteria. In *I. ricinus*, adult males do not need a blood meal to mate, while females are obliged to have a blood meal for completing oogenesis. It must be noted that *M. mitochondrii* shows the most remarkable population increase in this step of the host life cycle (i.e., oogenesis).

In addition to the above speculation, we emphasize three main points: (1) *M. mitochondrii* multiplies within mitochondria in *I. ricinus*, apparently causing a degeneration of these parasitized organelles; (2) the load of *M. mitochondrii* in tick ovaries and oocytes is dramatically high, considering that a single egg of *I. ricinus* can contain up to 1 million genomes of these bacteria; and (3) as evidenced by TEM, a high proportion of mitochondria in oocytes harbor *M. mitochondrii*, in agreement with the above estimation of the bacterial number in eggs.

It is clear that *M. mitochondrii* represents a nonnegligible component of the tick biomass, particularly in the ovaries and eggs. How can a tick manage to reproduce successfully and develop while harboring this huge bacterial population? How can a massive bacterial colonization of mitochondria be compatible with normal cell functioning? The interaction between *M. mitochondrii*, the tick, and its mitochondria is probably different from any kind of host-symbiont interaction so far described.

MIDICHLORIA-LIKE ORGANISMS IN ARTHROPODS AND ELSEWHERE

As previously discussed, the phylogeny of *M. mitochondrii* does not match that of the host ticks. This suggests that the bacterium can be transmitted horizontally even between ticks belonging to different species. For example, the distantly related tick species *Rhipicephalus turanicus* and *Hyalomma truncatum* harbor *M. mitochondrii* with identical 16S rDNA sequences (Epis et al. 2008). How could *M. mitochondrii* move horizontally between different ticks species? A possible hypothesis is that horizontal transmission occurs through a passage in a vertebrate host. PCR evidence for the presence of *M. mitochondrii* in the salivary glands of *I. ricinus* is consistent with this hypothesis (C. Bandi unpublished).

Is there any evidence for the capacity of *M. mitochondrii* to infect vertebrate hosts? Bacteria showing high sequence similarity with *M. mitochondrii* have been detected by PCR in the blood of roe deer, in the context of a screening for pathogens vectored by ticks (Skarphédinsson et al. 2005). Of course, this bacterial presence revealed by PCR does not provide any evidence for the capacity of *M. mitochondrii* to infect vertebrates. It is possible that PCR revealed bacteria, or even just bacterial DNA, as simply inoculated by ticks. However, considering the three lines of evidence so far acquired (horizontal transmission between ticks, DNA presence in salivary glands, and DNA presence in blood of tick hosts), we can hypothesize that *M. mitochondrii* is able to circulate among ticks, passing through the infection of a vertebrate host. There is no need to suppose that *M. mitochondrii* causes pathology in vertebrates, or a long-term infection, since horizontal transmission between ticks could occur through a transient infection of the host, or when ticks are aggregated on the host to get their blood meal (cofeeding; Randolph et al. 1996).

There is, however, circumstantial evidence that MLOs could cause pathological alterations in humans. In 2004 a novel microorganism, unofficially called Montezuma, was detected by PCR in ticks collected in Far East Russia (*Ixodes persulcatus* and *Haemaphysalis concinna*) and in samples from human patients presenting acute febrile symptoms (Mediannikov et al. 2004). Based on the 16S rDNA sequences deposited in the databases, Montezuma is a relative of *M. mitochondrii* (10% nucleotide divergence; Figure 6.1). Thus, based on available information, Montezuma can be regarded as a bacterium belonging to the same clade of *M. mitochondrii*, even though probably too distant to be assigned to the same species.

Besides the possible role of the MLO Montezuma in a human disease, the strawberry disease (SD) of the rainbow trout *Oncorhynchus mykiss* has recently been proposed as being caused by a MLO (Lloyd et al. 2008). SD is a fish skin disorder of unknown etiology, characterized by bright red inflammatory lesions. Analysis using conserved bacterial 16S rDNA primers consistently revealed the presence of MLOs in the skin lesions of fish specimens affected by SD. These bacteria have not been isolated, and there is no clear evidence that this microorganism is actually the etiologic agent of SD. These bacteria display 5% nucleotide divergence to the 16S rDNA of *M. mitochondrii* and, among the MLOs from hosts of different ticks, are the most closely related to *M. mitochondrii* (Figure 6.1). The studies performed on SD have thus provided significant evidence that MLOs can infect vertebrates.

M. mitochondrii and MLOs have also been detected by PCR analysis in other ectoparasitic/hematophagous arthropods: in two tabanid flies *Tabanus bovinus* and *Tabanus tergestinus* (Hornok et al. 2008), in the bed bug *Cimex lectularius* (Richard et al. 2009), in mites of the species *Spelaeorhynchus praecursor* (Acari: Dermanyssoidea) infecting bats (Reeves et al. 2006), and in *Xenopsylla cheopis* (Siphonaptera: Pulicidae), a flea that infests rats (Erickson et al. 2009).

Literature and database searches revealed that there are other 16S rDNA sequences, from different sources, that could be referred to as *M. mitochondrii* or MLOs (Figure 6.1). Among these MLOs, the best characterized are those harbored by amoebae of the genus *Acanthamoeba* (Fritsche et al. 1999). In addition, sequences from MLOs have been retrieved from ciliate protista, other amoebae, cnidarians, marine spongiae, and environmental microbial mats. In the case of the fresh water cnidarian *Hydra oligactis*, in addition to molecular sequence information (87% identity in 16S rDNA with *M. mitochondrii*), the candidate MLO was observed in electron microscopy. It appears as a rod-shaped bacterium located in the cytoplasm of ectodermal epithelial cells of the host (Fraune and Bosch 2007). In the case of ciliate protista, a study has recently been published, describing two different MLO endosymbionts located in the cytoplasm of the ciliate *Euplotes harpa* (Vannini et al. 2010). These MLOs, named *Candidatus* Anadelfobacter veles and *Candidatus* Cyrtobacter comes, represent the two first described species closely related to the *Midichloria mitochondrii* family cluster.

In summary, there is evidence that *M. mitochondrii* and MLOs are quite widespread symbionts, as evidenced by the variety of hosts in which their DNA was detected. Is the capacity of invading the mitochondria a peculiarity of *M. mitochondrii* of ticks, or is the intramitochondrial location a common feature of MLOs? It must be emphasized that bacteria with an intramitochondrial location have been observed in two different species of ciliates: *Halteria geleiana* (Yamataka and Hayashi 1970) and *Urotricha ovata* (de Puytorac and Grain 1972). In addition, bacteria strictly associated with mitochondria have been observed in other ciliate species: *Spirostomum* sp. (Fokin et al. 2005) and *Cyclidium* sp. (Beams and Kessel 1973). However, these bacteria have not yet been identified. It is thus not known whether these bacteria are phylogenetically related to *M. mitochondrii*. Considering that 16S rDNA sequences of MLOs have been detected in amoebae and ciliates, as well as in marine and fresh water metazoans (Figure 6.1), we could hypothesize that aquatic protists might have played an important role in MLO success, constituting a sort of reservoir from which these bacteria could reach or have reached their animal hosts.

LIVING INSIDE MITOCHONDRIA

The cytoplasm of eukaryotic cells represents a suitable niche for a variety of bacteria. In the order Rickettsiales, to which *M. mitochondrii* belongs, all known species share the common feature of multiplying only inside the eukaryotic cell. Different host types have been described for these microorganisms, ranging from vertebrates to nematodes. Several species in the two main families (Rickettsiaceae and Anaplasmataceae) of the order Rickettsiales are associated with arthropods. Bacteria of the genera *Rickettsia* and *Orientia* (Rickettsiaceae), often associated with ticks, are normally encountered free in the cytoplasm of the host cell: they are not surrounded by a host-derived membrane, and their cell wall is directly immersed into the cytosol (Ray et al. 2009). There are also species of *Rickettsia* that have been observed inside the nucleus, again without a surrounding membrane (Ogata et al. 2006). The lack of a host-derived membrane in symbionts is atypical, in that intracellular microorganisms are normally surrounded by this kind of membrane; i.e., they are normally located in a vacuole, which is frequently the result of a process of phagocytosis. This is indeed the case for members of the Anaplasmataceae (genera: *Anaplasma, Ehrlichia, Wolbachia,* and *Neorickettsia*). In the case of *M. mitochondrii*, three different types of locations have been reported: free in the cytosol, in the cytosol surrounded by a membrane, and in the intermembrane space of mitochondria (Figure 6.3). The second location should not necessarily be interpreted as the bacterium being in a vacuole; rather, TEM pictures suggest that the bacteria that are observed within a surrounding membrane are possibly located within a bulge of the mitochondrial membrane (although this interpretation would require further support). In summary, if we exclude the

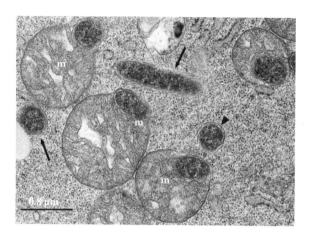

FIGURE 6.3   Detail of the cytoplasm of an oocyte from *Ixodes ricinus*, showing the presence of *Midichloria mitochondrii* either free in the cytoplasm (arrows), in the cytoplasm surrounded by a membrane (arrowhead), or inside mitochondria (m).

intramitochondrial location, *M. mitochondrii* seems to behave in a way similar to the Rickettsiaceae, in agreement with its phylogenetic position.

Intracellular bacteria inside host-derived vacuoles can take advantage of this location, gaining protection in terms of environmental stability, but at the same time they are confined and concentrated in a dangerous area, where they could be under the deleterious action of host lysosomes. Indeed, to avoid this defensive system of the host cell, intracellular bacteria of the genus *Chlamydia*, not phylogenetically related to Rickettsiales, prevent the fusion of the phagosome with the lysosome, and thus the formation of a phagolysosome (Eissenberg and Wyrick 1981). *Rickettsia* has clearly adopted a different strategy, consisting in the early escape from the vacuole soon after cell invasion (Ray et al. 2009).

The strategy adopted by *M. mitochondrii*, based on the invasion and multiplication inside mitochondria, could be interpreted as a third way to escape from the attack of the host cell. While the programmed, partial death of mitochondria (mitoptosis) is known to occur in particular phases of metazoan development (Meier et al. 2000), these organelles are obviously fundamental for the metabolic activity and long-term survival of the cell. We might thus expect that they represent a protected environment, where a bacterium could find a specific niche. Why, then, are intramitochondrial bacteria not widespread? Mitochondria play an important role in triggering the apoptosis of animal cells, both indirectly (decrease of mitochondrial activity results in an energetic crisis and acidification of the cell) and directly (mitochondrial pathway of apoptosis). Alteration of the mitochondrial membranes leads to the release of cytochrome C, triggering the apoptotic pathway (Garrido et al. 2006). Integrity of the mitochondrial membranes is thus crucial to avoid the death of the host cell. It is possible that intramitochondrial parasitism/symbiosis is rare not because of the unsuitability of the organelles for the symbionts, but for the delicate equilibrium between cell life and apoptosis, where mitochondria play a central role. We might hypothesize either that *M. mitochondrii* is capable of multiplying within mitochondria without triggering apoptosis, or that the apoptotic pathway in ticks is triggered differently than in other metazoans. We could go as far as to hypothesize that the mitochondrial pathway of apoptosis was selected as a system to eliminate cells parasitized by intramitochondrial pathogens.

The mitochondrial intermembrane space might represent for *M. mitochondrii* not only a protected shelter, but also an advantageous niche very rich in ATP actively pumped out of the mitochondrial matrix through the action of the mitochondrial ADP/ATP carrier. Could *M. mitochondrii* take advantage of the ATP molecules that flow from the mitochondrial matrix toward the cytoplasm? It is interesting to note that both members of the genus *Rickettsia* and the MLOs of *Acanthamoeba* sp. possess ATP/ADP translocases (Schmitz-Esser et al. 2004) that are believed to import ATP from the host cell cytoplasm. It is thus reasonable to expect that *M. mitochondrii* could also possess these import/export proteins. Should the *M. mitochondrii* genome really encode for an ATP/ADP translocase, this would represent a perfect example of specialization from the point of view of both function and intracellular location.

Mitochondrial ADP/ATP carrier and the proteins involved in ATP acquisition in bacteria of the orders Rickettsiales (including some MLOs) and Chlamydiales have different origin and are not considered homologous (Klingenberg 2008). Bacterial translocases act as importers of ATP, while the mithochondrial version works in the opposite direction, exporting ATP toward the cytoplasm. An additional difference between these two classes of transporters is their origin (Amiri et al. 2003). It is generally assumed that the current mitochondrial ADP/ATP carrier was not present in the free-living ancestor of mitochondria, and it has been proposed that this protein derived from the eukaryotic cell (Embley 2006), representing an important evolutionary step in the process of nuclear control over mitochondria.

MITOCHONDRIAL PATHOGENS

Mitochondria are extremely dynamic organelles that migrate, divide, and fuse inside eukaryotic cells, with repeated cycles of mitochondrial fission and fusion that guarantee mixing of metabolites and mitochondrial DNA (Chen and Chan 2009). This process might also guarantee the repair of mitochondria damaged by physical or chemical stresses. All these dynamic processes modify shape, number, and bioenergetic functionality of mitochondria, thus allowing the cell to face changes in energy demand and environmental conditions. There is mounting evidence that mitochondrial dysfunction has deleterious consequences on the functionality of cells, tissues, and organs. The origin of mitochondria from symbiotic bacteria raises the possibility that mitochondria themselves, like free-living bacteria, have their own pathogens, like predatory/parasitic bacteria and phages (Bongaerts and van den Heuvel 2008). It is thus worth considering whether pathogens of mitochondria do exist, and whether these pathogens might cause any disease in eukaryotes. The discovery of *M. mitochondrii* clearly demonstrates that microorganisms capable of living inside mitochondria do exist. Are there other examples? Is it possible that MLOs in organisms other than ticks cause pathologies? Or, is there any MLO-tick interaction that could represent a form of parasitism?

There is limited knowledge of pathogenic microorganisms for mitochondria. The previously mentioned review article by Bongaerts and van den Heuvel (2008) proposed the word *mitopathogens* to indicate infectious agents capable of determining alterations of mitochondria. In addition to *M. mitochondrii*, we can mention the mitochondrial viruses of the genus *Mitovirus* (family Narnaviridae), for example, the one infecting the mold *Ophiostoma novo-ulmi* (a pathogen of elm trees). Both a bacterium (*M. mitochondrii*) and a group of viruses (*Mitovirus*) have thus been shown to invade mitochondria. Whether *M. mitochondrii* is to be regarded as a pathogen, a commensal, or a beneficial symbiont is still unclear, and certainly this bacterium is not pathogenic for *I. ricinus*. However, from the point of view of the single mitochondrion that is consumed by *M. mitochondrii*, this bacterium is certainly a mitopathogen.

ACKNOWLEDGMENTS

Networking activity and missions of young scientists supported by COST Action FA0701. C.B. research is supported by Ministero della Salute and ISPESLS and by a MIUR-PRIN grant to Claudio Genchi.

REFERENCES

Amiri, H., Karlberg, O., and Andersson, S.G.E. 2003. Deep origin of plastid/parasite ATP/ADP translocases. *Journal of Molecular Evolution* 56: 137–150.

Baldridge, G.D., Burkhardt, N.J., Simser, J.A., et al. 2004. Sequence and expression analysis of the *ompA* gene of *Rickettsia peacockii*, an endosymbiont of the Rocky Mountain wood tick, *Dermacentor andersoni*. *Applied and Environmental Microbiology* 70: 6628–6636.

Beams, H.W., and Kessel, R.G. 1973. Studies on the fine structure of the protozoan *Cyclidium*, with special reference to the mitochondria, pellicle and surface-associated bacteria. *Zeitschrift für Zellforschung und Mikroskopische Anatomie* 139: 303–310.

Beninati, T., Lo, N., Sacchi, L., et al. 2004. A novel alpha-proteobacterium resides in the mitochondria of ovarian cells of the tick *Ixodes ricinus*. *Applied and Environmental Microbiology* 70: 2596–2602.

Beninati, T., Riegler, M., Vilcins, I.M., et al. 2009. Absence of the symbiont *Candidatus* Midichloria mitochondrii in the mitochondria of the tick *Ixodes holocyclus*. *FEMS Microbiology Letters* 299: 241–247.

Bongaerts, G.P., and van den Heuvel, L.P. 2008. Microbial mito-pathogens: fact or fiction? *Medical Hypotheses* 70: 1051–1053.

Chen, H., and Chan, D.C. 2009. Mitochondrial dynamics—fusion, fission, movement, and mitophagy—in neurodegenerative diseases. *Human Molecular Genetics* 18(R2): R169–R176.

Cotté, V., Bonnet, S., Cote, M., et al. 2009. Prevalence of five pathogenic agents in questing *Ixodes ricinus* ticks from western France. *Vector-Borne and Zoonotic Diseases* 10: 1–8.

Dale, C., and Moran, N.A. 2006. Molecular interactions between bacterial symbionts and their hosts. *Cell* 126: 453–465.

Davidov, Y., and Jurkevitch, E. 2009. Predation between prokaryotes and the origin of eukaryotes. *Bioessays* 31: 748–757.

de Puytorac, P., and Grain, J. 1972. Intramitochondrial bacteria and pecularities of cytostomopharyngeal ultrastructure in the ciliate, *Urotricha ovata* Kahl (Ciliata). *Comptes Rendus des Séances de la Société de Biologie et de Ses Filiales* 166: 604–607.

Eissenberg, L.G., and Wyrick, P.B. 1981. Inhibition of phagolysosome fusion is localized to *Chlamydia psittaci*-laden vacuoles. *Infection and Immunity* 32: 889–896.

Embley, T.M. 2006. Multiple secondary origins of the anaerobic lifestyle in eukaryotes. *Philosophical Transactions of the Royal Society B: Biological Sciences* 361: 1055–1067.

Epis, S., Sassera, D., Beninati, T., et al. 2008. *Midichloria mitochondrii* is widespread in hard ticks (Ixodidae) and resides in the mitochondria of phylogenetically diverse species. *Parasitology* 135: 485–494.

Erickson, D.L., Anderson, N.E., Cromar, L.M., et al. 2009. Bacterial communities associated with flea vectors of plague. *Journal of Medical Entomology* 46: 1532–1536.

Fokin, S.I., Schweikert, M., Brümmer, F., et al. 2005. *Spirostomum* spp. (Ciliophora, Protista), a suitable system for endocytobiosis research. *Protoplasma* 225: 93–102.

Fraune, S., and Bosch, T.C. 2007. Long-term maintenance of species-specific bacterial microbiota in the basal metazoan *Hydra*. *Proceedings of the National Academy of Sciences of the United States of America* 104: 13146–13151.

Fritsche, T.R., Horn, M., Seyedirashti, S., et al. 1999. *In situ* detection of novel bacterial endosymbionts of *Acanthamoeba* spp. phylogenetically related to members of the order Rickettsiales. *Applied and Environmental Microbiology* 65: 206–612.

Funk, D.J., Helbling, L., Wernegreen, J.J., et al. 2000. Intraspecific phylogenetic congruence among multiple symbiont genomes. *Proceedings of the Royal Society B: Biological Sciences* 267: 2517–2521.

Garrido, C., Galluzzi, L., Brunet, M., et al. 2006. Mechanisms of cytochrome C release from mitochondria. *Cell Death and Differentiation* 13: 1423–1433.

Graça-Souza, A.V., Maya-Monteiro, C., Paiva-Silva G.O., et al. 2006. Adaptations against heme toxicity in blood-feeding arthropods. *Insect Biochemistry and Molecular Biology* 36: 322–335.

Hornok, S., Földvári, G., Elek, V., et al. 2008. Molecular identification of *Anaplasma marginale* and rickettsial endosymbionts in blood-sucking flies (Diptera: Tabanidae, Muscidae) and hard ticks (Acari: Ixodidae). *Veterinary Parasitology* 154: 354–359.

Klingenberg, M. 2008. The ADP and ATP transport in mitochondria and its carrier. *Biochimica et Biophysica Acta* 1778: 1978–2021.

Lee, K.B., Liu, C.T., Anzai, Y., et al. 2005. The hierarchical system of the 'Alphaproteobacteria': description of Hyphomonadaceae fam. nov., Xanthobacteraceae fam. nov. and Erythrobacteraceae fam. nov. *International Journal of Systematic and Evolutionary Microbiology* 55: 1907–1919.

Lewis, D. 1979. The detection of *Rickettsia*-like microorganisms within the ovaries of female *Ixodes ricinus*. *Zeitschrift für Parasitenkunde* 59: 295–298.

Lloyd, S.J., LaPatra, S.E., Snekvik, K.R., et al. 2008. Strawberry disease lesions in rainbow trout from southern Idaho are associated with DNA from a *Rickettsia*-like organism. *Diseases of Aquatic Organisms* 82: 111–118.

Lo, N., Bandi, C., Watanabe, H., et al. 2003. Evidence for cocladogenesis between diverse dictyopteran lineages and their intracellular endosymbionts. *Molecular Biology and Evolution* 20: 907–913.

Lo, N., Beninati, T., Sassera, D., et al. 2006. Widespread distribution and high prevalence of an alpha-proteobacterial symbiont in the tick *Ixodes ricinus*. *Environmental Microbiology* 8: 1280–1287.

Mediannikov, O., Ivanov, L.I., Nishikawa, M., et al. 2004. Microorganism 'Montezuma' of the order Rickettsiales: the potential causative agent of tick-borne disease in the Far East of Russia. *Zhurnal Mikrobiologii, Epidemiologii i Immunobiologii* 1: 7–13.

Meier, P., Finch, A., and Evan, G. 2000. Apoptosis in development. *Nature* 407: 796–801.

Moran, N.A., Degnan, P.H., Santos, S.R., et al. 2005. The players in a mutualistic symbiosis: insects, bacteria, viruses, and virulence genes. *Proceedings of the National Academy of Sciences of the United States of America* 102: 16919–16926.

Ogata, H., La Scola, B., Audic, S., et al. 2006. Genome sequence of *Rickettsia bellii* illuminates the role of amoebae in gene exchanges between intracellular pathogens. *PLoS Genetics* 2: 733–744.

Randolph, S.E., Gern, I., and Nuttall, P.A. 1996. Co-feeding ticks: epidemiological significance for tick-borne pathogen transmission. *Parasitology Today* 12: 472–479.

Ray, K., Marteyn, B., Sansonetti P.J., et al. 2009. Life on the inside: the intracellular lifestyle of cytosolic bacteria. *Nature Reviews Microbiology* 7: 333–340.

Reeves, W.K., Dowling, A.P., and Dasch, G.A. 2006. Rickettsial agents from parasitic dermanyssoidea (Acari: Mesostigmata). *Experimental and Applied Acarology* 38: 181–188.

Richard, S., Seng, P., Parola, P., et al. 2009. Detection of a new bacterium related to '*Candidatus* Midichloria mitochondrii' in bed bugs. *Clinical Microbiology and Infection* 15(Suppl. 2): 84–85.

Sacchi, L., Bigliardi, E., Corona, S., et al. 2004. A symbiont of the tick *Ixodes ricinus* invades and consumes mitochondria in a mode similar to that of the parasitic bacterium *Bdellovibrio bacteriovorus*. *Tissue Cell* 36: 43–53.

Sassera, D., Beninati, T., Bandi, C., et al. 2006. '*Candidatus* Midichloria mitochondrii', an endosymbiont of the tick *Ixodes ricinus* with a unique intramitochondrial lifestyle. *International Journal of Systematic and Evolutionary Microbiology* 56: 2535–2540.

Sassera, D., Lo, N., Bouman, E.A., et al. 2008. '*Candidatus* Midichloria' endosymbionts bloom after the blood meal of the host, the hard tick *Ixodes ricinus*. *Applied and Environmental Microbiology* 74: 6138–6140.

Schmitz-Esser, S., Linka, N., Collingro, A., et al. 2004. ATP/ADP translocases: a common feature of obligate intracellular amoebal symbionts related to Chlamydiae and Rickettsiae. *Journal of Bacteriology* 186: 683–691.

Skarphédinsson, S., Jensen, P.M., and Kristiansen, K. 2005. Survey of tick-borne infections in Denmark. *Emerging Infectious Diseases* 11: 1055–1061.

Vannini, C., Ferrantini, F., Schleifer, K.H., et al. 2010. '*Candidatus* Anadelfobacter veles' and '*Candidatus* Cyrtobacter comes,' two new Rickettsiales species hosted by the protist ciliate *Euplotes harpa* (Ciliophora, Spirotrichea). *Applied and Environmental Microbiology* 76: 4047–4054.

Venzal, J.M., Estrada-Peña, A., Portillo, A., et al. 2008. Detection of alpha and gamma-proteobacteria in *Amblyomma triste* (Acari: Ixodidae) from Uruguay. *Experimental and Applied Acarology* 44: 49–56.

Weinert, L.A., Werren, J.H., Aebi, A., et al. 2009. Evolution and diversity of *Rickettsia* bacteria. *BMC Biology* 27: 6.

Yamataka, S., and Hayashi, R. 1970. Electron microscopic studies on the mitochondria and intramitochondrial microorganisms of *Halteria geleiana*. *Journal of Electron Microscopy* 19: 50–62.

Zhu, Z., Aeschlimann, A., and Gern, L. 1992. *Rickettsia*-like microrganisms in the ovarian primordia of molting *Ixodes ricinus* (Acari: Ixodidae) larvae and nymphs. *Annales de Parasitologie Humaine et Comparée* 67: 99–110.

7 Rickettsiella, Intracellular Pathogens of Arthropods

Didier Bouchon, Richard Cordaux, and Pierre Grève

CONTENTS

INTRODUCTION

Members of the genus *Rickettsiella* are responsible for a number of diseases in invertebrates. The first *Rickettsiella* disease was reported from larvae of the Japanese beetle *Popillia japonica* (Dutky and Gooden 1952). This deadly disease was called blue disease due to the discoloration of the insect larvae. Since then *Rickettsiella* diseases have been described from insects of the orders Lepidoptera, Orthoptera, Diptera, Dictyoptera, Coleoptera, Hymenoptera, and Hemiptera, as well as from arachnids and crustaceans (Table 7.1).

According to the currently accepted taxonomy (Fournier and Raoult 2005; Weiss et al. 1984), three main *Rickettsiella* species are recognized, each with a

Genus	*Rickettsiella*
Species	*R. popilliae, R. chironomi, R. grylli*
Family	Coxiellaceae
Order	Legionellales
Description year	1952
Origin of name	Small *Rickettsia* (Philip 1956)
Description	
Reference	Philip, C. B. 1956. Comments on the classification of the order *Rickettsiales. Can. J. Microbiol.* 2:261–270.
Anecdote	First assigned to *Coxiella* in 1952, then formally names *Rickettsiella* in 1956.

C. B. Philip (Photo provided courtesy of the University of Kansas Natural History Museum, Division of Entomology, Lawrence, Kansas).

list of synonyms: *R. popilliae, R. grylli,* and *R. chironomi. R. popilliae* constitutes the type species (Dutky and Gooden 1952). The wide geographical distribution of *Rickettsiella* species and their various host taxa suggest a high infectivity and adaptability, which may be facilitated by the fact that *Rickettsiella* transmission to arthropods occurs via environmental contamination rather than vertical transmission (Fournier and Raoult 2005).

The *Rickettsiella* diseases are relatively easy to diagnose. *R. popilliae* causes blue disease in insect larvae. In crickets and locusts, *R. grylli* causes swelling of the abdomen and at later stages turgidity of the intersegmentary membranes and ventral inclination of the head, leading to death. Crickets infected by *R. grylli* increase their preferred temperature from 26°C to 32°C (Adamo 1998; Louis et al. 1986). This behavioral fever is an adaptive process optimizing the probability of surviving the disease (Adamo 1998). In isopods infected with *Rickettsiella*, opaque white masses of infected connected tissue develop throughout hemocoel and are easily observed in advanced stages of the disease by examining the ventral surface of the isopod (Federici 1984; Vago et al. 1970).

The potentially devastating consequences that *Rickettsiella* infections may have on laboratory insectaries and other arthropod collections should drive our attention toward this pathogen. Indeed, even if *Rickettsiella* are not as virulent as viral and mycotic pathogens, the possible use of these bacteria as biological control agents of agricultural pests has been proposed, considering that they can survive in the soil for years (Hurpin and Robert 1977). Conservation of *R. popilliae* in the soil has been demonstrated for up to 2 years without any alteration of their pathogenicity (Hurpin and Robert 1976). A mesocosm experiment has been conducted in natural grassland infected by cockchafer larvae (Hurpin and Robert 1977). One year after, a suspension of *Rickettsiella* was sprayed, resulting in ~15% of the white grubs exhibiting *Rickettsiella* disease. This level of infection was maintained during the 3-year life cycle of the cockchafer (Hurpin and Robert 1977).

Virulence in vertebrate hosts has been demonstrated in mice after massive inoculation of *R. popilliae* (Giroud et al. 1958). *R. grylli* multiplied in mice after inhalation, but virulence depended on the inoculum size (Giroud et al. 1958). In heavy contaminations, the lungs underwent hepatization and bronchopneumopathy, evolving to chronic disease in half of the mice (Delmas and Timon-David 1985). No infection has been reported by ingestion, although *R. popilliae* (*R. armadillidii*) maintained virulence in their natural host after passage through the mice digestive tract (Yousfi 1976).

In this review, we present the state of the art and discuss recent results on *Rickettsiella* research, on issues as diverse as evolution, biology, host diversity, and interactions with the host at the molecular level.

RICKETTSIELLA EVOLUTION

RICKETTSIELLA TAXONOMIC CLASSIFICATION WITHIN THE CLASS GAMMA-PROTEOBACTERIA

On the basis of ultrastructural observations and their close association with arthropods, bacteria of the genus *Rickettsiella* were originally assigned to the order Rickettsiales, a group of intracellular bacteria belonging to the class alpha-Proteobacteria (Weiss et al. 1984). However, an alternative classification within the order Chlamydiales (class Chlamydiae) has also been considered, because of analogies in replication cycles of *Rickettsiella* and *Chlamydia* bacteria (Federici 1980). Molecular genetic evidence accumulated within the past few years has shed new light on the taxonomic position of *Rickettsiella* bacteria. In 1997, Roux et al. reported the first *Rickettsiella* nucleotide sequence, specifically the 16S rRNA sequence of a *R. grylli* strain isolated from the cricket *G. bimaculatus*. Unexpectedly, *R. grylli* was found to be most closely related to *Coxiella burnetii* and *Legionella* species, within the bacterial class gamma-Proteobacteria (Roux et al. 1997). This recently prompted the removal of the entire genus *Rickettsiella* from the Rickettsiales and assignment to the family Coxiellaceae, within the Legionellales (Fournier and Raoult 2005). Additional nucleotide sequence data from *R. armadillidii* strains (Cordaux et al. 2007) and subsequently from *R. grylli*, *R. melolonthae*, and *R. tipulae* strains

TABLE 7.1

Ecology of the *Rickettsiella* (updated from Fournier and Raoult 2005)

Rickettsiella Species	Host Class	Order	Species	Common Name	Location	References
R. popilliae	Insecta	Coleoptera	*Popillia japonica*	Japanese beetle	United States	Dutky and Gooden 1952; Philip 1956
				Charabid beetles	United States, UK	Sutter and Kirk 1968; Carter and Luff 1977
			Melolontha melolontha	Cockchafer	Germany	Krieg 1955; Philip 1956; Leclercque and Kleespies 2008a
			Tenebrio molitor	Mealworm	Germany	Krieg 1965
			Cetonia sp.	Cetonid beetle	Madagascar	Meynadier and Monsarrat 1969
		Diptera	*Tipula paludosa*	Crane fly	Germany	Müller-Kogler 1958; Leclercque and Kleespies 2008b
		Dictyoptera	*Blatta orientalis*	Cockroach	Germany	Huger 1964
	Crustacea	Isopoda	*Armadillidium vulgare*	Pill bug	France, United States	Vago et al. 1970; Cordaux et al. 2007; Federici 1984
			Porcellio sp.		United States	Federici 1984; Abd and Holdich 1987
			Oniscus asellus		UK	Abd and Holdich 1987
			Helleria brevicornis		France	Cordaux et al. 2007
			Philoscia muscorum		Spain	Cordaux et al. 2007
			Asellus aquaticus		Germany	Wang et al. 2007
			Armadillo officinalis		France	This study (Figure 7.2)
R. grylli	Insecta	Orthoptera	*Gryllus bimaculatus*	Cricket	France	Vago and Martoja 1963
			Schistocerca gregaria	Desert locust	Jordan	Vago and Meynadier 1965
	Crustacea	Amphipoda	*Crangonyx floridanus*		United States	Federici et al. 1974

R. chironomi	Insecta	Diptera	*Chironomus tentans*	Midge	Germany	Weiser 1963
	Arachnida	Aranea	*Argyrodes gibbosus*	Spider	Spain	Meynadier et al. 1974
			Pisaura mirabilis	Spider	France	Morel 1977
Not assigned	Insecta	Coleoptera	*Stethorus punctum*	Stethorus beetle	Morocco	Hall and Badgley 1957
		Lepidoptera	*Samia cynthia*	Sturnid moth	UK	Entwistle et al. 1968
			Paramyelois transitella	Navel orange worm	United States	Kellen et al. 1972
		Hemiptera	*Cecidotrioza sozanica*	Psyllid	United States	Spaulding and von Dohlen 2001
			Acyrthosiphon pisum	Pea aphid	Europe	Tsuchida et al. 2010
		Hymenoptera	*Vespa germanica*	Wasp	Australia	Reeson et al. 2003
	Arachnida	Acarina	*Phytoseiulus persimilis*	Mite	Ukraine	Sutakova and Ruttgen 1978
			Dermanyssus gallinae	Mite	UK, France	De Luna et al. 2009; Moro et al. 2009
			Ixodes woodi	Tick	United States	Kurtti et al. 2002
			Ixodes tasmani	Tick	Australia	Vilcins et al. 2009
		Scorpionida	*Buthus occitanus*	Scorpion	France	Morel 1976
	Crustacea	Decapoda	*Carcinus mediterraneus*	Crab	France	Bonami and Papparlardo 1980
			Cherax quadricarinatus	Redclaw crayfish	UK	Romero et al. 2000
	Collembola		*Folsomia candida*	Springtail	Germany	Czarnetski and Tebbe 2004

(Leclerque 2008; Leclerque and Kleespies 2008a, 2008b, 2008c) have confirmed the taxonomic classification of *Rickettsiella* bacteria within the gamma-proteobacterial order Legionellales.

Although the assignment of the genus *Rickettsiella* to the order Legionellales is well supported by molecular evidence, there is more uncertainty regarding the relationships of *Rickettsiella* relative to *Coxiella* (family Coxiellaceae) and *Legionella* (family Legionellaceae), the other two genera within the Legionellales. 16S rRNA sequence data generally suggest that *Rickettsiella* bacteria are more closely related to *Coxiella* than to *Legionella* (Cordaux et al. 2007; Leclerque and Kleespies 2008b; Roux et al. 1997), which was the basis for assigning the *Rickettsiella* genus to the family Coxiellaceae in the current classification (Fournier and Raoult 2005). However, other analyses based on 16S rRNA as well as *GroEL*, *MucZ*, and three genes encoding type IV secretion system (T4SS) components instead suggested that *Rickettsiella* bacteria may be more closely related to *Legionella* than to *Coxiella*, or that *Rickettsiella* is ancestral to both *Legionella* and *Coxiella* (Leclerque and Kleespies 2008a, 2008c). To further investigate these conflicting results, the draft genome sequence of *R. grilly* available in GenBank (see below) was used to perform a phylogenomic analysis of Legionellales. However, analysis of tens of genes at both nucleotide and amino acid levels failed to resolve the *Rickettsiella* phylogenetic position within the Legionellales (Leclerque 2008). These results suggest that ancestors of the three genera *Rickettsiella*, *Coxiella*, and *Legionella* may have diverged from each other within a relatively short period of time, early during Legionellales evolution. Consequently, it has been proposed that *Rickettsiella* bacteria should be assigned to a distinct family within the Legionellales (Leclerque 2008). Additional sequence data and analyses will be necessary before firm conclusions on Legionellales taxonomic organization can be drawn.

GENETIC DIVERSITY WITHIN THE GENUS *RICKETTSIELLA*

Numerous *Rickettsiella* pathotypes have been described, which fall into one of the three currently recognized species: *R. grylli*, *R. popilliae*, and *R. chironomi* (Fournier and Raoult 2005). At the time of writing, 16S rRNA nucleotide sequences are available for various strains encompassing the *R. grylli* and *R. popilliae* species. These sequences are closely related, with <5% average divergence between strains, compared with ~14% average divergence between *Rickettsiella* strains and closely related non-*Rickettsiella* species (Cordaux et al. 2007; Leclerque and Kleespies 2008b). Thus, the available molecular evidence is consistent with monophyly of the *Rickettsiella* genus (Cordaux et al. 2007; Leclerque and Kleespies 2008a). However, sequence information from the third species *R. chironomi* is currently lacking; resolving this pitfall will definitely settle the question of the origin of *Rickettsiella* bacteria. The available molecular data suggest that diversification within the genus *Rickettsiella* started ~100 million years ago (Cordaux et al. 2007). Thus, the distribution of *Rickettsiella* bacteria in a wide range of arthropods implies extensive horizontal transfers, which may be facilitated by the ability of these bacteria to survive in the soil outside of their arthropod hosts for years (Cordaux et al. 2007; Hurpin and Robert 1976).

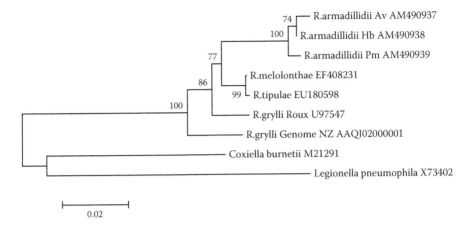

FIGURE 7.1 Neighbor-joining tree of *Rickettsiella* strains based on 16S rRNA sequences. Only sequences derived from diagnosed *Rickettsiella* pathotypes are included. Several additional *Rickettsiella*-like sequences are available in GenBank, but they were omitted because *Rickettsiella* infection has not been ascertained independently of molecular data. Consequently, pathotype information is lacking for these sequences. The tree was built by using the software MEGA 4, based on the Kimura two-parameter substitution model and rooted with *Coxiella burnetii* and *Legionella pneumophila.* Bootstrap values (based on 1,000 replicates) are shown on branches as percentages. GenBank accession numbers are indicated in brackets. Av, Hb, Pm, and Gb indicate strains isolated from *Armadillidium vulgare, Helleria brevicornis, Philoscia muscorum,* and *Gryllus bimaculatus,* respectively.

Species delineation and relationships within the genus *Rickettsiella* are also controversial. For example, 16S rRNA sequences from two *R. grylli* strains are available: the authentic pathotype isolated from the cricket *Gryllus bimaculatus* (Roux et al. 1997) and the isopod-derived *R. grylli* strain used for genome sequencing. The two sequences exhibit 3.7% nucleotide divergence, which is unusually high at this taxonomic level (Leclerque and Kleespies 2008b). In addition, the two sequences suggest a polyphyletic origin of *R. grylli* (Figure 7.1), thus casting doubt on the taxonomic assignment of the *Rickettsiella* strain used for genome sequencing (Leclerque and Kleespies 2008b).

Regarding the species *R. popilliae,* 16S rRNA sequences are available for three pathotypes: *R. armadillidii* from the isopods *Armadillidium vulgare, Philoscia muscorum,* and *Helleria brevicornis* (Cordaux et al. 2007); *R. melolonthae* from the coleopteran *Melolontha hippocastani* (Leclerque and Kleespies 2008a); and *R. tipulae* from the dipteran *Tipula paludosa* (Leclerque and Kleespies 2008b). The sequences from these three pathotypes form a monophyletic group in the *Rickettsiella* phylogeny (Figure 7.1). This is consistent with their current recognition as synonyms of the type species *R. popilliae* (Fournier and Raoult 2005). Nevertheless, the closely related *R. armadillidii* sequences (0.3% average divergence) exhibit ~3% divergence with the very closely related *R. melolonthae* and *R. tipulae* sequences (0.1% divergence). This strongly suggests that the two groups of strains may actually correspond to two distinct species (Cordaux et al. 2007; Leclerque and Kleespies 2008b).

Molecular evidence from the type species *R. popilliae* would undoubtedly be useful to resolve the taxonomic status of *R. popilliae* strains.

There is thus a great deal of uncertainty regarding numerous aspects of *Rickettsiella* classification and evolution. Although the phylogenetic affinities of *Rickettsiella* bacteria to gamma-Proteobacteria are now largely accepted, important questions remain. They are related to the phylogenetic position of the *Rickettsiella* genus relative to other bacterial groups within the Legionellales, as well as the organization of the internal structure of the *Rickettsiella* genus. While molecular data have been instrumental in solving some questions, they have also highlighted the complex evolutionary history of *Rickettsiella* bacteria. Acquisition of additional data from a wide range of pathotypes along with the completion of the *Rickettsiella* genome sequence will hopefully provide new insight into these open questions.

BIOLOGY OF *RICKETTSIELLA*

REPRODUCTION AND MORPHOGENESIS

Rickettsiella bacteria may be present in all host tissues, including the fat tissues and the hepatopancreas. *Rickettsiella* have been observed particularly in hemocytes as well as free in the hemolymph, indicating a systemic infection (Devauchelle et al. 1972; Federici et al. 1974; Romero et al. 2000; Yousfi 1976). Isolated bacteria were also observed in the lumen of the hepatopancreas (Romero et al. 2000).

Rickettsiella bacteria have a complex intravacuolar cycle similar in many aspects to the characteristic cycle of the Chlamydiales. They appear as oval or disk-shaped forms that can be present in all tissues of the host. The bacteria are pleomorphic and contained in membrane-bound vacuoles in the host cells. During their reproduction, *Rickettsiella* bacteria typically undergo a cycle of development that can be divided into two main stages. In the first stage, small, dense infectious particles referred to as elementary bodies (EBs) enter the host cells by phagocytosis. Being enclosed in host cell vacuoles, they gradually transform into larger forms of much lesser electron density. In the second stage, these larger forms, referred to as reticulate bodies (RBs), multiply by binary fission. In advanced phases of the disease, accumulation of numerous *Rickettsiella*-filled vesicles in host cells results in a typical opaque white coloration of the tissues. As vacuoles become crowded, RBs condense and form new EBs. In some host species, vacuoles may also contain large refractive crystalline bodies.

Electron micrographs of nervous tissues of infected isopods illustrate these subcellular structures (Figure 7.2). EBs appear as uniformly electron-dense rod shapes surrounded by bilayered envelopes. They measure from 0.5 to 0.6 μm in length and 0.3 μm in diameter (Figure 7.2). In heavily infected cells, no organelles can be observed. The replicative forms of *Rickettsiella* are constituted by RBs, larger forms of medium electron density measuring ~1 μm in length and ~0.5 μm in diameter (Figure 7.2). RBs are the only cells exhibiting elongations and constrictions in the center. Electron-dense material is concentrated in the RB center, whereas the RB periphery appears granular (Figure 7.2). Some large forms are giant, round cells from which crystalline bodies arise, whose origin (host or bacteria) is not clearly determined (Fournier and Raoult 2005) (Figure 7.2).

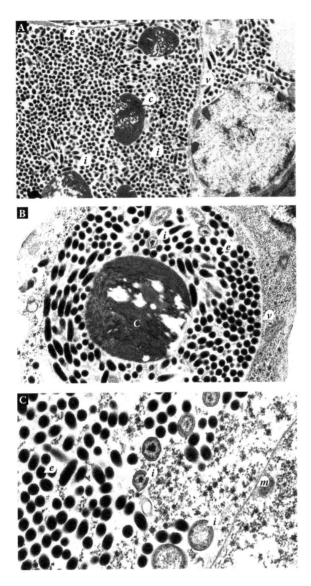

FIGURE 7.2 Ultrastructure (TEM) of various stages of *Rickettsiella armadillidii* in (a) the nervous chord of an *Armadillidium vulgare* female and in (b,c) the ovary of an *Armadillo officinalis* female. e, electron-dense elementary bodies; i, intermediate or reticulate bodies with three-layered cell boundary and condensed cytoplasm in the center; c, crystalline bodies; v, vacuolar membrane; N, host cell nucleus; m, cell membrane. (a) Details of *Rickettsiella* infecting a nerve cell. Bacterial vacuole containing all three stages. ×3,300. (b) Details of *Rickettsiella* infecting an oocyte. Bacterial vacuole containing all the three stages. ×13,000. (c) Details of *Rickettsiella* infecting an oocyte. ×20,000.

However, there are some significant differences to this general cycle, depending on the three recognized *Rickettsiella* species. In *R. popilliae* and *R. grylli*, cells remain rod shaped during all their reproductive cycle. The rods may be oval, curved, or kidney shaped with rounded edges. The giant forms (or RBs) contain crystals (Huger and Krieg 1967). In *R. chironomi*, the infectious and reproductive forms are disk shaped and crystalline bodies are never observed (Federici 1980; Fournier and Raoult 2005; Weiser 1963).

HOST DIVERSITY

Most of the knowledge on *Rickettsiella* bacteria is based on light and electron microscopy. However, the similarity of the *Rickettsiella* cycle with that of the order Chlamydiales may lead to misidentifications. Therefore, many intracellular bacteria were usually referred to as *Rickettsia*-like organisms (RLOs), leading to uncertainty or false taxonomic assignation in the absence of more definitive data. Hence, the bacteria of the isopod *Porcellio scaber* (Drobne et al. 1999) and of the cockroach *Blatta orientalis* (Radek 2000) that had been initially described as *Rickettsiella* based on morphologic criteria were recently assigned to the "*Candidatus* Rhabdochlamydia" genus of the order Chlamydiales, based on 16S rRNA sequence data (Corsaro et al. 2007; Kostanjsek et al. 2004). *Rickettsiella* species have been identified to date in hosts belonging to the three major arthropod groups: Hexapoda, Arachnida, and Crustacea (Table 7.1).

Hexapods

Rickettsiella infection has been reported in species from diverse insect orders, including Coleoptera, Diptera, Dictyoptera, Orthoptera, Lepidoptera, Hymenoptera, and Hemiptera (Table 7.1).

Coleopterans were the first insects in which *Rickettsiella* bacteria were identified. The type species, *R. popilliae*, was identified in the Japanese beetle *P. japonica* (Dutky and Gooden 1952). A virulent disease observed in insectary populations of five species of *Stethorus* beetles was attributed to a different *Rickettsiella* (*R. stethorae*) (Hall and Badgley 1957). A few years later, RLO diseases were reported in different European cockchafer species of the genus *Melolontha* (Hurpin and Vago 1958; Krieg 1958; Wille and Martignoni 1952). As cockchafer larvae can be serious pests of grasses and cereals (they live in the soil and feed roots), extensive work has been conducted on *Rickettsiella* disease diagnosis, developmental cycle, and virulence. Laboratory tests and field experiments have been performed to study the virulence of *R. melolonthae* in its natural host (Hurpin 1971a; Hurpin and Robert 1976, 1977). The three larval stages of *M. melolontha* are equally susceptible to *Rickettsiella* injection into the hemocoel or esophagus, as well as after free ingestion by food contamination (Hurpin 1971b). Bacteria multiply and death occurs within ~80 days, depending on larval stage and environment (temperature and soil composition) (Hurpin 1971b) (Figure 7.3). In the yellow mealworm *Tenebrio molitor*, *Rickettsiella* bacteria were detected in fat tissues of larvae, pupae, and adults, but no crystalline inclusion was observed (Krieg 1965). *Rickettsiella* particles have also been reported in the fat body of carabid and cetonid beetles (Carter and Luff 1977; Meynadier and

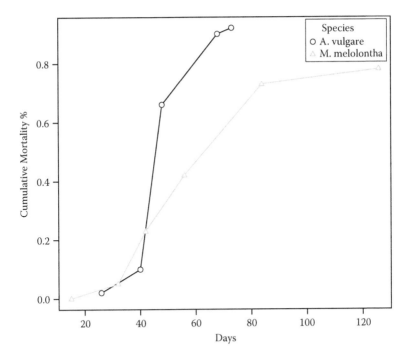

FIGURE 7.3 Mortality curves in *Rickettsiella* hosts. (a) *Armadillidium vulgare* specimens after *Rickettsiella* inoculation (n = 50). (Modified from Yousfi 1976.) (b) *Melolontha melolontha* larvae (L3) after *Rickettsiella* ingestion (n = 25). (Modified from Hurpin 1971b.)

Monsarrat 1969; Sutter and Kirk 1968). A putative *Rickettsiella* infection was also identified in third-instar white grub (Scarabaeidae) collected in Mexico (Sanchez-Pena 2000).

Rickettsiella infection has also been documented in Dipterans, an insect order containing many species that are particularly important as vectors of diseases in humans, animals, and plants. The first evidence concerned the midge *Chironomus tentans*, in which a rickettsiosis was initially described (Weiser 1949). The disease primarily affects the last two instars and pupal stages. Experimental infection results in death within 7 to 14 days (Weiser and Zizka 1968). The pathotype was described as *R. chironomi*, from which flat disks are formed instead of rods (Weiser 1963). A second *Rickettsiella*, related to *R. popilliae* rather than *R. chironomi*, was reported in the crane fly *T. paludosa* (Leclerque and Kleespies 2008b; Müller-Kogler 1958). The fat tissue was the main infected tissue exhibiting pleomorphic bacteria with rhombohedric crystalline inclusions, which were later assigned to host metabolic perturbations (Huger 1959).

The first *Rickettsiella* infection in Orthopterans was recorded by Vago and Martoja (1963), who observed bacteria in the fat tissue of the cricket *G. bimaculatus*. The scientific name of *R. grylli* initially proposed for this pathogen was later considered as a valid species of *Rickettsiella* (Fournier and Raoult 2005). The separation of *R. popilliae* and *R. grylli* was initially based on serological cross-reaction between the cricket and isopod pathogens (Croizier and Meynadier 1971) and later confirmed by DNA hybrization (Frutos et al. 1994). *Rickettsiella* found in fat tissue and hypodermic tissue of larvae and adults of the desert locust *Schistocerca gregaria* were less studied (Vago and Meynadier 1965).

In the cockroach *Blatta germanica,* Huger (1964) observed pleomorphic bacteria with two forms exhibiting crystalline inclusions in both larvae and adults. This constitutes the unique example of *Rickettsiella* infection reported in Dictyoptera. The pathotype of *R. blattae,* a synonym of *R. popilliae,* was proposed by Huger (1964).

Rickettsiella were also observed in two Lepidoptera species: (1) the sturnid moth *Samia cynthia,* in which rod-shaped bacteria were found in the digestive tract, fat tissue, and hypodermis of the larvae (Entwistle and Robertson 1968), and (2) the navel orange worm *Paramyelois transitella,* in which pleomorphic bacteria showing crystalline inclusions were observed (Kellen et al. 1972). However, insufficient data exist to support further assignment of these two lepidopteran pathogens to one of the three *Rickettsiella* species.

Rickettsiella infection has only recently been described in Hymenopterans and Hemipterans. The microbial ecology of the eusocial wasp *Vespa germanica* was examined through 16S rDNA-DGGE (denaturing gradient gel electrophoresis) (Reeson et al. 2003). Based on DGGE profiles of larval gut microflora, two *Rickettsiella* genotypes were identified and sequenced, showing 98% similarity with the *R. grylli* 16S rRNA sequence from at least two nests. Corresponding bands in DGGE profiles were present in many other samples, indicating a widespread *Rickettsiella* distribution in this eusocial wasp. In Hemiptera, eubacterial 16S rRNA was sequenced from endosymbionts of seven psyllids (Psylloidea) (Spaulding and von Dohlen 2001). A sequence showing 94% similarity to *R. grylli* 16S rRNA gene was identified in *Cecidotrioza sozanica.* Approximately 8% of pea aphids collected in Western Europe carried *Rickettsiella* infection (Tsuchida et al. 2010). Most of these individuals were coinfected with other facultative symbionts (*Hamiltonella* or *Serratia*-like organism) that confer protection to their aphid hosts against parasitoid wasps. Fitness measurements revealed that infection status did not affect growth rate and body size. However, infection with *Rickettsiella* changes insect body color from red to green and is expected to influence prey-predator interactions, as well as interactions with other symbionts.

Collembola (springtails) are soil arthropod species that are closely related to insects (Timmermans et al. 2008). Intracellular bacteria were first described by transmission electron microscopy (TEM) in the fat body and ovaries of *Folsomia candida* (Palevody 1972). *Wolbachia* endosymbionts have been detected in this species (Czarnetzki and Tebbe 2004; Vandekerckhove et al. 1999), and broader bacterial communities were characterized by generating clone libraries of 16S rRNA amplified from total DNA, revealing a 16S rRNA sequence related to *Rickettsiella* sp. in a breeding stock of *F. candida* (Czarnetzki and Tebbe 2004). Interestingly, the host did not seem affected by *Rickettsiella,* although dominant in this breeding stock.

Arachnids

Bacteria found in the spider *Argyrones gibbosus* (Meynadier et al. 1974) and the scorpion *Buthus occitanus* (Morel et al. 1974) are the first records of *Rickettsiella* in arachnids. The latter bacteria were subsequently described as *R. buthi* (Morel 1976). Based on their morphological characteristics, they were assigned to *R. chironomi* (Fournier and Raoult 2005). However, molecular data suggested no relationship with *R. chironomi* (Frutos et al. 1989, 1994), and further molecular analyses will be required to determine their correct assignment. In 1977, Morel described *Rickettsiella* in the spider *Pisaura mirabilis* (Morel 1977).

Ticks are also hosts to *Rickettsiella* bacteria. A parthenogenetic strain of *Ixodes woodi* was found to be infected by *Rickettsiella* (Kurtti et al. 2002). Pleomorphic bacteria were observed by TEM in ovaries and malphigian tubules. The 16S rRNA gene was amplified and sequenced from *I. woodi* female tissues. All clones exhibited the same sequence, indicating that the females were all infected with a single bacterial strain. Overall, TEM and phylogenetic analyses indicated that this bacterium is closely related to *R. grylli*. Molecular evidence for *Rickettsiella* bacteria has also recently been reported from field samples of three native Australian ticks, *I. tasmani*, *Bothriocroton concolor*, and *B. auriginans* (Vilcins et al. 2009). Interestingly, different *Rickettsiella* sequences were identified in samples of *I. tasmani* collected from three koalas (Vilcins et al. 2009).

In mites, research has largely focused on intracellular reproductive parasites such as *Wolbachia* (Breeuwer and Jacobs 1996; Gotoh et al. 2003), and few bacteria have been reported as pathogens of the Acari. *Rickettsiella* microorganisms were identified by TEM in the predatory mite *Phytoseiulus persimilis* (Sutakova 1988; Sutakova and Ruttgen 1978) from a laboratory mass rearing but not from other sources (Sutakova and Arutunyan 1990). Pleomorphic bacteria were found in all individuals examined; however, the effects of these bacteria upon host fitness were not reported. *R. phytoseiuli* cultivated in adult females of *Dermacentor reticulatus* (Ixodidae) were able to multiply as in the typical host mite (Sutakova and Rehacek 1989). The poultry red mite *Dermanyssus gallinae* is considered the most important hematophagous ectoparasite of laying hens in Europe, causing reduction of egg production, blood staining of eggs, anemia, and even death (Chauve 1998). A survey of intracellular bacteria of *D. gallinae* sampled from poultry farms in the UK and France by 16S rRNA amplification, temperature gradient gel electrophoresis (TGGE), fingerprinting, and sequencing identified *Rickettsiella* bacteria in one sampled farm out 16 (De Luna et al. 2009). Another TGGE study (Moro et al. 2009) conducted on poultry farms in Britanny, France, identified one 16S rRNA sequence showing 91% similarity to the *Rickettsiella* previously identified in *D. gallinae* (De Luna et al. 2009) and one showing 97% similarity to the *Rickettsiella* isolated from the springtail *F. candida* (Czarnetzki and Tebbe 2004).

Crustaceans

The first observed *Rickettsiella* infection in terrestrial isopods (woodlice) was described in *A. vulgare* (Vago et al. 1970). *R. armadillidii*, a subjective synonym of *R. popilliae*, was therefore proposed as the pathotype designation (Vago et al. 1970). The common transmission route was through ingestion. In terrestrial isopod species, *Rickettsiella* diseases were reported in *A. vulgare, A. nasatum, A. granulatum, Eluma purpurascens, Oniscus asellus, P. laevis*, and *P. gallicus* (Yousfi 1976) and *H. brevicornis* and *P. muscorum* (Cordaux et al. 2007) from France; *P. dilatatus* from the United States (Federici 1984); and *P. scaber* and *O. asellus* from the UK (Abd and Holdich 1987). *Rickettsiella* was also detected in the freshwater isopod *Asellus aquaticus* (Wang et al. 2007). It is noteworthy that intracellular bacteria of the genus *Wolbachia* are commonly found in isopods, including the aforementioned species (Bouchon et al. 2008).

An extensive study of virulence and transmission routes of *R. armadillidii* in woodlice has been conducted by Yousfi (1976), which showed that virulence is conserved through various interspecies inoculations. In *A. vulgare* death occurred ~60 days after inoculation (Figure 7.3), whereas death occurred 5 to 6 months after ingestion. Results were similar when *R. armadillidii* was ingested or inoculated in diverse *Porcellio* species. However, inoculations were unsuccessful in the freshwater isopod *A. aquaticus*. This result is in agreement with the distant phylogenetic relationships of *Rickettsiella* isolated from *A. vulgare* and *A. aquaticus* (Cordaux et al. 2007). Interestingly, injections of *R. armadillidii* were also unsuccessful in the diplopod *Glomeris marginata*, the greater wax moth *Galleria mellonella* (Lepidoptera), and the field cricket *G. bimaculatus* (Orthoptera) (Yousfi 1976).

In amphipods, *Rickettsiella* was identified in the grangonid freshwater amphipod *Crangonyx floridanus* (Federici et al. 1974). Forty-three percent of the 270 individuals sampled were naturally infected as ascertained by diagnostic iridescence in the legs. In the freshwater amphipod, *Rivulogammarus pulex*, no iridescence was observed, but bacteria in the fat tissue exhibited apparent crystal formation similar to that described for *R. popiliiae* (Larsson 1982).

Rickettsiella was also detected in decapods. In the crab *Carcinus mediterraneus*, bacteria were observed in the connective tissue between pancreatic tubules (Bonami and Pappalardo 1980). Healthy crabs died 15 days after experimental inoculation of a suspension of diseased hepatopancreatic tissue. This study represented the first description of a *Rickettsiella* infection in a marine species. Bacteria were detected in the hepatopancreatic epithelial cells of the crab *Paralithodes platypus* and tentatively identified as *Rickettsiella* (Johnson 1984), as well as in the freshwater decapod *Cherax quadricarinatus*, a farmed redclaw crayfish (Romero et al. 2000). Affected juveniles presented a greenish blue coloration and were generally lethargic. Intracellular bacteria were found in almost all host tissues, including hemocytes. Although the morphogenesis has ultrastructural characteristics similar to those of *Rickettsiella*, no information is available regarding its genetic affinities.

Environmental Samples

Apart from the well-described *Rickettsiella* hosts discussed above, many other species might harbor these bacteria. Based on disease diagnosis and morphology using TEM, bacteria were described as RLOs in many studies, but their precise identification is still pending. RLOs are known to cause a variety of diseases in vertebrates and in free-living or cultured invertebrates. In particular, the development of shellfish-based aquaculture around the world, along with an expanding market demand, has prompted attention to the infectious diseases and parasites of shellfish (Bower et al. 1994). Many RLOs were described in Mollusca (oysters, mussels, clams and cockles, scallops) and in Crustacea (shrimps and prawns, crabs and crayfish). Even if no formal assignment was proposed, some cases of *Rickettsiella* infection might be included in a number of described RLO diseases.

More recently, 16S rRNA fingerprinting has been used to identify large bacterial communities in hosts or habitats. For example, molecular assessment of bacterial communities in the Salton Sea was performed by sequencing of 16S rRNA clone libraries obtained from the water column and sediments (Dillon et al. 2009). Various sequences most closely related to *R. grylli* were found in both the water column and the sediments. Interestingly, this result represents the only exception to the nonoverlapping bacterioplankton diversity between water and sediment.

Another study of pathogen infection in wild Atlantic salmon populations from six rivers in Québec (Canada) was performed recently (Dionne et al. 2009). Bacterial pathogenic infections were evaluated by total DNA extraction from kidney tissue of juvenile salmons, followed by 16S rRNA sequencing using eubacterial universal primers. A total of 13 pathogens were identified, including *Rickettsiella* sp. (96% sequence similarity) with a prevalence of 1.7% in one river. Taken together, these data underline the probably underestimated prevalence of *Rickettsiella* in different habitats and hosts. The development of metagenomics approaches using next-generation sequencing together with the completion of the *Rickettsiella* genome sequence (see below) will certainly offer new insight into the diversity and extent of distribution of this pathogen.

RICKETTSIELLA-HOST INTERACTIONS

Although *Rickettsiella* was defined as a pathogenic agent of invertebrates, including insects, crustaceans, and arachnids, little is known on virulence factors secreted by the pathogen to cause disease. Characteristic steps of the *Rickettsiella* life cycle are multiplication in acidic vacuoles within hosts cells and the formation of large crystalline bodies, which require *Rickettsiella*-host interactions and communication at the molecular level. The specific *R. grylli* genes involved in host cell manipulation are unknown. Nonetheless, analysis of the publicly available draft genome sequence of *R. grylli* (GenBank accession no. NZ_AAQJ00000000) is expected to offer new insights into the mechanisms that allow the bacteria to infect their hosts. The *R. grylli* genome size has been estimated to 2.1 Mb by pulse-field gel electrophoresis, which is in good agreement with the sizes of other *Rickettsiella* genomes, which range from 1.55 to 2.65 Mb (Frutos et al. 1989). As expected, genomes of these obligate intracellular bacteria are reduced in size compared to the ones of facultative intracellular bacteria, which show an evolutionary tendency toward genomic reduction (Merhej et al. 2009; Moran et al. 2008). Similarly, the G + C nucleotide content of *Rickettsiella* genomes ranges from 36.3 to 41%, as estimated from melting curves (Frutos et al. 1994). Such nucleotide bias is typically expected from the genomes of obligate intracellular bacteria (Merhej et al. 2009). The partially sequenced *R. grylli* genome contains 1,581,239 nucleotides, of which 86% are coding sequences. According to the GenBank annotation, the genome contains 1,557 genes, including 1,410 protein encoding genes, 50 RNA genes, and 97 disrupted genes.

Many symbiotic and pathogenic bacteria are able to export DNA or proteins into the cytoplasm of eukaryotic host cells. Such bacteria use a wide range of mechanisms, including a T4SS, which forms a pore that spans the periplasm and the two bacterial membranes (Christie et al. 2005). In the genome of *R. grylli*, homologous genes of the T4SS (*dot/Icm*) of *Legionella pneumophila* were found (Zusman et al. 2007). It belongs to the type IVB and was shown to be essential for *L. pneumophila* pathogenesis (Segal et al. 1998). The *dot/Icm* system is present in another pathogenic bacterium, *C. burnetii* (Seshadri and Samuel 2005), and it seems to function similarly to *L. pneumophila* T4SS (Zamboni et al. 2003; Zusman et al. 2003). These two pathogens are evolutionarily closely related bacteria that also replicate in vacuoles when infecting human immune cells (Segal et al. 2005). In *C. burnetii*, 23 of the 26 Dot/Icm components of *L. pneumophila* have been identified (Zamboni et al. 2003), and at least 19 of them are present in the genome of *R. grylli* (Voth and Heinzen 2009; Zusman et al. 2007). To date, it is unclear whether this system constitutes a functional T4SS in *R. grylli* given its apparently incomplete nature. Nevertheless, this T4SS is thought to deliver macromolecules into the vacuole and the host cell cytoplasm to ensure necessary dialogs between the bacteria and the host (Christie and Cascales 2005).

It is expected that bacterial effectors will have in common known eukaryotic-like motifs that may facilitate interactions with host cell proteins (Stebbins and Galan 2001). In *L. pneumophila*, about 140 Dot/Icm effector proteins have been identified using various approaches, including bioinformatics analyses, which are considered characteristics of known T4SS effector proteins (Burstein et al. 2009). These effectors include proteins with ankyrin repeat domains (Anks), tetratricopeptide repeats (TRPs), coiled-coil domains (CCDs), leucine-rich repeats (LRRs), GTPase domains, or ubiquitination-related motifs and multiple kinases and phosphatases (de Felipe et al. 2005). Very few *L. pneumophila* effector homologs have been identified in *C. burnetii* and *R. grylli*.

Ank repeats, which are tandem motifs of around 33 amino acids, are found mainly in eukaryotic proteins, where they are known to mediate protein-protein interactions (Mosavi et al. 2004). The increasing amount of bacterial genome sequences is revealing that these proteins can represent an unexpectedly high proportion of bacterial genomes, as in *Wolbachia* genomes, which contain several tens of Ank domain-containing proteins (Iturbe-Ormaetxe et al. 2005). In the genome of *R. grylli*, nine Ank proteins have been identified so far as putative T4SS effectors (Voth and Heinzen 2009). The secretion of three of them, including the *L. pneumophila* Ank K homolog, has been tested without success using the Dot/Icm system of *Legionella* as surrogate T4SS (Voth et al. 2009). The Ank K protein of *L. pneumophila* is not secreted either. Surprisingly, it has been shown that the Ank K homolog of *C. burnetii* is efficiently secreted by the *Legionella* Dot/Icm system (Voth et al. 2009). *C. burnetti* Ank K is 41 and 39% identical at the amino acid level to Ank K homologs of *L. pneumophila* and *R. grylli*, respectively. These results suggest that, in *R. grylli*, the function of the three Ank proteins tested is not associated with secretion. According to GenBank annotation, several other candidate effector proteins have been identified in the *R. grylli* genome, including TRP repeat proteins, putative effector protein B, and a structural toxin protein, but no translocation assay has been performed to test their secretion. Identifying the *R. grylli* Dot/Icm substrates may provide a list of potential effector proteins whose functional characterization will substantially aid our ability to decipher *Rickettsiella*-host cell interactions and the pathophysiology of induced diseases. It should be noted that even in *Legionella*, the cellular function of most effectors is still unknown (Burstein et al. 2009).

More recently, it has been suggested that the Dot/Icm-independent secretion system of *C. burnetii* may also participate in the secretion of bacterial effectors (Voth and Heinzen 2009). Among these other secretion systems, the *Coxiella* genome contains genes for a T1SS and nine *Pil* genes that are involved in type IV pilus biogenesis and evolutionarily related to type II secretion genes (Peabody et al. 2003). Draft annotation of the genome of *R. grylli* also revealed the presence of a type I secretion system, including several ABC-type transporters, as well as *Pil* genes and other putative outer membrane autotransporters. According to the annotated genome, *R. grylli*, like other gamma-Proteobacteria, possesses the Sec system that could deliver macromolecules into the periplasm, which could then be exported in the host cell cytoplasm through the T1SS or the T4SS. Overall, the *R. grylli* genome may be used for comparative analysis with the related *C. burnetii* and *L. pneumophila* to get insight into the specific genes of obligate intracellular bacteria.

CONCLUSION

Rickettsiella are likely more widespread than expected. Indeed, *Rickettsiella* are found in many arthropods, including the largest groups of insects. Furthermore, *Rickettsiella* are present in arthropods that have colonized diverse habitats from terrestrial to aquatic environments. *Rickettsiella* are most commonly considered virulent for their hosts, leading to death. *Rickettsiella* also potentially affects a host trait of ecological importance: prey-predator relationships (Tsuchida et al. 2010). Consequently, it has been proposed that *Rickettsiella* may be used to control arthropod pest populations or to reduce vector competence. Such biocontrol agents would be environmentally friendly and could replace chemical control methods. However, as mammals may be susceptible to *Rickettsiella* (Giroud et al. 1958), such a control strategy needs further investigation to characterize *Rickettsiella*-host interactions. These approaches are being developed using *Wolbachia* endosymbionts, which are the most abundant intracellular bacteria so far discovered in arthropods (Bourtzis 2008). Intracellular bacteria therefore are invaluable models for fundamental research, and they may also be excellent assets of agricultural and biomedical relevance.

ACKNOWLEDGMENTS

We are very grateful to Maryline Raimond for technical assistance and TEM micrographs. Our research is funded by the Centre National de la Recherche Scientifique (CNRS), the French Ministère de l'Education Nationale, de l'Enseignement Supérieur et de la Recherche, the Agence Nationale de la Recherche (ANR-06-BLAN-0316 EndoSymbArt project coordinated by D.B.). R.C. is supported by a CNRS Young Investigator ATIP award.

REFERENCES

Abd, E. M., and Holdich, D. M. 1987. The occurrence of a Rickettsial disease in British wood-lice (Crustacea, Isopoda, Oniscidea) populations. *J Invertebr Pathol* 49, 252–58.

Adamo, S. A. 1998. The specificity of behavioral fever in the cricket *Acheta domesticus*. *J Parasitol* 84, 529–33.

Bonami, J. R., and Pappalardo, R. 1980. Rickettsial infection in marine crustacea. *Experientia* 36, 180–81.

Bouchon, D., Cordaux, R., and Grève, P. 2008. Feminizing *Wolbachia* and the evolution of sex determination in isopod. In *Insect symbiosis*, ed. K. Bourtzis and T. A. Miller, 273–94. Vol. 3. Boca Raton, FL: CRC Press.

Bourtzis, K. 2008. *Wolbachia*-based technologies for insect pest population control. *Adv Exp Med Biol* 627, 104–13.

Bower, S. M., McGladdery, S. E., and Price, I. M. 1994. Synopsis of infectious diseases and parasites of commercially exploited shellfish. *Ann Rev Fish Dis* 4, 1–199.

Breeuwer, J. A., and Jacobs, G. 1996. *Wolbachia*: intracellular manipulators of mite reproduction. *Exp Appl Acarol* 20, 421–34.

Burstein, D., Zusman, T., Degtyar, E., Viner, R., Segal, G., and Pupko, T. 2009. Genome-scale identification of *Legionella pneumophila* effectors using a machine learning approach. *PLoS Pathog* 5, e1000508.

Carter, J. B., and Luff, L. F. 1977. Rickettsia-like organisms infecting *Harpalus rufipes* (Coleoptera: Carabidae). *J Invertebr Pathol* 30, 99–101.

Chauve, C. 1998. The poultry red mite *Dermanyssus gallinae* (De Geer, 1778): current situation and future prospects for control. *Vet Parasitol* 79, 239–45.

Christie, P. J., Atmakuri, K., Krishnamoorthy, V., Jakubowski, S., and Cascales, E. 2005. Biogenesis, architecture, and function of bacterial type IV secretion systems. *Annu Rev Microbiol* 59, 451–85.

Christie, P. J., and Cascales, E. 2005. Structural and dynamic properties of bacterial type IV secretion systems (review). *Mol Membr Biol* 22, 51–61.

Cordaux, R., Paces-Fessy, M., Raimond, M., Michel-Salzat, A., Zimmer, M., and Bouchon, D. 2007. Molecular characterization and evolution of arthropod-pathogenic *Rickettsiella* bacteria. *Appl Environ Microbiol* 73, 5045–47.

Corsaro, D., Thomas, V., Goy, G., Venditti, D., Radek, R., and Greub, G. 2007. '*Candidatus Rhabdochlamydia crassificans*', an intracellular bacterial pathogen of the cockroach *Blatta orientalis* (Insecta: Blattodea). *Syst Appl Microbiol* 30, 221–28.

Croizier, G., and Meynadier, G. 1971. [Demonstration and comparison of antigens extracted from giant forms of 3 *Rickettsiella* Philip]. *Ann Inst Pasteur (Paris)* 121, 87–92.

Czarnetzki, A. B., and Tebbe, C. C. 2004. Diversity of bacteria associated with *Collembola*—a cultivation-independent survey based on PCR-amplified 16S rRNA genes. *FEMS Microbiol Ecol* 49, 217–27.

de Felipe, K. S., Pampou, S., Jovanovic, O. S., Pericone, C. D., Ye, S. F., Kalachikov, S., and Shuman, H. A. 2005. Evidence for acquisition of *Legionella* type IV secretion substrates via interdomain horizontal gene transfer. *J Bacteriol* 187, 7716–26.

Delmas, F., and Timon-David, P. 1985. Effect of invertebrate *Rickettsia* on vertebrates: experimental infection of mice by *Rickettsiella grylli*. *C R Acad Sci Hebd Seances Acad Sci D* 300, 115–17.

De Luna, C. J., Moro, C. V., Guy, J. H., Zenner, L., and Sparagano, O. A. E. 2009. Endosymbiotic bacteria living inside the poultry red mite (*Dermanyssus gallinae*). *Exp Appl Acarol* 48, 105–13.

Devauchelle, G., Meynadier, G., and Vago, C. 1972. [Ultrastructure of the multiplication cycle of *Ricketsiella melolonthae* (Krieg), Philip, in blood cells of its host]. *J Ultrastruct Res* 38, 134–48.

Dillon, J. G., Miller, S., Bebout, B., Hullar, M., Pinel, N., and Stahl, D. A. 2009. Spatial and temporal variability in a stratified hypersaline microbial mat community. *FEMS Microbiol Ecol* 68, 46–58.

Dionne, M., Miller, K. M., Dodson, J. J., and Bernatchez, L. 2009. MHC standing genetic variation and pathogen resistance in wild Atlantic salmon. *Philos Trans R Soc Lond B Biol Sci* 364, 1555–65.

Drobne, D., Strus, J., Znidarsic, N., and Zidar, P. 1999. Morphological description of bacterial infection of digestive glands in the terrestrial isopod *Porcellio scaber* (Isopoda, crustacea). *J Invertebr Pathol* 73, 113–19.

Dutky, S. R., and Gooden, E. L. 1952. *Coxiella popilliae*, n. sp., a *Rickettsia* causing blue disease of Japanese beetle larvae. *J Bacteriol* 63, 743–50.

Entwistle, P. F., and Robertson, J. S. 1968. The ultrastructure of a *Rickettsia* pathogenic to a saturnid moth. *J Gen Microbiol* 54, 97–104.

Federici, B. A. 1980. Reproduction and morphogenesis of *Rickettsiella chironomi*, an unusual intracellular procaryotic parasite of midge larvae. *J Bacteriol* 143, 995–1002.

Federici, B. A. 1984. Diseases of terrestrial isopods. *Symp. Zool. Soc. London* 53, 233–45.

Federici, B. A., Hazard, E. I., and Anthony, D. W. 1974. *Rickettsia*-like organism causing disease in a crangonid amphipod from Florida. *Appl Microbiol* 28, 885–86.

Fournier, P. E., and Raoult, D. 2005. Genus II. *Rickettsiella* Philip 1956, 267AL. In *Bergey's manual of systematic bacteriology*, ed. D. J. Brenner, N. R. Krieg, and J. T. Staley, 241–47. 2nd ed. New York: Springer.

Frutos, R., Federici, B. A., Revet, B., and Bergoin, M. 1994. Taxonomic studies of *Rickettsiella*, *Rickettsia*, and *Chlamydia* using genomic DNA. *J Invertebr Pathol* 63, 294–300.

Frutos, R., Pages, M., Bellis, M., Roizes, G., and Bergoin, M. 1989. Pulsed-field gel electrophoresis determination of the genome size of obligate intracellular bacteria belonging to the genera *Chlamydia, Rickettsiella*, and *Porochlamydia. J Bacteriol* 171, 4511–13.

Giroud, P., Dumas, N., and Hurpin, B. 1958. [Attempted adaptation of white mice to the rickettsial agent causing blue disease from *Melolontha melolontha* L.: adaptation by pulmonary and by oral routes]. *C R Hebd Seances Acad Sci* 247, 2499–501.

Gotoh, T., Noda, H., and Hong, X. Y. 2003. *Wolbachia* distribution and cytoplasmic incompatibility based on a survey of 42 spider mite species (Acari: Tetranychidae) in Japan. *Heredity* 91, 208–16.

Hall, I. M., and Badgley, M. E. 1957. A rickettsial disease of larvae of species of *Stethorus* caused by *Rickettsiella stethorae*, n. sp. *J Bacteriol* 74, 452–55.

Huger, A. 1959. Histological observations on the development of crystalline inclusions of the rickettsial disease of *Tipula paludosa* Meigen. *J Insect Pathol* 1, 60:66.

Huger, A. 1964. Eine rickettsiose der orientalischen schabe Blatta orientalis l verursacht durch Rickettsiella blattae nov spec. *Naturwissenschaften* 51, 22.

Huger, A. M., and Krieg, A. 1967. [A new mode of multiplication of rickettsiae in insects]. *Naturwissenschaften* 54, 475.

Hurpin, B. 1971a. [Principles of microbiological control in agriculture]. *Ann Parasitol Hum Comp* 46(Suppl.), 243–76.

Hurpin, B. 1971b. Specificity of *Rickettsiella melolonthae* and pathogenicity for vertebrates. *Ann Soc Entomol France* 7, 439.

Hurpin, B., and Robert, P. H. 1976. Conservation dans le sol de trois germes pathogènes pour les larves de *Melolontha melolontha* [Col.: Scarabaeidae]. *Entomophaga* 21, 73–80.

Hurpin, B., and Robert, P. H. 1977. Effets en population naturelle de *Melolontha melolontha* (col.: Scarabaeidae) d'une introduction de *Rickettsiella melolonthae* et de *Entomopoxvirus melolonthae. Entomophaga* 22, 85–91.

Hurpin, B., and Vago, C. 1958. Les maladies du hanneton commun (*Melolontha melolontha* L.) (Col. Scarabaeidae). *Entomophaga* 4, 285–330.

Iturbe-Ormaetxe, I., Burke, G. R., Riegler, M., and O'Neill, S. L. 2005. Distribution, expression, and motif variability of ankyrin domain genes in *Wolbachia pipientis*. *J Bacteriol* 187, 5136–45.

Johnson, P. T. 1984. A *Rickettsia* of the blue king crab, *Paralithodes platypus*. *J Invertebr Pathol* 44, 112–13.

Kellen, W. R., Lindegren, J. E., and Hoffmann, D. F. 1972. Developmental stages and structure of a *Rickettsiella* in the navel orangeworm, *Poramyelois transitella* (Lepidoptera: Phycitidae). *J Invertebr Pathol* 20, 193–99.

Kostanjsek, R., Strus, J., Drobne, D., and Avgustin, G. 2004. '*Candidatus Rhabdochlamydia porcellionis*', an intracellular bacterium from the hepatopancreas of the terrestrial isopod *Porcellio scaber* (Crustacea: Isopoda). *Int J Syst Evol Microbiol* 54, 543–49.

Krieg, A. 1955. Licht- und elektronenmikroskopische Untersuchungen zur Pathologie der "Lorscher Erkrankung" von Engerlingen und zur Zytologie der *Rickettsia melolonthae* nov. spec. *Naturforsch* 10b, 34–37.

Krieg, A. 1958. [Comparative taxonomic, morphological and serological research on Rickettsiae pathogenic for insects]. *Z Naturforsch B* 13B, 555–57.

Krieg, A. 1965. Über eine neue Rickettsie aus Colepoteren, Rickettsiella tenebrionis nov. spec. *Naturwissenschaften* 52, 144–45.

Kurtti, T. J., Palmer, A. T., and Oliver, J. H. 2002. *Rickettsiella*-like bacteria in *Ixodes woodi* (Acari : Ixodidae). *J Med Entomol* 39, 534–40.

Larsson, R. 1982. A *Rickettsial* pathogen of the amphipod *Rivulogammarus pulex*. *J Invertebr Pathol* 40, 28–35.

Leclerque, A. 2008. Whole genome-based assessment of the taxonomic position of the arthropod pathogenic bacterium *Rickettsiella grylli*. *FEMS Microbiol Lett* 283, 117–27.

Leclerque, A., and Kleespies, R. G. 2008a. 16S ribosomal RNA, GroEL, and MucZ based assessment of the taxonomic position of *Rickettsiella melolonthae* and its implications for the organization of the genus *Rickettsiella*. *Int J Syst Evol Microbiol* 58, 749–755.

Leclerque, A., and Kleespies, R. G. 2008b. Genetic and electron-microscopic characterization of *Rickettsiella tipulae*, an intracellular bacterial pathogen of the crane fly, *Tipula paludosa*. *J Invertebr Pathol* 98, 329–34.

Leclerque, A., and Kleespies, R. G. 2008c. Type IV secretion system components as phylogenetic markers of entomopathogenic bacteria of the genus *Rickettsiella*. *FEMS Microbiol Lett* 279, 167–73.

Louis, C., Jourdan, M., and Cabanac, M. 1986. Behavioral fever and therapy in a *Rickettsia*-infected Orthoptera. *Am J Physiol* 250, R991–95.

Merhej, V., Royer-Carenzi, M., Pontarotti, P., and Raoult, D. 2009. Massive comparative genomic analysis reveals convergent evolution of specialized bacteria. *Biol Direct* 4, 13.

Meynadier, G., Lopez, B., and Duthoit, J. L. 1974. Mise en évidence de Rickettsiales chez une araignée (Argyrodes gibbosus Lucas), Ananeae, Theridiidae. *C R Acad Sci Hebd Seances Acad Sci D* 278, 2365–67.

Meynadier, G., and Monsarrat, P. 1969. Une rickettsiose chez une cétoine de Madagascar. *Entomophaga* 14, 401–6.

Moran, N. A., McCutcheon, J. P., and Nakabachi, A. 2008. Genomics and evolution of heritable bacterial symbionts. *Annu Rev Genet* 42, 165–90.

Morel, G. 1976. Studies on *Porochlamydia buthi* g. n., sp. n., an intracellular pathogen of the scorpion *Buthus occitanus*. *J Invertebr Pathol* 28, 167–75.

Morel, G. 1977. [Study of a "*Rickettsiella*" (Rickettsia) pathogen of the spider "*Pisaura mirabilis*" (author's transl.)]. *Ann Microbiol (Paris)* 128A, 49–59.

Morel, G., Veyrunes, J. C., and Vago, C. 1974. [Viral infection in rickettsia of the scorpion *Buthus occitanus* Amoreux]. *C R Acad Sci Hebd Seances Acad Sci D* 279, 1365–71.

Moro, C. V., Thioulouse, J., Chauve, C., Normand, P., and Zenner, L. 2009. Bacterial taxa associated with the hematophagous mite *Dermanyssus gallinae* detected by 16S rRNA PCR amplification and TTGE fingerprinting. *Res Microbiol* 160, 63–70.

Mosavi, L. K., Cammett, T. J., Desrosiers, D. C., and Peng, Z. Y. 2004. The ankyrin repeat as molecular architecture for protein recognition. *Protein Sci* 13, 1435–48.

Müller-Kogler, E. 1958. Eine Rickettsiose von *Tipula paludosa* Meig. durch *Rickettsiella tipulae* nov. spec. *Naturwissenschaften* 45, 248–50.

Palevody, C. 1972. [Intracellular microorganisms in the ovary of the isotomide collembole *Folsomia candida*]. *C R Acad Sci Hebd Seances Acad Sci D* 275, 401–4.

Peabody, C. R., Chung, Y. J., Yen, M. R., Vidal-Ingigliardi, D., Pugsley, A. P., and Saier, M. H., Jr. 2003. Type II protein secretion and its relationship to bacterial type IV pili and archaeal flagella. *Microbiology* 149, 3051–72.

Philip, C. B. 1956. Comments on the classification of the order *Rickettsiales. Can J Microbiol* 2, 261–270.

Radek, R. 2000. Light and electron microscopic study of a *Rickettsiella* species from the cockroach *Blatta orientalis. J Invertebr Pathol* 76, 249–56.

Reeson, A. F., Jankovic, T., Kasper, M. L., Rogers, S., and Austin, A. D. 2003. Application of 16S rDNA-DGGE to examine the microbial ecology associated with a social wasp *Vespula germanica. Insect Mol Biol* 12, 85–91.

Romero, X., Turnbull, J. F., and Jimenez, R. 2000. Ultrastructure and cytopathology of a *Rickettsia*-like organism causing systemic infection in the redclaw crayfish, *Cherax quadricarinatus* (Crustacea: decapoda), in Ecuador. *J Invertebr Pathol* 76, 95–104.

Roux, V., Bergoin, M., Lamaze, N., and Raoult, D. 1997. Reassessment of the taxonomic position of *Rickettsiella grylli. Int J Syst Bacteriol* 47, 1255–57.

Sanchez-Pena, S. R. 2000. Entomopathogens from two Chihuahuan desert localities in Mexico. *BioControl* 45, 63–78.

Segal, G., Feldman, M., and Zusman, T. 2005. The Icm/Dot type-IV secretion systems of *Legionella pneumophila* and *Coxiella burnetii. FEMS Microbiol Rev* 29, 65–81.

Segal, G., Purcell, M., and Shuman, H. A. 1998. Host cell killing and bacterial conjugation require overlapping sets of genes within a 22-kb region of the *Legionella pneumophila* genome. *Proc Natl Acad Sci USA* 95, 1669–74.

Seshadri, R., and Samuel, J. 2005. Genome analysis of *Coxiella burnetii* species: insights into pathogenesis and evolution and implications for biodefense. *Ann NY Acad Sci* 1063, 442–50.

Spaulding, A. W., and von Dohlen, C. D. 2001. Psyllid endosymbionts exhibit patterns of co-speciation with hosts and destabilizing substitutions in ribosomal RNA. *Insect Mol Biol* 10, 57–67.

Stebbins, C. E., and Galan, J. E. 2001. Structural mimicry in bacterial virulence. *Nature* 412, 701–5.

Sutakova, G. 1988. Electron microscopic study of developmental stages of *Rickettsiella phytoseiuli* in *Phytoseiulus persimilis* Athias-Henriot (Gamasoidea:Phytoseiidae) mites. *Acta Virol* 32, 50–54.

Sutakova, G., and Arutunyan, E. S. 1990. The spider-mite predator *Phytoseiulus persimilis* and its association with microorganisms—an electron-microscope study. *Acta Entomol Bohemoslovaca* 87, 431.

Sutakova, G., and Rehacek, J. 1989. Experimental infection with *Rickettsiella phytoseiuli* in adult female *Dermacentor reticulatus* (Ixodidae): an electron microscopy study. *Exp Appl Acarol* 7, 299–311.

Sutakova, G., and Ruttgen, F. 1978. *Rickettsiella phytoseiuli* and virus-like particles in *Phytosfiulus persimilis* (Gamasoidea: Phytoseiidae) mites. *Acta Virol* 22, 333–36.

Sutter, G. R., and Kirk, V. M. 1968. *Rickettsia*-like particles in fat-body cells of carabid beetles. *J Invertebr Pathol* 10, 445–449.

Timmermans, M. J., Roelofs, D., Marien, J., and van Straalen, N. M. 2008. Revealing pan-crustacean relationships: phylogenetic analysis of ribosomal protein genes places Collembola (springtails) in a monophyletic Hexapoda and reinforces the discrepancy between mitochondrial and nuclear DNA markers. *BMC Evol Biol* 8, 83.

Tsuchida, T., Koga, R., Horikawa, M., Tsunoda, T., Maoka, T., Matsumoto, S., Simon, J. C., and Fukatsu, T. 2010. Symbiotic bacterium modifies aphid body color. *Science* 330, 1102–4.

Vago, C., and Martoja, R. 1963. [Rickettsiosis in the Gryllidae (Orthoptera)]. *C R Hebd Seances Acad Sci* 256, 1045–47.

Vago, C., and Meynadier, G. 1965. Une rickettsiose chez le criquet pélerin (Schistocerca gregaria Forsk.) *Entomophaga* 10, 307–10.

Vago, C., Meynadier, G., Juchault, P., Legrand, J. J., Amargier, A., and Duthoit, J. L. 1970. [A rickettsial disease of isopod crustaceans]. *C R Acad Sci Hebd Seances Acad Sci D* 271, 2061–63.

Vandekerckhove, T. T., Watteyne, S., Willems, A., Swings, J. G., Mertens, J., and Gillis, M. 1999. Phylogenetic analysis of the 16S rDNA of the cytoplasmic bacterium *Wolbachia* from the novel host *Folsomia candida* (Hexapoda, Collembola) and its implications for wolbachial taxonomy. *FEMS Microbiol Lett* 180, 279–86.

Vilcins, I. M. E., Old, J. M., and Deane, E. 2009. Molecular detection of *Rickettsia*, *Coxiella* and *Rickettsiella* DNA in three native Australian tick species. *Exp Appl Acarol* 49, 229–42.

Voth, D. E., and Heinzen, R. A. 2009. Coxiella type IV secretion and cellular microbiology. *Curr Opin Microbiol* 12, 74–80.

Voth, D. E., Howe, D., Beare, P. A., Vogel, J. P., Unsworth, N., Samuel, J. E., and Heinzen, R. A. 2009. The *Coxiella burnetii* ankyrin repeat domain-containing protein family is heterogeneous, with C-terminal truncations that influence Dot/Icm-mediated secretion. *J Bacteriol* 191, 4232–42.

Wang, Y., Brune, A., and Zimmer, M. 2007. Bacterial symbionts in the hepatopancreas of isopods: diversity and environmental transmission. *FEMS Microbiol Ecol* 61, 141–52.

Weiser, J. 1949. Deux nouvelles infections à virus des insectes. *Ann Parasitol* 24, 259–64.

Weiser, J. 1963. Diseases of insects of medical importance in Europe. *Bull World Health Organ* 28, 121–27.

Weiser, J., and Zizka, Z. 1968. [Electron microscopic study of old viral material]. *Mikroskopie* 22, 336–40.

Weiss, E., Dasch, G. A., and Chang, K.-P. 1984. Genus VIII. *Rickettsiella* Philip 1956, 267AL. In *Bergey's manual of systematic bacteriology*, ed. N. R. Krieg and J. G. Holt, 713–17. Vol. 1. Baltimore: Williams and Wilkins.

Wille, H., and Martignoni, M. E. 1952. [Preliminary report on a new disease in larvae of the cockchafer]. *Schweiz Z Pathol Bakteriol* 15, 470–74.

Yousfi, A. 1976. *Recherches sur la pathologie des Crustacés Isopodes Oniscoïdes*. Ph.D. thesis. University of Montpellier.

Zamboni, D. S., McGrath, S., Rabinovitch, M., and Roy, C. R. 2003. *Coxiella burnetii* express type IV secretion system proteins that function similarly to components of the *Legionella pneumophila* Dot/Icm system. *Mol Microbiol* 49, 965–76.

Zusman, T., Aloni, G., Halperin, E., Kotzer, H., Degtyar, E., Feldman, M., and Segal, G. 2007. The response regulator PmrA is a major regulator of the icm/dot type IV secretion system in *Legionella pneumophila* and *Coxiella burnetii*. *Mol Microbiol* 63, 1508–23.

Zusman, T., Yerushalmi, G., and Segal, G. 2003. Functional similarities between the icm/dot pathogenesis systems of *Coxiella burnetii* and *Legionella pneumophila*. *Infect Immun* 71, 3714–23.

8 Arthropods Shopping for *Wolbachia*

*Daniela Schneider, Wolfgang J. Miller,
and Markus Riegler*

CONTENTS

Genus	*Wolbachia*
Species	*pipientis*
Family	Anaplasmataceae
Order	Rickettsiales
Description year	1924
Origin of name	After Prof. Dr. S. Burt Wolbach (1880–1954), an American pathologist who described the rickettsial agent of Rocky Mountain spotted fever, and worked on epidemic typhus and in the fields of radiation and vitamin research, in particular the relationships of vitamins to tissue structure and the pathology of scurvy and other diseases. From 1922 to 1947 he held the position as Shattuck professor of pathology at Harvard University.
Description	
References	Hertig and Wolbach (1924), Hertig (1936), Shields (1954)

S. Burt Wolbach. (Photo from Woodward et al. 1992. With permission from ACCA.)

INTRODUCTION

Within the large group of arthropod-associated bacteria, one genus of the α-Proteobacteria, *Wolbachia*, has attracted particular attention by serving as a microbial model system for deciphering the complexities of arthropod symbiosis, ranging from manipulation of host reproduction to interactions in nutritional and metabolic pathways, interferences in development and life span, and protection from pathogens and parasites (reviewed in Moran et al. 2008; Brownlie and Johnson 2009; Gross et al. 2009; Cook and McGraw 2010). *Wolbachia* were first described as intracellular *Rickettsia*-like organisms (RLOs) in the gonad cells of the mosquito *Culex pipiens* (Hertig and Wolbach 1924) and named *Wolbachia pipientis* (Hertig 1936). Since their first description in the early twentieth century, *Wolbachia* have been found in a uniquely wide range of host species mostly belonging to the arthropod phylum. *Wolbachia* infect up to two-thirds of all insect species (Jeyaprakash and Hoy 2000; Hilgenboecker et al. 2008), as well as a variety of mites, spiders, scorpions, and terrestrial crustaceans (Rowley et al. 2004; Bordenstein and Rosengaus 2005; Baldo et al. 2007, 2008; Wiwatanaratanabutr et al. 2009). In addition to arthropods, *Wolbachia* are also present in filarial nematodes (see Taylor et al. 2010 for review), and have been detected in a plant-associated nematode (Haegeman et al. 2009). Over the last two decades *Wolbachia* have attracted major foci within arthropod research and evince the largest growth in publications compared to any other bacterial arthropod symbiont. The reasons for this are multiple: *Wolbachia* are extremely common, have a wide host range, and have manifold ways of influencing host fitness and behavior. Host species appear to experience recurring *Wolbachia* epidemics, or occasionally accumulate *Wolbachia* strains, perhaps as a strategy to extend their metabolic pathways with gene networks that they could otherwise not as quickly exploit. This "shopping for *Wolbachia*" could be seen as an analogy to the stepwise acquisition of different types of plastids by photosynthetic organisms, previously introduced as "shopping for plastids" (Larkum et al. 2007) in the light of the endosymbiotic theory (Sagan 1967).

This book chapter will add to extensive reviews about *Wolbachia* biology and ecology (O'Neill et al. 1997; Werren 1997; Stouthamer et al. 1999; Riegler and O'Neill 2006; Serbus et al. 2008, Werren et al. 2008) by incorporating recent progress in this fast-moving research field. We will discuss potential impacts of recent findings on future research directions in the comprehensive biology of these exhilarating symbionts, including some novel and unconventional thoughts.

BIOLOGY OF *WOLBACHIA*

PHYLOGENY

The genus *Wolbachia* belongs to the family Anaplasmataceae, order Rickettsiales, class α-Proteobacteria (Dumler et al. 2001). Within the family Anaplasmataceae, *Wolbachia* are the closest phylogenetic relatives of the genera *Anaplasma*, *Ehrlichia*, and *Neorickettsia* (Figure 8.1). *Wolbachia pipientis* is the only species within the genus (reviewed by Riegler and O'Neill 2006), and researchers often refer to the bacteria just by the genus name.

On the basis of sequence information obtained from the bacterial genes 16S rRNA, *ftsZ* (expressed in host cell division), and the gene coding for the outer surface protein WSP, *Wolbachia* infections have been characterized and subdivided into eight supergroups: A–H (O'Neill et al. 1992; Werren et al. 1995; Zhou et al. 1998; Bordenstein and Rosengaus 2005; Lo et al. 2002, 2007) (see Figure 8.2). Infections classified into supergroups A, B, and E are primarily associated with arthropods (Vandekerckhove et al. 1999; Lo et al. 2002, 2007), whereas C and D infections can be found in filarial nematodes (Bandi et al. 1998; Lo et al. 2002, 2007). Three supergroups have been more recently proposed: Type F *Wolbachia* infections are associated with both arthropods and filarial nematodes (Campbell et al. 1992; Lo et al. 2002, 2007; Rasgon and Scott 2004; Casiraghi et al. 2005). Type G *Wolbachia* infections can be found in spiders (Rowley et al. 2004), whereas type H infections are associated with termites (Bordenstein and Rosengaus 2005). Recently, Ros and colleagues suggested the existence of three more supergroups: I (Siphonaptera), J (Spirurida, Nematoda), and K (Prostigmata, Acarina) (Ros et al. 2009).

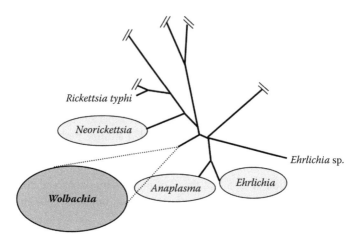

FIGURE 8.1 Schematic phylogenetic tree (maximum likelihood) inferred from small subunit 16S rRNA gene sequences of *Ehrlichia*, *Anaplasma*, *Neorickettsia*, and *Wolbachia* species. (Modified from Dumler, J. S., Barbet, A. F., Bekker, C. P., et al., *Int. J. Syst. Evol. Microbiol.*, 51, 2145–65, 2001.)

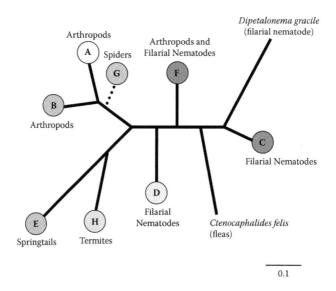

FIGURE 8.2 See color insert. Schematic diagram of *Wolbachia pipientis* phylogeny based on various phylogenetic studies of the genes *ftsZ*, *groEL*, *gltA*, and *dnaA*. Letters represent supergroups that have been confirmed on the basis of these four genes. The position of supergroup G is tentative since it was estimated using the *wsp* and 16S rRNA genes. Host species are indicated next to each clade or lineage. Two lineages, from *Dipetalonema gracile* and *Ctenocaphalides felis*, have not yet been classified into supergroups. The status of supergroup G is currently being debated. Bar indicates 0.1 substitution per site and is an approximation based on a concatenated gene analysis of these four genes. (Modified from Lo et al. 2007.)

It has been debated whether some of the latter supergroups are indeed supergroups, as analyses based on more loci established in a multilocus sequence typing (MLST) system (Baldo et al. 2006; Paraskevopoulos et al. 2006) have shown that the *wsp* locus is not suitable for supergroup assignment, and it was even suggested that supergroup G should be "decommissioned" (Baldo and Werren 2007). These considerations are based on recent findings that demonstrate that the *wsp* gene locus is affected by high recombination frequencies (Jiggins et al. 2002; Jiggins 2002) and strong diversifying selection (Baldo et al. 2006), mistaking extreme polymorphism in *wsp* sequence data as a novel supergroup. Furthermore, single-gene phylogenies are not considered reliable for resolving close *Wolbachia* relationships, or indeed other ancestral relationships (Jiggins et al. 2001; Baldo et al. 2005, 2006, 2007).

In order to avoid false supergroup assignments, as well as to access a reliable strain typing tool, a multi-locus sequence typing (MLST) system was introduced. MLST was originally established for applications in epidemiology as well as for surveillance of recombinant pathogenic bacteria (Maiden et al. 1998). In *Wolbachia* research, MLST has proven a reliable tool for the differentiation of diverse strains and the determination of supergroups (Paraskevopoulos et al. 2006; Baldo et al. 2006). *Wolbachia*-MLST comprises a set of conserved single-copy housekeeping genes, widely distributed throughout the *w*Mel genome of *Drosophila melanogaster* (Wu et al. 2004). The inclusion of higly polymorphic markers such as variable number tandem repeats (VNTRs; Riegler et al. 2005), and genes encoding for ankyrin domain repeats (Iturbe-Ormaetxe et al. 2005) further facilitate the differentiation between closely related isolates (e.g., Miller and Riegler 2006).

CELL BIOLOGY AND DISTRIBUTION

Wolbachia are intracellular bacteria and totally dependent on the cytoplasmic environment of their hosts. They are coccoid or bacilliform in morphology with an average size of 0.8–1.5 μm in length (Hertig 1936). Inside their hosts, *Wolbachia* are located within vacuoles, presumably of host origin. The bacteria themselves are surrounded by at least two cell membranes (Figure 8.3A).

Although *Wolbachia* are maternally transmitted through the germline, infection is not restricted to reproductive organs. *Wolbachia* colonize various somatic tissues such as muscle, digestive system, head/brain, fat body, and the hemolymph (Min and Benzer 1997; Cheng and Aksoy 1999; Dobson et al. 1999; Cheng et al. 2000; Serbus et al. 2008; Albertson et al. 2009; summarized in Figure 8.3B,C). Within assorted host tissues, *Wolbachia* can be identified by using diverse techniques. Hertig showed that the bacteria are easily visualized using Giemsa stain (Hertig 1936), and the fluorescent dye DAPI (4',6-diamidino-2-phenylindole) has been extensively used for detection (O'Neill and Karr 1990). A study by Albertson et al. (2009) has recently used the fluorescent nucleic acid dye Syto-11 to consistently label *Wolbachia*. *In situ* hybridization techniques with *Wolbachia*-specific DNA probes (Heddi et al. 1999) and immuno stainings using *Wolbachia* antibodies (Kose and Karr 1995; Dobson et al. 1999; Masui et al. 2001; Miller and Riegler 2006; Moreira et al. 2009b) are well established (Figure 8.3). Application of highly sensitive polymerase chain reaction (PCR) methods (Jeyaprakash and Hoy 2000) has revealed that *Wolbachia* infection

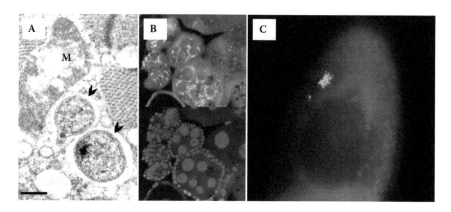

FIGURE 8.3   See color insert. Localization of *Wolbachia* within various host tissues. (A) Electron micrograph showing *Wolbachia* in ultra-thin sections of *Rhagoletis cerasi* thorax; scale bar = 0.5 μm, M = Mitochondria; *Wolbachia* are indicated by black arrowheads. (B) Fluorescent *in situ* hybridization performed on 5 μm paraffin sections of *R. cerasi* ovaries. *Wolbachia* (green) localize with nuclei of nurse cells (blue) in egg chambers. (C) Immunostainings performed on blastodermal (stage 9) embryos of *Drosophila willistoni*. *Wolbachia* (*w*Wil), indicated in green, localize in the primordial germ cells (PGCs).

is present in up to 76% of all arthropod species. These highly sensitive PCR protocols, however, need to be accompanied by other techniques, such as DNA hybridization and sequence analysis (Arthofer et al. 2009a, 2009b; Miller et al. 2010) in order to avoid PCR artifacts or contamination.

MANIPULATION OF HOST BIOLOGY—INDUCTION OF REPRODUCTIVE PHENOTYPES AND TRANSMISSION

Vertical maternal inheritance is the primary mode for *Wolbachia* transmission from one host generation to another. The bacteria are passed from females to offspring transovarially via the egg cytoplasm. Paternal transmission is unlikely since *Wolbachia* and sperm cytoplasm are stripped off during late spermatogenesis and egg fertilization (Tokuyasu et al. 1972a, 1972b; Fabrizio et al. 1998; Clark et al. 2003). Conclusive evidence for paternal inheritance in insects is still needed, as it has been reported in *Drosophila simulans* at an extremely low frequency (Hoffmann and Turelli 1988), and recent findings indicate potential transfer of *Wolbachia* from males to females in some *Nasonia* species hybrids (Chafee et al. 2010). In order to be transmitted efficiently, *Wolbachia* target the female germline of their hosts, while somatic tissue can also be colonized, but such somatic colonization is not mandatory. *Wolbachia* profoundly influence host reproductive biology in order to secure their transgenerational transmission. *Wolbachia* can induce cytoplasmic incompatibility (CI) in matings between infected males and uninfected females, leading to embryonic mortality. Maternally inherited *Wolbachia* in the early embryo can rescue such incompatibility. This selection fosters high maternal transmission rates. In some instances, *Wolbachia* favor female host offspring and can shift the sex ratio toward females by inducing reproductive phenotypes such as thelytokous parthenogenesis, male killing, and feminization (reviewed in Stouthamer et al. 1999).

Although rare, horizontal *Wolbachia* transfer from infected to uninfected individuals within (intraspecific) and between (interspecific) host species has been observed over evolutionary timescales with the outcome of closely related strains in distantly related hosts. Horizontal acquisition allows these bacteria to move between host species and contribute to their vast host range. Intraspecific horizontal transmission has been reported in wasps of the genus *Trichogramma* (Huigens et al. 2004). Raychoudhury and colleagues (2009) showed that out of the 11 different *Wolbachia* strains found in 4 species of the parasitoid wasps *Nasonia*, 5 were acquired by interspecific horizontal transmission from other taxa. Repeated horizontal transmission was also observed in the wood ants *Formica rufa* (Viljakainen et al. 2008) and in the North American funnel-web spider genus *Agelenopsis* (Baldo et al. 2008).

Cytoplasmic Incompatibility (CI)

Cytoplasmic incompatibility (CI) is the most commonly described *Wolbachia*-induced reproductive phenotype. Already in the 1970s, Yen and Barr (1971) had uncovered *Wolbachia*'s role in causing CI between populations of the mosquito *Culex pipiens* by antibiotic treatment.

Unidirectional CI arises whenever *Wolbachia*-infected males mate with uninfected females. Bidirectional CI occurs between populations when individuals carry different strains that are incompatible with each other (Figure 8.4A; and recently reviewed by Merçot and Poinsot 2009). Both modes of CI can result in high embryonic mortality and low numbers of viable offspring (Hoffmann et al. 1986; Tram and Sullivan 2002; Vavre et al. 2000, 2002; Bordenstein and Werren 2007); however, many *Wolbachia* associations have limited CI (Yamada et al. 2007) or totally lack CI expression (Hoffmann et al. 1996).

Functionally, CI can be explained by a two-component model that proposes the existence of two different antagonistic *Wolbachia* functions (Werren 1997; Poinsot et al. 2003; Mercot and Poinsot 2009). Sperm of infected males is modified during spermatogenesis, i.e., the modifier function (mod), resulting in early embryonic death unless a compatible *Wolbachia* is present in the egg to restore viability, i.e., the rescue function (resc). Accordingly, *Wolbachia* can be divided into the following four mod/resc classes:

1. $mod^+/resc^+$: *Wolbachia* can induce and then rescue CI (invasive).
2. $mod^-/resc^-$: *Wolbachia* are not capable of inducing CI, nor do they rescue (helpless or indifferent).
3. $mod^-/resc^+$: *Wolbachia* do not induce CI but can rescue (defensive).
4. $mod^+/resc^-$: The fourth class, the suicide phenotype, is in contrast to the other three, a theoretical one that could be maintained in a population only with a balancer strain; however, such a suicide strain has so far not been observed in nature (reviewed by Mercot and Poinsot 2009).

Zabalou et al. (2008) further elucidated the mod/resc model by analyzing modification and rescue properties of the closely related *Wolbachia* strains *w*Yak, *w*Tei, and

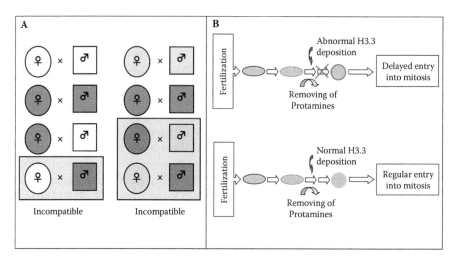

FIGURE 8.4   See color insert. (A) Schematic presentation of cytoplasmic incompatibility. Depending on *Wolbachia* strain and host species, unidirectional CI may result when *Wolbachia*-infected males (red) mate with uninfected females (white). These matings result in few viable offspring; all other crosses are compatible. Bidirectional CI may result when *Wolbachia*-infected males (red) mate with females harboring a different *Wolbachia* strain (green). (B) Proposed model of key events in the transformation of sperm to male pronucleus in embryos from normal and CI crosses: in normal crosses proper H3.3 deposition is not inhibited; in CI crosses abnormal H3.3 deposition leads to formation of a ring of histone H3.3 encompassing the paternal pronucleus (indicated in red). (Modified from Landmann, F., Orsi, G. A., Loppin, B., Sullivan, W., *PLoS Pathog.*, 5, e1000343, 2009.)

*w*San from the *Drosophila yakuba* complex upon transfer into the genomic background of an uninfected *Drosophila simulans* line. In the case of *w*Tei transfer from its natural host into an uninfected recipient, the authors uncovered a clear shift from mod⁻ to mod⁺ phenotypes, resulting in expression of high CI levels. To explain this observation, the existence of multiple rescue factors carried by a single *Wolbachia* strain was suggested. Each factor can partially or fully rescue the imprint of another *Wolbachia* strain, leading to asymmetric CI expression patterns, and multiple modification and rescuing factors exist (Zabalou et al. 2008).

The molecular mechanisms of CI are not yet fully understood, but some recent cell biological studies have contributed significantly to our current understanding. Cytological analyses suggested that *Wolbachia* induce CI by altering host cell cycle timing. Male pronuclei from infected males are delayed when entering first mitotic divisions, resulting in aneuploid nuclei and death in early stages of embryonic development of uninfected egg cells. The fact that this asynchrony between pronuclei of both sexes is compensated in the presence of *Wolbachia* in eggs suggests that the bacteria target host cell cycle proteins (Tram and Sullivan 2002; Tram et al. 2003, 2006; Ferree and Sullivan 2006). A recent study showed that in *Drosophila*, *Wolbachia* interfere with the host by delaying recruitment of the replication-independent histone H3.3/H4 complex to the male pronucleus (Landmann et al. 2009). Sperm chromatin is packaged with protamines, small arginine-rich proteins (Balhorn 2007), like in mammals. These proteins are immediately eliminated after fertilization and maternally supplied core histones initiate *de novo* nucleosome assembly (Loppin et al. 2005). The histone variant H3.3 and its specific chaperone HIRA are responsible for such assembly, and with CI, proper protamine removals, as well as H3.3/H4 deposits, are inhibited (Loppin et al. 2005, Landmann et al. 2009; and summarized in Figure 8.4B).

Genome sequence analyses have revealed a series of *Wolbachia* gene candidates that might be involved in coding for this intricate interference, including genes encoding for ankyrin repeat domains (Iturbe-Ormaetxe et al. 2005) as well as bacteriophage-associated genes (Sinkins et al. 2005). It is also possible that *Wolbachia* do not directly interact through expressed and secreted proteins but manipulate host gene expression in ways similarly seen in the viral and transposon interference of the host immune systems (reviewed by Aravin et al. 2007; Halic and Moazed 2009).

Thelytokous Parthenogenesis

Sex determination systems in insects comprise diplodiploidy (diploid males and females), haplodiploidy (haploid males and diploid females), and thelytoky (diploid females without males), reviewed in Normark (2003). Parthenogenesis induction (PI) by *Wolbachia* has been reported in mites and Hymenoptera (Huigens and Stouthamer 2003). In these orders, *Wolbachia* shift the sex ratio toward female offspring by utilizing the haplodiploid sex determination system. In uninfected haplodiploids, unfertilized haploid eggs normally develop into haploid males, whereas females develop from fertilized diploid eggs—a reproductive system that is known as arrhenotoky. Thelytoky occurs when diploid females are produced parthenogenetically without the need of males. In some host species thelytoky is induced by the presence of *Wolbachia*, leading to unfertilized diploidized eggs during the first mitotic division, which then develop into females (Stouthamer et al. 1990). In this case, *Wolbachia*-infected mothers produce more female offspring than uninfected ones, and hence contribute to the expansion of the infection. Other bacteria, such as *Cardinium hertigii* (Giorgini et al. 2009), and *Rickettsia* have also been reported to induce thelytoky (Hagimori et al. 2006).

Male Killing

A diverse range of bacteria, such as *Rickettsia*, *Flavobacteria*, *Spiroplasma*, *Arsenophonus*, and *Wolbachia*, has been associated with induction of male killing (Werren et al. 1986, 1994; Hurst et al. 1997, 1999; Williamson et al. 1999; Jiggins et al. 2000; Anbutsu and Fukatsu 2003; Bentley et al. 2007; Engelstädter and Hurst 2007; Perotti et al. 2007; Ferree et al. 2008; Kageyama et al. 2009). This phenotype leads to severe sex ratio distortions for the host by bacterial-induced killing of male embryos. Although male-killing bacteria are widespread among arthropods, male-killing *Wolbachia* strains are rare (Hornett et al. 2006; Sheeley and McAllister 2009). *Wolbachia*-induced male killing has been observed in three different orders of arthropods: in Coleoptera (Majerus et al. 2000), in Lepidoptera (Jiggins et al. 1998), and in Diptera (Hurst et al. 2000). Recent research with butterflies demonstrated that resistance against the male-killing phenotype can emerge rapidly. The nymphalid butterfly *Hypolimnas bolina* is capable of suppressing the *Wolbachia*-induced male-killing phenotype in Southeast Asian populations, leading to a rapid expansion of the resistance gene throughout wild populations of these butterflies in the Pacific region (Hornett et al. 2006, 2008; Jaenike 2007; Charlat et al. 2011).

Feminization

Feminization causes an imbalance in host sex ratios by turning genetic males into functional females, and thereby accelerating symbiont distribution throughout their host range. Manipulation of host reproduction has been mainly reported in terrestrial crustaceans (Bouchon et al. 1998), but also occurs in Lepidoptera, as well as in Hemiptera (Kageyama et al. 2002; Negri et al. 2006). In crustaceans and butterflies feminization acts on the unique ZW/ZZ sex chromosome system, with females being the heterogametic sex (Rigaud 1997; Hiroki et al. 2002). In one species of the Hemiptera order, however, *Wolbachia* cause feminization in an XX/X0 sex determination system. The leafhopper *Zyginidia pullula* was shown to produce females having a male genotype when infected with *Wolbachia* (Negri et al. 2006). In a more recent study, Negri et al. (2009) demonstrated that *Wolbachia* are capable of inducing transgenerational epigenetic changes that might be responsible for the feminization phenotype. Analyses of host imprinting patterns, via methylation-sensitive random amplification of polymorphic DNA, revealed female genomic imprints in feminized males, suggesting disruption of male phenotypic development by *Wolbachia* (Negri et al. 2009).

LOW-TITER *WOLBACHIA* INFECTIONS

Wolbachia infections are extremely common in nature. Species infection frequencies within the class Insecta were estimated at approximately 20% after sampling and PCR screening of specimens from tropical (Werren et al. 1997) and temperate (Werren and Windsor 2000; West et al. 1998) habitats. The application of a highly sensitive detection method, long PCR, increased estimated infection frequencies to over 70% (Jeyaprakash and Hoy 2000). Recent work indicates that even highly sensitive techniques such as long PCR can overlook low-titer *Wolbachia* infections, leading to false negatives (Arthofer et al. 2009a, 2009b). In the European cherry fruit fly *Rhagoletis cerasi*, multiple low-titer *Wolbachia* infections were uncovered by using a novel detection strategy, augmenting *Wolbachia* detection limits by combining a classical PCR approach with hybridization techniques by three orders of magnitude (Arthofer et al. 2009b). Similarly, low-titer infections were recently uncovered in *Drosophila* species (Miller et al. 2010) that were earlier diagnosed as uninfected by using standard PCR approaches (Mateos et al. 2006). All members of the neotropical *Drosophila paulistorum* complex (Burla et al. 1949), a superspecies group in *statu nascendi* comprising at least six semispecies (Dobzhansky and Spassky 1959), are naturally infected with *Wolbachia*. Out of these, only one semispecies harbors high-titer infections, while the five others exhibit low-titer infections. The low-titer infections in the five semispecies were verified by applying the improved two-level detection strategy of first PCR, amplifying multicopy *Wolbachia* genes, and then Southern hybridization for final detection (Miller et al. 2010).

Detection difficulties can arise in *Wolbachia* strains with extreme tissue tropism and at densities of less than one bacterial cell per host cell when using whole individual DNA extracts. Under such circumstances quality control for DNA extraction becomes an essential analysis component and should involve single-copy host genes. Mitochondrial markers, host ribosomal DNA, and multicopy genes will be less adequate to use because of their high copy numbers in host cells.

Besides technical detection issues, *Wolbachia* frequencies can still be underestimated through sampling errors. Field population studies of infection frequencies often reveal that *Wolbachia* infections can be patchy, regionally restricted, variable in frequency, and at different densities throughout the host's life cycle, as well as in their natural distribution range (Arthofer et al. 2009a, 2009b). Localized studies based on low sample numbers per species will hence underestimate the actual infection frequencies of infected species. Estimates of infection frequencies in habitat surveys (Werren et al. 1997; Werren and Windsor 2000) build on the assumption that *Wolbachia* are fixed within and between populations of a species. A recent meta-analysis compiled data from 20 individual studies and concluded that 65% is a more realistic estimate of species infection frequencies (Hilgenboecker et al. 2008). However, this meta-analysis was based on results that were obtained from standard PCR screening techniques. It can be expected that the inclusion of highly sensitive techniques (Jeyaprakash and Hoy 2000; Arthofer et al. 2009b) will further increase *Wolbachia* frequencies in species communities. This raises the question as to why low-titer *Wolbachia* infections can persist in host species, and hence will require more research on their functions and their evolution in host species.

EXTREME TISSUE TROPISM

Because of their vertical transmission, it can be expected that *Wolbachia* selectively target germline cells and supportive tissues in order to facilitate host colonization (Figure 8.3) (Frydman et al. 2006; Miller and Riegler 2006; Serbus and Sullivan 2007). However, many surveys have demonstrated that *Wolbachia* are not restricted to reproductive tissues. To the contrary, the bacteria occupy a broad repertoire of somatic tissue types (see the "Cell Biology and Distribution" section). Although *Wolbachia* distribution throughout host bodies is common, it is not possible to generalize distribution patterns, as they seem to vary between host species (Ijichi et al. 2002; Dobson 2003; Clark et al. 2003; McGarry et al. 2004; Veneti et al. 2004; Miller and Riegler 2006) and developmental stages of the hosts (Arthofer et al. 2009a; Schneider 2008). This suggests that *Wolbachia* can play diverse roles in different host species. Somatic tissue tropism may arise from the functional interactions, as well as adaptative strategies of both *Wolbachia* and host species. A recent study elucidated the localization dynamics of *Wolbachia* during *Drosophila* development (Albertson et al. 2009). The authors demonstrated that the bacteria are equally distributed throughout early syncytial embryonic development, but exhibit an asymmetrical segregation pattern in later stages of embryonic and larval development. *Wolbachia* asymmetrically colocalize with embryonic neuroblasts, cells that later will give rise to a variety of specialized neural cells in the third instar larval brain (Ceron et al. 2001). These cells divide along an apical-basal axis to regenerate a large self-renewing apical neuroblast and a small ganglion mother cell from the basal cell (Kaltschmidt et al. 2000; Albertson and Doe 2003; Wu et al. 2008). In the adult fly brain, highest *Wolbachia* densities were determined in specific areas of the central brain, including subesophageal ganglia, antennal lobes, and the superior protocerebrum (Albertson et al. 2009). This kind of extreme tissue tropism is of great interest in the context of insect behavior. Localization of *Wolbachia* in certain areas of the adult brain and head (Alberston et al. 2009) could be directly or indirectly associated with interference of altered olfactory (Peng et al. 2008), mosquito probing behavior (Moreira et al. 2009a), and feeding success (Turley et al. 2009), plus sex-specific insect behavior (de Crespigny et al. 2006; Koukou et al. 2006; Villella and Hall 2008). In the *D. paulistorum* species complex, obligate *Wolbachia* can direct assortative mating behavior of females by triggering sexual isolation against closely related semispecies males that are infected with a different strain type of the symbiont (Miller et al. 2010). Strict somatic *Wolbachia* tissue tropism, similar to the one observed in *Drosophila* neural tissue during embryonic and larval development by Albertson et al. (2009), could potentially be extended to other tissue types. We have recently uncovered the pervasiveness of *Wolbachia* in insect flight muscle of *R. cerasi*, suggesting potential significance regarding development and function of flight muscles (Schneider et al. unpublished).

MULTIPLE INFECTIONS WITH *WOLBACHIA* AND OTHER SYMBIONTS

Superinfections with multiple *Wolbachia* strains that coexist within host individuals have been reported in various host species (Frank 1998; Malloch et al. 2000; Malloch and Fenton 2005; Vautrin et al. 2007; Arthofer et al. 2009a). Individuals of *Acromyrmex* leafcutter ants were diagnosed with four different *Wolbachia* strains (Van Borm et al. 2003). Some individuals of tephritid fruit fly species carry even more than four *Wolbachia* strains (Jamnongluk et al. 2002; Arthofer et al. 2009b). The frequency of observed multiple infections with A and B supergroup *Wolbachia* is higher than expected randomly (Werren et al. 1997), indicating that some insect hosts might be highly permissive for acquiring and accumulating *Wolbachia* strains. This shopping for *Wolbachia* hypothesis is contrary to earlier host-parasite theories that suggest symbionts and parasites would compete for resources in hosts and eventually outcompete each other, leading to individual or clonal infections (reviewed by Van Baalen and Sabelis 1995). *Wolbachia* are certainly becoming an interesting model for deciphering the evolutionary origin of multiple infections. A major question is whether the persistence of more than one *Wolbachia* strain within a host individual can lead to interstrain competition or synergies (Mouton et al. 2004). It can be expected that concurring organisms such as host and parasite follow Red Queen dynamics, resulting in a constant arms race (Van Valen 1973; Decaestecker et al. 2007). In the case of multiple *Wolbachia* infections, host-parasite interactions need to be extended to also include parasite-parasite interactions. Cross-interactions between various *Wolbachia* strains can be either synergistic or antagonistic, meaning that competition for survival and stable persistence within the host can occur.

Empirical studies on these types of *Wolbachia* interactions between strains are still lacking, and more research on the interactive dynamics of multiple infections is needed. In the parasitic haplodiploid wasp *Leptopilina heterotoma*, the density of different *Wolbachia* strains remained constant over time, regardless of the presence of other strains (Mouton et al. 2003). However, in another wasp, *Asobara tabida*, naturally infected with two facultative and one mutualistic *Wolbachia* (Dedeine et al. 2001), the titer of the mutualistic strain was significantly lower in singly infected lines that were generated artificially (Mouton et al. 2004).

Apart from multiple *Wolbachia* infections, combinations with obligate (primary) and facultative (secondary) symbionts within host species and individuals should be taken into account. Recent research illustrates that *Wolbachia* can share the intracellular environment with many other bacteria. Duron et al. (2008) analyzed 136 arthropod species for combined infections between *Wolbachia* and other bacteria. Seven out of 44 infected species were determined to harbor multiple bacterial infections with *Wolbachia* and other bacteria. Individual *Bemisia tabaci* whiteflies harbor five symbionts comprising *Hamiltonella*, *Arsenophonus*, *Cardinium*, *Wolbachia*, and *Rickettsia* (Gottlieb et al. 2008). Trypanosome-transmitting tsetse flies, *Glossina* spp., are infected with three different symbiotic organisms: *Wigglesworthia glossinida*, *Sodalis*, and *Wolbachia* (Cheng and Aksoy 1999). Some of these bacterial associations colocalize with *Wolbachia*; others are restricted to bacteriocytes, specific tissues, or the digestive system. It will be important to analyze not only host-bacteria interactions in isolation, but also the interactions within the bacterial communities. These dynamics are expectedly complex and could result in a constant arms race for persistence and resources within hosts (Van Valen 1973; Decaestecker et al. 2007). Alternatively, bacterial symbionts could as well facilitate their joint existence via interactive cooperation (Wu et al. 2006; Gottlieb et al. 2008; Gosalbes et al. 2008).

CRYPTIC *WOLBACHIA* DIVERSITY

Wolbachia can occur in host species at low densities as single or multiple infections. Depending on host species, *Wolbachia* can display extreme tissue tropism with changes throughout the host's development. All of these *Wolbachia* characteristics unavoidably contribute to underestimating *Wolbachia* abundance and diversity. It can be expected that many hidden *Wolbachia* infections are yet to be isolated from hosts. In the European cherry fruit fly *R. cerasi*, for example, an originally determined double infection (*w*Cer1 and *w*Cer2, Rieger, and Stauffer 2002) turned out to be a multiple infection, comprising five different *Wolbachia* strains, once highly sensitive detection techniques were applied to different developmental host stages (Arthofer et al. 2009a). Some *Wolbachia* strains seem to hide within their original hosts system, and only become apparent once transferred into novel host systems. This has been observed at least twice, in artificial *Wolbachia* transfer from *R. cerasi* to the medfly *Ceratitis capitata*, and to *D. simulans*. *Wolbachia* transfer via microinjection of cytoplasm from infected, field-collected *R. cerasi* into uninfected medfly resulted in the establishment of a novel strain in medfly (Zabalou et al. 2004) that had previously not been isolated from its original host *R. cerasi* (Arthofer et al. 2009a). Similarly, two *Wolbachia* strains derived from *R. cerasi* (*w*Cer1 and *w*Cer2) were transferred into uninfected *D. simulans* via microinjection. Based on standard PCR detection techniques, it was assumed that wCer1 was lost from this system in the first-generation postinjection, whereas wCer2 was stably maintained (Riegler et al. 2004). However, after a time period of over 150 generations posttransfer, the application of highly sensitive detection methods clearly uncovered the presence of wCer1 in *D. simulans* (Schneider 2008). The only reason why these fly lines were maintained was because of the presence of wCer2. The fly lines would probably have been discarded if they were negative for *Wolbachia*. Our findings have two major consequences: First, the detection of effective horizontal transfer, in ecological terms, might not be apparent for many generations after the initial transfer event. Second, in an applied context, artificially transferred *Wolbachia* might not arrive solely, but accompanied by unexpected bacterial diversity that can interfere with the stability and persistence of the desired symbiont-induced phenotype.

THE EVOLUTIONARY LIFE CYCLE OF *WOLBACHIA* IN HOST SPECIES

Wolbachia infections were originally discovered in host species because of their remarkable capacity of causing cytoplasmic incompatibilities between natural populations (Yen and Barr 1971), or by triggering other pronounced reproductive phenotypes (see the "Manipulation of Host Biology—Induction of Reproductive Phenotypes and Transmission" section). Since then, however, numerous *Wolbachia* strains were isolated from arthropods that do not exhibit reproductive phenotypes in the laboratory (Hoffmann et al. 1996; Merçot and Charlat 2004; Miller and Riegler 2006), or that only express the CI rescue but not the CI modifier phenotype (Bourtzis et al. 1998; Merçot and Poinsot 1998, 2009; Zabalou et al. 2004a, b, 2008). Presence of *Wolbachia* can hence be associated with measurable phenotypes, or phenotypes that are not detectable. Based on this, it has been concluded that *Wolbachia*-host associations are highly dynamic (Riegler and O'Neill 2007), and can evolve toward mutualism in insects (Dedeine et al. 2001, 2005; Pannebakker et al. 2007; Miller et al. 2010), even within short evolutionary periods (Weeks et al. 2007). This is in concordance with earlier theories that vertically transmitted symbiont will evolve toward lower virulence levels and mutualism in hosts (Ewald 1987). In the course of this review we present an evolutionary life history model of this ubiquitous symbiont of arthropods by linking the three symbiotic states—parasitism, commensalism, and mutualism—in a unified symbiotic life cycle. A similar model has been developed for evolutionary dynamics of transposable elements colonizing host genomes where they can be invasive, suppressed and reactivated, domesticated, or horizontally transferred into other genomes (Miller et al. 1999; Pinsker et al. 2001). The different stages in the evolutionary life history of *Wolbachia* are summarized in Figure 8.5. After successful *Wolbachia* establishment in the germline of a novel host B, via horizontal transfer from an original host A, the newly acquired symbiont colonizes the germline of both sexes and spreads throughout a naive population by reproductive parasitism, in general via favoring the fitness of infected females. Depending on a range of factors, the infection will then reach an equilibrium frequency or fixation in the *de novo* host population (Hoffmann and Turelli 1997). In the process of *Wolbachia* colonization of novel hosts it is expected that host-encoded

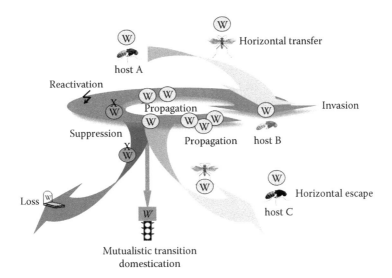

FIGURE 8.5 Evolutionary life history of *Wolbachia* in arthropods. This model starts with a *Wolbachia* (W) horizontal transfer from host A to a naive gene pool (host B) mediated by still unknown vectors (here symbolized by a parasitoid wasp). The initial infection burst leads to rapid propagation and adaptation of the endosymbiont in the germline of both sexes, as well as in the population via reproductive parasitism such as CI, male killing, feminization, or parthenogenesis. The reproductive phenotypes can be suppressed by various mechanisms encoded by both partners, the host and the symbiont. Such attenuated *Wolbachia* are either reactivated by mutations to start a new period of reproductive parasitism, replacing the original infection type, or subjected to molecular erosion by random mutations. Eventual symbiont loss can then occur through incomplete maternal transmission in the absence of selective pressure such as CI or fitness benefits. Symbiont extinction can be avoided by horizontal escape into an uninfected gene pool (host C), where a new parasitic cycle is then initiated. A sidetrack from this cycle leads to symbiont domestication, i.e., transition into obligate mutualism, by the acquisition of novel host function, or compensating loss of host-encoded factors with essential regulatory functions (symbolized by the traffic light).

suppressor systems may evolve against the intruder. Once fixation is reached, the presence of *Wolbachia* no longer provides any advantage in infected females if it coincides with a universal loss of the mod+ function in the symbiont or resistance within the host. Therefore, the symbiont may get lost from its host population over time by natural selection (Koehncke et al. 2009). Complete extinction of attenuated *Wolbachia* seems likely unless reactivation within the host is initiated by recombination between different *Wolbachia* strains (intersymbiotic) or variants (intrasymbiotic), mainly triggered by the activity of mobile DNAs in the *Wolbachia* genome, the accumulation of point mutations, or suspension/suppression of the repressing host regime. Another possible escape route is the invasion of an uninfected species via horizontal transmission, and the beginning of a new cycle in another permissive host environment. Finally, long-term survival as a germline-associated fixed host compound can be achieved by symbiont domestication, the transition toward obligate mutualistic interactions stably integrated into cellular host machinery to provide novel functions benefiting their hosts. There is a growing body of empirical evidence in support of the view that *Wolbachia* can provide significant fitness benefits by contributing to the fecundity, immunity, odorant perception, and mating success of their hosts under field and laboratory conditions (Dedeine et al. 2005; de Crespigny et al. 2006; Pannebakker et al. 2007; Weeks et al. 2007; Peng et al. 2008; Kambris et al. 2009; Miller et al. 2010).

WHAT IS IT THAT *WOLBACHIA* DON'T ACTUALLY DO?

Recent studies demonstrated that *Wolbachia* are involved in more host pathways than it was earlier anticipated. Brownlie et al. (2009) and Kremer et al. (2009) showed that *Wolbachia* may generally interfere with iron metabolism, and protect host cells from oxidative stress and apoptosis in *D. melanogaster* and parasitic wasps, respectively. Furthermore, the insulin/IGF-like signaling (IIS) pathway, which has pleiotropic functions in multicellular organisms (Skorokhod et al. 1999), was shown to be influenced by this omnipresent symbiont in *Drosophila*. *Wolbachia* potentially increases IIS, and thereby exerts influence on host aging, and the removal of the bacteria via antibiotic treatment results in reduced life span (Ikeya et al. 2009).

Independently it was shown that *Wolbachia* protect *D. melanogaster* from infections of RNA viruses (Hedges et al. 2008; Teixeira et al. 2008), whereas similar studies on the *Wolbachia* protection from entomopathogenic fungi in *D. melanogaster* are still inconclusive (Fytrou et al. 2006; Panteleev et al. 2007). Overall, the metabolic and protective processes may have contributed to the rapid selective *Wolbachia* sweep detected as the global replacement of one *Wolbachia* variant by another in *D. melanogaster* within the last century (Riegler et al. 2005). The analyses of *Wolbachia*-mediated protective effects against microbes have recently been extended from the *Drosophila* host system to the mosquito *Aedes aegypti*, where a *Wolbachia* strain, originating from *D. melanogaster*, interferes with the establishment of viral infections, as well as *Plasmodium* (Moreira et al. 2009b), plus filarial nematodes (Kambris et al. 2009). The *Wolbachia* strain of *Drosophila* had originally been transferred into *Ae. aegypti*, because of its virulent life span reduction in the original host (Min and Benzer 1997). As expected from the phenotype in its original host, *Wolbachia*-infected *Ae. aegypti* have reduced life spans (McMeniman et al. 2009). *Wolbachia*-infected *Ae. aegypti* are less likely to transmit viral diseases such as dengue, that require specific incubation periods in the mosquito vectors before transmission to humans can occur. The application of *Wolbachia* is therefore a novel strategy for fighting vector-borne diseases, such as dengue, malaria, and filariasis, by interfering with the vector biology (Sinkins 2004; Kambris et al. 2009; McMeniman et al. 2009, Moreira et al. 2009b). Because of these seminal discoveries, *Wolbachia* will continue to attract an applied research interest as a novel biocontrol agent for arthropod pests and vectors (Laven 1967; Zabalou et al. 2004; McMeniman et al. 2009).

Considering the variety of manipulations of a host's life history traits, including reproduction, longevity, and behavior, the question about the extent and diversity of *Wolbachia* phenotypes will continue to be explored and debated (Weeks et al.

2002). At this stage it looks like *Wolbachia* are super "bugs" or master manipulators, capable of shaping host biology in every conceivable way.

ACKNOWLEDGMENTS

We thank Traude Kehrer and Monika Imb for excellent technical support and fly work; we thank Lee Ehrman and two anonymous reviewers for very helpful comments on the manuscript. W.J.M. and D.S. were partly supported by the research grant FWF P19206-B17, and P22634-B17 from the Austrian Science Fund and the COST Action FA0701. M.R. is supported by grants from the University of Western Sydney (UWS) and the Australian Research Council (ARC).

REFERENCES

Albertson, R., Casper-Lindley, C., Cao, J., Tram, U., Sullivan, W. 2009. Symmetric and asymmetric mitotic segregation patterns influence *Wolbachia* distribution in host somatic tissue. *J. Cell. Sci.* 122: 4570–83.

Albertson, R., Doe, C. Q. 2003. Dlg, Scrib and Lgl regulate neuroblast cell size and mitotic spindle asymmetry. *Nat. Cell. Biol.* 5: 166–70.

Anbutsu, H., Fukatsu, T. 2003. Population dynamics of male-killing and non-male-killing spiroplasmas in *Drosophila melanogaster*. *Appl. Environ. Microbiol.* 69: 1428–34.

Aravin, A. A., Hannon, G. J., Brennecke, J. 2007. The Piwi-piRNA pathway provides an adaptive defense in the transposon arms race. *Science* 318: 761–64.

Arthofer, W., Riegler, M., Avtzis, D., Stauffer, C. 2009b. Evidence for low-titre infections in insect symbiosis: *Wolbachia* in the bark beetle *Pityogenes chalcographus* (Coleoptera, Scolytinae). *Environ. Microbiol.* 11: 1923–33.

Arthofer, W., Riegler, M., Schneider, D., Krammer, M., Miller, W. J., Stauffer, C. 2009a. Hidden *Wolbachia* diversity in field populations of the European cherry fruit fly, *Rhagoletis cerasi* (Diptera, Tephritidae). *Mol. Ecol.* 18: 3816–30.

Baldo, L., Ayoub, N. A., Hayashi, C. Y., Russell, J. A., J. K., Werren, J. H. 2008. Insight into the routes of *Wolbachia* invasion: high levels of horizontal transfer in the spider genus *Agelenopsis* revealed by *Wolbachia* strain and mitochondrial DNA diversity. *Mol. Ecol.* 17: 557–69.

Baldo, L., Dunning Hotopp, J. C. Jolley, K. A., et al. 2006. Multilocus sequence typing system for the endosymbiont *Wolbachia pipientis*. *Appl. Environ. Microbiol.* 72: 7098–110.

Baldo, L., Lo, N., Werren, J. H. 2005. Mosaic nature of the *Wolbachia* surface protein. *J. Bacteriol.* 187: 5406–18.

Baldo, L., Prendini, L., Corthals, A., Werren, J. H. 2007. *Wolbachia* are present in Southern African scorpions and cluster with supergroup F. *Curr. Microbiol.* 55: 367–73.

Baldo, L., Werren, J. H. 2007. Revisiting *Wolbachia* supergroup typing based on WSP: spurious lineages and discordance with MLST. *Curr. Microbiol.* 55: 81–87.

Balhorn, R. 2007. The protamine family of sperm nuclear proteins. *Genome Biol.* 8: 227.

Bandi, C., Anderson, T. J., Genchi, C., Blaxter, M. L. 1998. Phylogeny of *Wolbachia* in filarial nematodes. *Proc. Biol. Sci.* 265: 2407–13.

Bentley, J. K., Veneti, Z., Heraty, J., Hurst, G. D. 2007. The pathology of embryo death caused by the male-killing *Spiroplasma* bacterium in *Drosophila nebulosa*. *BMC Biol.* 5: 9.

Bordenstein, S., Rosengaus, R. B. 2005. Discovery of a novel *Wolbachia* super group in Isoptera. *Curr. Microbiol.* 51: 393–98.

Bordenstein, S. R., Werren, J. H. 2007. Bidirectional incompatibility among divergent *Wolbachia* and incompatibility level differences among closely related *Wolbachia* in *Nasonia*. *Heredity* 99: 278–87.

Bouchon, D., Rigaud, T., Juchault, P. 1998. Evidence for widespread *Wolbachia* infection in isopod crustaceans: molecular identification and host feminization. *Proc. Biol. Sci.* 265: 1081–90.

Bourtzis, K., Dobson, S. L., Braig, H. R., O'Neill, S. L. 1998. Rescuing *Wolbachia* have been overlooked. *Nature* 391: 852–53.

Brownlie, J. C., Cass, B. N., Riegler, M. 2009. Evidence for metabolic provisioning by a common invertebrate endosymbiont, *Wolbachia pipientis*, during periods of nutritional stress. *PLoS Pathog.* 5: e1000368.

Brownlie, J. C., Johnson, K. N. 2009. Symbiont-mediated protection in insect hosts. *Trends Microbiol.* 17(8):348–54.

Burla, H., Da Cunha, A. B., Cordeiro, A. R., Dobzhansky, T., Malogolowkin, C., Pavan, C. 1949. The *willistoni* group of sibling species *of Drosophila*. *Evolution* 3: 300–14.

Campbell, B. C., Bragg, T. S., Turner, C. E. 1992. Phylogeny of symbiotic bacteria of four weevil species (Coleoptera: Curculionidae) based on analysis of 16S ribosomal DNA. *Insect Biochem. Mol. Biol.* 22: 415–421.

Casiraghi, M., Bordenstein, S. R., Baldo, L., et al. 2005. Phylogeny of *Wolbachia pipientis* based on *gltA*, *groEL* and *ftsZ* gene sequences: clustering of arthropod and nematode symbionts in the F supergroup, and evidence for further diversity in the *Wolbachia* tree. *Microbiology* 151: 4015–22.

Ceron, J., González, C., Tejedor, F. J. 2001. Patterns of cell division and expression of asymmetric cell fate determinants in postembryonic neuroblast lineages of *Drosophila*. *Dev. Biol.* 230: 125–38.

Chafee, M. E., Zecher, C. N., Gourley, M. L., et al. 2011. Decoupling of host-symbiont-phage coadaptations following transfer between insect species. *Genetics*, 187: 203–15.

Charlat, S., Duplouy, A., Hornett, E., et al. 2009. The joint evolutionary histories of *Wolbachia* and mitochondria in *Hypolimnas bolina*. *BMC Evol. Biol.* 9: 64.

Cheng, O., Ruel, T. D., Zhou, W., et al. 2000. Tissue distribution and prevalence of *Wolbachia* infections in tsetse flies, *Glossina* spp. *Med. Vet. Entomol.* 14: 44–50.

Cheng, Q., Aksoy, S. 1999. Tissue tropism, transmission and expression of foreign genes *in vivo* in midgut symbionts of tsetse flies. *Insect Mol. Biol.* 8: 125–32.

Clark, M. E., Veneti, Z., Bourtzis, K., Karr, T. L. 2003. *Wolbachia* distribution and cytoplasmic incompatibility during sperm development: the cyst as the basic cellular unit of CI expression. *Mech. Dev.* 120: 185–98.

Cook, P. E., McGraw, E. A. 2010. *Wolbachia pipientis*: an expanding bag of tricks to explore for disease control. *Trends Parasitol.* 26(8):373–75.

Decaestecker, E., Gaba, S., Raeymaekers, J. A., et al. 2007. Host-parasite 'Red Queen' dynamics archived in pond sediment. *Nature* 450: 870–73.

de Crespigny, F. E., Pitt, T. D., Wedell, N. 2006. Increased male mating rate in *Drosophila* is associated with *Wolbachia* infection. *J. Evol. Biol.* 19: 1964–72.

Dedeine, F., Boulétreau, M., Vavre, F. 2005. *Wolbachia* requirement for oogenesis: occurrence within the genus *Asobara* (Hymenoptera, Braconidae) and evidence for intraspecific variation in *A. tabida*. *Heredity* 95: 394–400.

Dedeine, F., Vavre, F., Fleury, F., Loppin, B., Hochberg, M. E., Bouletreau, M. 2001. Removing symbiotic *Wolbachia* bacteria specifically inhibits oogenesis in a parasitic wasp. *Proc. Natl. Acad. Sci. USA* 98: 6247–52.

Dobson, S. L. 2003. Reversing *Wolbachia*-based population replacement. *Trends Parasitol.* 19: 128–33.

Dobson, S. L., Bourtzis, K., Braig, H. R., et al. 1999. *Wolbachia* infections are distributed throughout insect somatic and germ line tissues. *Insect Biochem. Mol. Biol.* 29: 153–60.

Dobzhansky, T., Spassky, B. 1959. *Drosophila paulistorum*, a cluster of species in *statu nascendi. Proc. Natl. Acad. Sci. USA* 45: 419–28.

Dumler, J. S., Barbet, A. F., Bekker, C. P., et al. 2001. Reorganization of genera in the families Rickettsiaceae and Anaplasmataceae in the order Rickettsiales: unification of some species of *Ehrlichia* with *Anaplasma*, *Cowdria* with *Ehrlichia* and *Ehrlichia* with *Neorickettsia*, descriptions of six new species combinations and designation of *Ehrlichia equi* and 'HGE agent' as subjective synonyms of *Ehrlichia phagocytophila. Int. J. Syst. Evol. Microbiol.* 51: 2145–65.

Duron, O., Bouchon, D., Boutin, S., et al. 2008. The diversity of reproductive parasites among arthropods: *Wolbachia* do not walk alone. *BMC Biol.* 6: 27.

Engelstädter, J., Hurst, G. D. 2007. The impact of male-killing bacteria on host evolutionary processes. *Genetics* 175: 245–54.

Ewald, P. W. 1987. Transmission modes and evolution of the parasitism-mutualism continuum. *Ann. NY Acad. Sci.* 503: 295–306.

Fabrizio, J. J., Hime, G., Lemmon, S. K., Bazinet, C. 1998. Genetic dissection of sperm individualization in *Drosophila melanogaster. Development* 125: 1833–43.

Ferree, P. M., Sullivan, W. 2006. A genetic test of the role of the maternal pronucleus in *Wolbachia*-induced cytoplasmic incompatibility in *Drosophila melanogaster. Genetics* 173: 839–47.

Frank, S. A. 1998. Dynamics of cytoplasmic incompatability with multiple *Wolbachia* infections. *J. Theor. Biol.* 192: 213–18.

Frydman, H. M., Li, J. M., Robson, D. N., Wieschaus, E. 2006. Somatic stem cell niche tropism in *Wolbachia. Nature* 441: 509–12.

Fytrou, A., Schofield, P. G., Kraaijeveld, A. R., Hubbard, S. F. 2006. *Wolbachia* infection suppresses both host defence and parasitoid counter-defence. *Proc. Biol. Sci.* 273: 791–96.

Giorgini, M., Monti, M. M., Caprio, E., Stouthamer, R., Hunter, M. S. 2009. Feminization and the collapse of haplodiploidy in an asexual parasitoid wasp harboring the bacterial symbiont *Cardinium. Heredity* 102: 365–71.

Gosalbes, M. J., Lamelas, A., Moya, A., Latorre, A. 2008. The striking case of tryptophan provision in the cedar aphid *Cinara cedri. J. Bacteriol.* 190: 6026–29.

Gottlieb, Y., Ghanim, M., Gueguen, G., et al. 2008. Inherited intracellular ecosystem: symbiotic bacteria share bacteriocytes in whiteflies. *FASEB J.* 22: 2591–99.

Gross, R., Vavre, F., Heddi, A., Hurst, G. D., Zchori-Fein, E., Bourtzis, K. 2009. Immunity and symbiosis. *Mol. Microbiol.* 73: 751–59.

Haegeman, A., Vanholme, B., Jacob, J., et al. 2009. An endosymbiotic bacterium in a plant-parasitic nematode: member of a new *Wolbachia* supergroup. *Int. J. Parasitol.* 39: 1045–54.

Hagimori, T., Abe, Y., Date, S., Miura, K. 2006. The first finding of a *Rickettsia* bacterium associated with parthenogenesis induction among insects. *Curr. Microbiol.* 52: 97–101.

Halic, M., Moazed, D. 2009. Transposon silencing by piRNAs. *Cell* 138: 1058–60.

Heddi, A., Grenier, A. M., Khatchadourian, C., Charles, H., Nardon, P. 1999. Four intracellular genomes direct weevil biology: nuclear, mitochondrial, principal endosymbiont, and *Wolbachia. Proc. Natl. Acad. Sci. USA* 96: 6814–19.

Hedges, L. M., Brownlie, J. C., O'Neill, S. L., Johnson, K. N. 2008. *Wolbachia* and virus protection in insects. *Science* 322: 702.

Hertig, M. 1936. The *Rickettsia, Wolbachia pipientis* and associated inclusions of the mosquito, *Culex pipiens. Parasitology* 28: 453–90.

Hertig, M., Wolbach, S. B. 1924. Studies on *Rickettsia*-like microorganisms in insects. *J. Med. Res.* 44: 329–74.

Hilgenboecker, K., Hammerstein, P., Schlattmann, P., Telschow, A., Werren, J. H. 2008. How many species are infected with *Wolbachia*? A statistical analysis of current data. *FEMS Microbiol. Lett.* 281: 215–20.

Hiroki, M., Kato, Y., Kamito, T., Miura, K. 2002. Feminization of genetic males by a symbiotic bacterium in a butterfly, *Eurema hecabe* (Lepidoptera: Pieridae). *Naturwissenschaften* 89: 167–70.

Hoffmann, A. A., Clancy, D., Duncan, J. 1996. Naturally-occurring *Wolbachia* infection in *Drosophila simulans* that does not cause cytoplasmic incompatibility. *Heredity* 76: 1–8.

Hoffmann, A. A., Turelli, M. 1988. Unidirectional incompatibility in *Drosophila simulans*: inheritance, geographic variation and fitness effects. *Genetics* 119: 435–44.

Hoffmann, A. A., Turelli, M. 1997. Cytoplasmic incompatibility. In *Influential passengers: inherited microorganisms and arthropod reproduction*, ed. S. L. O'Neill, A. A. Hoffmann, J. H. Werren, 42–80. Oxford University Press, Oxford.

Hoffmann, A. A., Turelli, M., Simmons, G. M. 1986. Unidirectional incompatibility between populations of *Drosophila simulans*. *Evolution* 40: 692–701.

Hornett, E. A., Charlat, S., Duplouy, A. M., et al. 2006. Evolution of male-killer suppression in a natural population. *PLoS Biol.* 4: e283.

Hornett, E. A., Duplouy, A. M., Davies, N., et al. 2008. You can't keep a good parasite down: evolution of a male-killer suppressor uncovers cytoplasmic incompatibility. *Evolution* 62: 1258–63.

Huigens, M. E., de Almeida, R. P., Boons, P. A., Luck, R. F., Stouthamer, R. 2004. Natural interspecific and intraspecific horizontal transfer of parthenogenesis-inducing *Wolbachia* in *Trichogramma* wasps. *Proc. Biol. Sci.* 271: 509–15.

Huigens, M. E., Stouthamer, R. 2003. Parthenogenesis associated with *Wolbachia*. In *Insect symbiosis*, ed. K. Bourtzis, T. A. Miller, 247–66. CRC, Boca Raton, FL.

Hurst, G. D. D., Graf von der Schulenburg, J. H., Majerus, T.M., et al. 1999. Invasion of one insect species, *Adalia bipunctata*, by two different male-killing bacteria. *Insect. Mol. Biol.* 8: 133–39.

Hurst, G. D. D., Hammarton, T. C., Bandi, C., Majerus, T. M. O., Bertrand, M. O. D., Majerus, M. E. N. 1997. The diversity of inherited parasites of insects: the male-killing agent of the ladybird beetle *Coleomegilla maculata* is a member of the Flavobacteria. *Genet. Res.* 70: 1–6.

Hurst, G. D., Johnson, A. P., Schulenburg, J. H., Fuyama, Y. 2000. Male-killing *Wolbachia* in *Drosophila*: a temperature-sensitive trait with a threshold bacterial density. *Genetics* 156: 699–709.

Ijichi, N., Kondo, N., Matsumoto, R., Shimada, M., Ishikawa, H., Fukatsu, T. 2002. Internal spatiotemporal population dynamics of infection with three *Wolbachia* strains in the adzuki bean beetle, *Callosobruchus chinensis* (Coleoptera: Bruchidae). *Appl. Environ. Microbiol.* 68: 4074–80.

Ikeya, T., Broughton, S., Alic, N., Grandison, R., Partridge, L. 2009. The endosymbiont *Wolbachia* increases insulin/IGF-like signalling in *Drosophila*. *Proc. Biol. Sci.* 276: 3799–807.

Iturbe-Ormaetxe, I., Burke, G. R., Riegler, M., O'Neill, S. L. 2005. Distribution, expression, and motif variability of ankyrin domain genes in *Wolbachia pipientis*. *J. Bacteriol.* 187: 5136–45.

Jaenike, J. 2007. Fighting back against male-killers. *Trends Ecol. Evol.* 22: 167–69.

Jamnongluk, W., Kittayapong, P., Baimai, V., O'Neill, S. L. 2002. *Wolbachia* infections of tephritid fruit flies: molecular evidence for five distinct strains in a single host species. *Curr. Microbiol.* 45: 255–60.

Jeyaprakash, A., Hoy, M. A. 2000. Long PCR improves *Wolbachia* DNA amplification: *wsp* sequences found in 76% of sixty-three arthropod species. *Insect. Mol. Biol.* 9: 393–405.

Jiggins F. M. 2002. The rate of recombination in *Wolbachia* bacteria. *Mol. Biol. Evol.* 19: 1640–43.

Jiggins, F. M., Hurst, G. D., Jiggins, C. D., Graf von der Schulenburg, J. H., Majerus, M. E. 2000. The butterfly *Danaus chrysippus* is infected by a male-killing *Spiroplasma* bacterium. *Parasitology* 120: 439–46.

Jiggins, F. M., Hurst, G. D. D., Majerus, M. E. N. 1998. Sex ratio distortion in *Acraea encedon* (Lepidoptera: Nymphalidae) is caused by a male-killing bacterium. *Heredity* 81: 87–91.

Jiggins, F. M., Hurst, G. D., Yang, Z. 2002. Host-symbiont conflicts: positive selection on an outer membrane protein of parasitic but not mutualistic Rickettsiaceae. *Mol. Biol. Evol.* 19: 1341–49.

Jiggins, F. M., von Der Schulenburg, J. H., Hurst, G. D., Majerus, M. E. 2001. Recombination confounds interpretations of *Wolbachia* evolution. *Proc. Biol. Sci.* 268: 1423–27.

Kageyama, D., Anbutsu, H., Shimada, M., Fukatsu, T. 2009. Effects of host genotype against the expression of spiroplasma-induced male killing in *Drosophila melanogaster*. *Heredity* 102: 475–82.

Kageyama, D., Nishimura, G., Hoshizaki, S., Ishikawa, Y. 2002. Feminizing *Wolbachia* in an insect, *Ostrinia furnacalis* (Lepidoptera: Crambidae). *Heredity* 88: 444–49.

Kaltschmidt, J. A., Davidson, C. M., Brown, N. H., Brand, A. H. 2000. Rotation and asymmetry of the mitotic spindle direct asymmetric cell division in the developing central nervous system. *Nat. Cell. Biol.* 1: 7–12.

Kambris, Z., Cook, P. E., Phuc, H. K., Sinkins, S. P. 2009. Immune activation by life-shortening *Wolbachia* and reduced filarial competence in mosquitoes. *Science* 326: 134–36.

Koehncke, A., Telschow, A., Werren, J. H., Hammerstein, P. 2009. Life and death of an influential passenger: *Wolbachia* and the evolution of CI-modifiers by their hosts. *PLoS One.* 4: e4425.

Kose, H., Karr, T. L. 1995. Organization of *Wolbachia pipientis* in the *Drosophila* fertilized egg and embryo revealed by an anti-*Wolbachia* monoclonal antibody. *Mech. Dev.* 51: 275–88.

Koukou, K., Pavlikaki, H., Kilias, G., Werren, J. H., Bourtzis, K., Alahiotis, S. N. 2006. Influence of antibiotic treatment and *Wolbachia* curing on sexual isolation among *Drosophila melanogaster* cage populations. *Evolution* 60: 87–96.

Kremer, N., Voronin, D., Charif, D., Mavingui, P., Mollereau, B., Vavre, F. 2009. *Wolbachia* interferes with ferritin expression and iron metabolism in insects. *PLoS Pathog.* 5: e1000630.

Landmann, F., Orsi, G. A., Loppin, B., Sullivan, W. 2009. *Wolbachia*-mediated cytoplasmic incompatibility is associated with impaired histone deposition in the male pronucleus. *PLoS Pathog.* 5: e1000343.

Larkum, A. W., Lockhart, P. J., Howe, C. J. 2007. Shopping for plastids. *Trends Plant. Sci.* 12: 189–95.

Laven, H. 1967. Speciation and evolution in *Culex pipiens*. In *Genetics of insect vectors of disease*, ed. J. W. Wright, R. Pal, 251–75. Elsevier, Amsterdam.

Lo, N., Casiraghi, M., Salati, E., Bazzocchi, C., Bandi, C. 2002. How many *Wolbachia* supergroups exist? *Mol. Biol. Evol.* 19: 341–46.

Lo, N., Paraskevopoulos, C., Bourtzis, K., et al. 2007. Taxonomic status of the intracellular bacterium *Wolbachia pipientis*. *Int. J. Syst. Evol. Microbiol.* 57: 654–57.

Loppin, B., Bonnefoy, E., Anselme, C., Laurencon, A., Karr, T. L., Couble, P. 2005. The histone H3.3 chaperone HIRA is essential for chromatin assembly in the male pronucleus. *Nature* 437: 1386–90.

Maiden, M. C., Bygraves, J. A., Feil, E., Morelli, G., et al. 1998. Multilocus sequence typing: a portable approach to the identification of clones within populations of pathogenic microorganisms. *Proc. Natl. Acad. Sci. USA* 95: 3140–45.

Majerus, M. E., Hinrich, J., Schulenburg, G. V., Zakharov, I. A. 2000. Multiple causes of male-killing in a single sample of the two-spot ladybird, *Adalia bipunctata* (Coleoptera: coccinellidae) from Moscow. *Heredity* 84: 605–9.

Malloch, G., Fenton, B. 2005. Super-infections of *Wolbachia* in byturid beetles and evidence for genetic transfer between A and B super-groups of *Wolbachia*. *Mol. Ecol.* 14: 627–37.

Malloch, G., Fenton, B., Butcher, R. D. 2000. Molecular evidence for multiple infections of a new subgroup of *Wolbachia* in the European raspberry beetle *Byturus tomentosus*. *Mol. Ecol.* 9: 77–90.

Masui, S., Kuroiwa, H., Sasaki, T., Inui, M., Kuroiwa, T., Ishikawa, H. 2001. Bacteriophage WO and virus-like particles in *Wolbachia*, an endosymbiont of arthropods. *Biochem. Biophys. Res. Commun.* 283: 1099–104.

Mateos, M., Castrezana, S. J., Nankivell, B. J., Estes A. M., Markov, T. A., Moran, N. A. 2006. Heritable endosymbionts of *Drosophila*. *Genetics*. 174: 363–76.

McGarry, H. F., Egerton, G. L., Taylor, M. J. 2004. Population dynamics of *Wolbachia* bacterial endosymbionts in *Brugia malayi*. *Mol. Biochem.* Parasitol. 135: 57–67.

McMeniman, C. J., Lane, R. V., Cass, B. N., et al. 2009. Stable introduction of a life-shortening *Wolbachia* infection into the mosquito *Aedes aegypti*. *Science* 323: 141–44.

Merçot, H., Charlat, S. 2004. *Wolbachia* infections in *Drosophila melanogaster* and *D. simulans*: polymorphism and levels of cytoplasmic incompatibility. *Genetica* 120: 51–59.

Merçot, H., Poinsot, D. 1998. ... and discovered on Mount Kilimanjaro. *Nature* 391: 853.

Merçot, H., Poinsot, D. 2009. Infection by *Wolbachia*: from passengers to residents. *C. R. Biol.* 332: 284–97.

Miller, W. J., Ehrman, L., Schneider, D. 2010. Infectious speciation revisited: impact of symbiont-depletion on female fitness and mating behavior of *Drosophila paulistorum*. *PLoS Pathog.* 6(12): e1001214. doi:10.1371/journal.ppat.1001214.

Miller, W. J., McDonald, J. F., Nouaud, D., Anxolabéhère, D. 1999. Molecular domestication—more than a sporadic episode in evolution. *Genetica* 107: 197–207.

Miller, W. J., Riegler, M. 2006. Evolutionary dynamics of *w*Au-like *Wolbachia* variants in neotropical *Drosophila* spp. *Appl. Environ. Microbiol.* 72: 826–35.

Min, K. T., Benzer, S. 1997. *Wolbachia*, normally a symbiont of *Drosophila*, can be virulent, causing degeneration and early death. *Proc. Natl. Acad. Sci. USA* 94: 10792–96.

Moran, N. A., McCutcheon, J. P., Nakabachi, A. 2008. Genomics and evolution of heritable bacterial symbionts. *Annu. Rev. Genet.* 42: 165–90.

Moreira, L. A., Iturbe-Ormaetxe, I., Jeffery, J. A., et al. 2009b. A *Wolbachia* symbiont in *Aedes aegypti* limits infection with dengue, chikungunya, and *Plasmodium*. *Cell* 139: 1268–78.

Moreira, L. A., Saig, E., Turley, A. 2009a. Human probing behavior of *Aedes aegypti* when infected with a life-shortening strain of *Wolbachia*. *PLoS Negl. Trop. Dis.* 3: e568.

Mouton, L., Dedeine, F., Henri, H., Boulétreau, M., Profizi, N., Vavre, F. 2004. Virulence, multiple infections and regulation of symbiotic population in the *Wolbachia-Asobara tabida* symbiosis. *Genetics* 168: 181–89.

Mouton, L., Henri, H., Bouletreau, M., Vavre, F. 2003. Strain-specific regulation of intracellular *Wolbachia* density in multiply infected insects. *Mol. Ecol.* 12: 3459–65.

Negri, I., Franchini, A., Gonella, E., et al. 2009. Unravelling the *Wolbachia* evolutionary role: the reprogramming of the host genomic imprinting. *Proc. Biol. Sci.* 276: 2485–91.

Negri, I., Pellecchia, M., Mazzoglio, P. J., Patetta, A., Alma, A. 2006. Feminizing *Wolbachia* in *Zyginidia pullula* (Insecta, Hemiptera), a leafhopper with an XX/X0 sex-determination system. *Proc. Biol. Sci.* 273: 2409–16.

Normark, B. B. 2003. The evolution of alternative genetic systems in insects. *Annu. Rev. Entomol.* 48: 397–423.

O'Neill, S. L., Giordano, R., Colbert, A. M., Karr, T. L., Robertson, H. M. 1992. 16S rRNA phylogenetic analysis of the bacterial endosymbionts associated with cytoplasmic incompatibility in insects. *Proc. Natl. Acad. Sci. USA* 89: 2699–702.

O'Neill, S. L., Hoffmann, A. A., Werren, J. H. 1997. *Influential passengers*. Oxford University Press, Oxford.

O'Neill, S. L., Karr, T. L. 1990. Bidirectional incompatibility between conspecific populations of *Drosophila simulans*. *Nature* 348: 178–180.

Pannebakker, B. A., Loppin, B., Elemans, C. P., Humblot, L., Vavre, F. 2007. Parasitic inhibition of cell death facilitates symbiosis. *Proc. Natl. Acad. Sci. USA* 104: 213–15.

Panteleev, D., Goriacheva, I. I., Andrianov, B. V., Reznik, N. L., Lazebny, O. E., Kulikov, A. M. 2007. The endosymbiotic bacterium *Wolbachia* enhances the nonspecific resistance to insect pathogens and alters behavior of *Drosophila melanogaster*. *Genetika* 43: 1277–80.

Paraskevopoulos, C., Bordenstein, S. R., Wernegreen, J. J., Werren, J. H., Bourtzis, K. 2006. Toward a *Wolbachia* multilocus sequence typing system: discrimination of *Wolbachia* strains present in *Drosophila* species. *Curr. Microbiol.* 53: 388–89.

Peng, Y., Nielsen, J. E., Cunningham, J. P., McGraw, E. A. 2008. *Wolbachia* infection alters olfactory-cued locomotion in *Drosophila* spp. *Appl. Environ. Microbiol.* 74: 3943–48.

Perotti, M. A., Allen, J. M., Reed, D. L., Braig, H. R. 2007. Host-symbiont interactions of the primary endosymbiont of human head and body lice. *FASEB J.* 21: 1058–66.

Pinsker, W., Haring, E., Hagemann, S., Miller, W. J. 2001. The evolutionary life history of P transposons: from horizontal invaders to domesticated neogenes. *Chromosoma* 110: 148–58.

Poinsot, D., Charlat, S., Merçot, H. 2003. On the mechanism of *Wolbachia*-induced cytoplasmic incompatibility: confronting the models with the facts. *Bioessays* 25: 259–65.

Rasgon, J. L., Scott, T. W. 2004. Phylogenetic characterization of *Wolbachia* symbionts infecting *Cimex lectularius* L. and *Oeciacus vicarius* Horvath (Hemiptera: Cimicidae). *J. Med. Entomol.* 41: 1175–78.

Raychoudhury, R., Baldo, L., Oliveira, D. C., Werren, J. H. 2009. Modes of acquisition of *Wolbachia*: horizontal transfer, hybrid introgression, and codivergence in the *Nasonia* species complex. *Evolution* 63: 165–83.

Riegler, M., Stauffer, C. 2002. *Wolbachia* infections and super infections in cytoplasmically incompatible populations of the European cherry fruit fly Rhagoletis cerasi (diptera, tephritidae) *Mol. Ecol.* 11: 2425–34.

Riegler, M., Charlat, S., Stauffer, C., Merçot, H. 2004. *Wolbachia* transfer from *Rhagoletis cerasi* to *Drosophila simulans*: investigating the outcomes of host-symbiont coevolution. *Appl. Environ. Microbiol.* 70: 273–79.

Riegler, M., O'Neill, S. L. 2006. The genus *Wolbachia*. In *The prokaryotes: a handbook on the biology of bacteria. Proteobacteria: alpha and beta subclass*, ed. M. Dworkin, S. Falkow, E. Rosenberg, K. H. Schleifer, E. Stackebrandt, 547–61. 3rd ed., vol. 5. Springer, New York.

Riegler, M., O'Neill, S. L. 2007. Evolutionary dynamics of insect symbiont associations. *Trends Ecol. Evol.* 22: 625–27.

Riegler, M., Sidhu, M., Miller, W. J., O'Neill, S. L. 2005. Evidence for a global *Wolbachia* replacement in *Drosophila melanogaster*. *Curr. Biol.* 15: 1428–33.

Rigaud, T., Antoine, D., Marcadé, I., Juchault, P. 1997. The effect of temperature on sex ratio in the isopod *Porcellionides pruinosus*: environmental sex determination or a by-product of cytoplasmic sex determination? *Evol. Ecol.* 11: 205–15.

Ros, V. I., Fleming, V. M., Feil, E. J., Breeuwer, J. A. 2009. How diverse is the genus *Wolbachia*? Multiple-gene sequencing reveals a putatively new *Wolbachia* supergroup recovered from spider mites (Acari: Tetranychidae). *Appl. Environ. Microbiol.* 75: 1036–43.

Rowley, S. M., Raven, R. J., McGraw, E. A., et al. 2004. *Wolbachia pipientis* in Australian spiders. *Curr. Microbiol.* 49: 208–14.

Sagan, L. 1967. On the origin of mitosing cells. *J. Theor. Biol.* 14: 255–74.

Schneider, D. 2008. Monitoring the infection and genome dynamics of artificially transferred *Wolbachia* in *Drosophila simulans*. Diploma thesis. Medical University of Vienna and University of Applied Life Sciences and Natural Resources Vienna.

Serbus, L., Sullivan, W. 2007. A cellular basis for *Wolbachia* recruitment to the host germline. *PLoS Pathog.* 3(12):e190.

Serbus, L. R., Casper-Lindley, C., Landmann, F., Sullivan, W. 2008. The genetics and cell biology of *Wolbachia*-host interactions. *Annu. Rev. Genet.* 42: 683–707.

Sheeley, S. L., McAllister, B. F. 2009. Mobile male-killer: similar *Wolbachia* strains kill males of divergent *Drosophila* hosts. *Heredity* 102: 286–92.

Shields, W. (1954). Simon Wolbach, 3rd July 1880–19th March 1954. *J. Path. Bacter.* 68: 656–658.

Sinkins, S. P. 2004. *Wolbachia* and cytoplasmic incompatibility in mosquitoes. *Insect Biochem. Mol. Biol.* 34: 723–29.

Sinkins S. P., Walker T., Lynd A. R., Steven A. R., Makepeace B. L., Godfray H. C. J., Parkhill J. 2005. *Wolbachia* variability and host effects associated with crossing type in *Culex* mosquitoes. *Nature* 436: 257–60.

Skorokhod, A., Gamulin, V., Gundacker, D., Kavsan, V., Muller, I. M., Muller, W. E. 1999. Origin of insulin receptor-like tyrosine kinases in marine sponges. *Biol. Bull.* 197: 198–206.

Stouthamer, R., Breeuwer, J. A., Hurst, G. D. 1999. *Wolbachia pipientis*: microbial manipulator of arthropod reproduction. *Annu. Rev. Microbiol.* 53: 71–102.

Stouthamer, R., Luck, R. F., Hamilton, W. D. 1990. Antibiotics cause parthenogenetic *Trichogramma* (Hymenoptera/Trichogrammatidae) to revert to sex. *Proc. Natl. Acad. Sci. USA* 87: 2424–27.

Taylor, M. J., Hoerauf, A., Bockarie, M. 2010. Lymphatic filariasis and onchocerciasis. *Lancet* 376: 1175–85.

Teixeira, L., Ferreira, A., Ashburner, M. 2008. The bacterial symbiont *Wolbachia* induces resistance to RNA viral infections in *Drosophila melanogaster*. *PLoS Biol.* 6: e2.

Tokuyasu, K. T., Peacock, W. J., Hardy, R. W. 1972a. Dynamics of spermiogenesis in *Drosophila melanogaster*. I. Individualization process. *Z. Zellforsch. Mikrosk. Anat.* 124: 479–506.

Tokuyasu, K. T., Peacock, W. J., Hardy, R. W. 1972b. Dynamics of spermiogenesis in *Drosophila melanogaster*. II. Coiling process. *Z. Zellforsch. Mikrosk. Anat.* 127: 492–525.

Tram, U., Sullivan, W. 2002. Role of delayed nuclear envelope breakdown and mitosis in *Wolbachia*-induced cytoplasmic incompatibility. *Science* 296: 1124–26.

Tram, U., Ferree, P. M., Sullivan, W. 2003. Identification of *Wolbachia*-host interacting factors through cytological analysis. *Microbes Infect.* 5: 999–1011.

Tram, U., Frederick, K., Werren, J. H., Sullivan, W. 2006. Paternal chromosome segregation during the first mitotic division determines *Wolbachia*-induced cytoplasmic incompatibility phenotype. *J. Cell Sci.* 119: 3655–63.

Turley, A. P., Moreira, L. A., O'Neill, S. L., McGraw, E. A. 2009. *Wolbachia* infection reduces blood-feeding success in the dengue fever mosquito, *Aedes aegypti*. *PLoS Negl. Trop. Dis.* 3: e516.

Van Baalen, M., Sabelis, M. W. 1995. The dynamics of multiple infection and the evolution of virulence. *Am. Naturalist* 146: 881–910.

Van Borm, S., Wenseleers, T., Billen, J., Boomsma, J. J. 2003. Cloning and sequencing of *wsp* encoding gene fragments reveals a diversity of co-infecting *Wolbachia* strains in *Acromyrmex* leafcutter ants. *Mol. Phylogenet. Evol.* 26: 102–9.

Vandekerckhove, T. T., Watteyne, S., Willems, A., Swings, J. G., Mertens, J., Gillis M. 1999. Phylogenetic analysis of the 16S rDNA of the cytoplasmic bacterium *Wolbachia* from the novel host *Folsomia candida* (Hexapoda, Collembola) and its implications for *Wolbachia* taxonomy. *FEMS Microbiol. Lett.* 180: 279–86.

Van Valen, L. 1973. A new evolutionary law. *Evol. Theory* 1: 1–30.

Vautrin, E., Charles, S., Genieys, S., Vavre, F. 2007. Evolution and invasion dynamics of multiple infections with *Wolbachia* investigated using matrix based models. *J. Theor. Biol.* 245: 197–209.

Vavre, F., Fleury, F., Varaldi, J., Fouillet, P., Boulétreau, M. 2000. Evidence for female mortality in *Wolbachia*-mediated cytoplasmic incompatibility in haplodiploid insects, epidemiologic and evolutionary consequences. *Evolution* 54: 191–200.

Vavre, F., Fleury, F., Varaldi, J., Fouillet, P., Boulétreau, M. 2002. Infection polymorphism and cytoplasmic incompatibility in *Hymenoptera-Wolbachia* associations. *Heredity* 88: 361–65.

Veneti, Z., Clark, M. E., Karr, T. L., Savakis, C., Bourtzis, K. 2004. Heads or tails: host-parasite interactions in the *Drosophila-Wolbachia* system. *Appl. Environ. Microbiol.* 70: 5366–72.

Viljakainen, L. Reuter, M., Pamilo, P. 2008. *Wolbachia* transmission dynamics in *Formica* wood ants. *BMC Evol. Biol.* 8: 55.

Villella, A., Hall, J. C. 2008. Neurogenetics of courtship and mating in *Drosophila*. *Adv Genet* 62: 67–184.

Weeks, A. R., Reynolds K. T., Hoffmann, A. A. 2002. *Wolbachia* dynamics and host effects: what has (and has not) been demonstrated? *Trends Ecol. Evol.* 17: 257–62.

Weeks, A. R., Turelli, M., Harcombe, W. R., Reynolds, K. T., Hoffmann, A. A. 2007. From parasite to mutualist: rapid evolution of *Wolbachia* in natural populations of *Drosophila*. *PLoS Biol.* 5: e114.

Werren, J. H. 1997. Biology of *Wolbachia*. *Ann. Rev. Entomol.* 42: 587–609.

Werren, J. H., Baldo, L., Clark, M. E. 2008. *Wolbachia*: master manipulators of invertebrate biology. *Nat. Rev. Microbiol.* 6: 741–51.

Werren, J. H., Hurst, G. D., Zhang, W., Breeuwer, J. A., Stouthamer, R., Majerus, M. E. 1994. Rickettsiales relative associated with male killing in the ladybird beetle (*Adalia bipunctata*). *J. Bacteriol.* 176: 388–94.

Werren, J. H., Skinner, S. W., Huger, A. M. 1986. Male-killing bacteria in a parasitic wasp. *Science* 231: 990–92.

Werren, J. H., Windsor, D. M. 2000. *Wolbachia* infection frequencies in insects: evidence of a global equilibrium? *Proc. Biol. Sci.* 267: 1277–85.

Werren, J. H., Zhang, W., Guo, L. R. 1995. Evolution and phylogeny of *Wolbachia* reproductive parasites of arthropods. *Proc. Biol. Sci.* 261: 55–63.

West, S. A., Cook, J. M., Werren, J. H., Godfrey, H. C. 1998. *Wolbachia* in two insect host-parasitoid communities. *Mol. Ecol.* 7: 1457–65.

Williamson, D. L., Sakaguchi, B., Hackett, K. J., Whitcomb, R. F., Tully, J. G., Carle, P., Bové, J. M., Adams, J. R., Konai, M., Henegar, R. B. 1999. *Spiroplasma poulsonii sp.* nov., a new species associated with male-lethality in *Drosophila willistoni*, a neotropical species of fruit fly. *Int. J. Syst. Bacteriol.* 2: 611–18.

Wiwatanaratanabutr, I., Kittayapong, P., Caubet, Y., Bouchon, D. 2009. Molecular phylogeny of *Wolbachia* strains in arthropod hosts based on *groE*-homologous gene sequences. *Zoolog. Sci.* 26: 171–77.

Woodward, T. E., Walker, D. H., and Dumler, J. S. 1992. The remarkable contributions of S. Burt Wolbach on rickettsial vasculitis updated. *Trans. Am. Clin. Climatol. Assoc.* 103: 78–94.

Wu, D., Daugherty, S. C., Van Aken, S. E., et al. 2006. Metabolic complementarity and genomics of the dual bacterial symbiosis of sharpshooters. *PLoS Biol.* 4: e188.

Wu, M., Sun, L. V., Vamathevan, J., et al. 2004. Phylogenomics of the reproductive parasite *Wolbachia pipientis* wMel: a streamlined genome overrun by mobile genetic elements. *PLoS Biol.* 2(3):E69.

Wu, P. S., Egger, B., Brand, A. H. 2008. Asymmetric stem cell division: lessons from *Drosophila*. *Semin. Cell. Dev. Biol.* 19: 283–93.

Yamada, R., Floate, K. D., Riegler, M., O'Neill, S. L. 2007. Male development time influences the strength of *Wolbachia*-induced cytoplasmic incompatibility expression in *Drosophila melanogaster*. *Genetics* 177: 801–8.

Yen, J. H., Barr, A. R. 1971. New hypothesis of the cause of cytoplasmic incompatibility in *Culex pipiens* L. *Nature*. 232: 657–58.

Zabalou, S., Apostolaki, A., Pattas, S., et al. 2008. Multiple rescue factors within a *Wolbachia* strain. *Genetics* 178: 2145–60.

Zabalou, S., Charlat, S., Nirgianaki, A., Lachaise, D., Merçot, H. et al. 2004a. Natural *wolbachia* infections in the drosophila yakuba species complex do not induce cytoplasmic incompatibility but fully rescue the WRI modification. *Genetics* 167: 827–834.

Zabalou, S., Riegler, M., Theodorakopoulou, M., Stauffer, C., Savakis, C., Bourtzis, K. 2004b. *Wolbachia*-induced cytoplasmic incompatibility as a means for insect pest population control. *Proc. Natl. Acad. Sci. USA* 101: 15042–45.

Zhou, W., Rousset, F., O'Neil, S. 1998. Phylogeny and PCR-based classification of *Wolbachia* strains using *wsp* gene sequences. *Proc. Biol. Sci.* 265: 509–15.

9 Host and Symbiont Adaptations Provide Tolerance to Beneficial Microbes
Sodalis and Wigglesworthia Symbioses in Tsetse Flies

*Brian L. Weiss, Jingwen Wang,
Geoffrey M. Attardo, and Serap Aksoy*

CONTENTS

Genus	*Sodalis*
Species	*glossinidius*
Family	Enterobacteriaceae
Order	Enterobacteriales
Description year	1999
Origin of name	Latin for "a companion"
Description	
Reference	Dale and Maudlin, 1999

Genus	*Wigglesworthia*
Species	*glossinidia*
Family	Enterobacteriaceae
Order	Enterobacteriales
Description year	1929
Origin of name	After Sir Vincent Brian Wigglesworth (1899–1994), a British entomologist who made significant contributions to the field of insect metamorphosis
Description	
Reference	Aksoy, 1995

INTRODUCTION

Human African trypanosomiasis (HAT) is a fatal disease caused by the protozoan *Trypanosoma brucei* spp. and transmitted by the tsetse fly (genus *Glossinidiae*). Both male and female adult flies are strict blood feeders during all developmental stages, and both sexes can transmit trypanosomes to a vertebrate host while feeding. Tsetse flies also harbor two enteric microbes, the primary symbiont *Wigglesworthia glossinidia* and the secondary symbiont *Sodalis glossinidius*. In addition, some tsetse populations harbor a third microorganism that is related to parasitic *Wolbachia pipientis* (Figure 9.1). Both *Wigglesworthia* and *Sodalis* are members of γ-Proteobacteria closely related to the enteric microbes (Figure 9.2), while *Wolbachia* belongs to α-Proteobacteria. All three symbionts are thought to be important for host physiological functions.

GENUS *WIGGLESWORTHIA* HIGHLIGHTS

To date every tsetse species examined has an obligate association with *Wigglesworthia*. Phylogenetic analysis of *Wigglesworthia* from different tsetse species shows concordant history with their host. This finding suggests that a tsetse ancestor had been infected with a bacterium some 50–80 million years ago, and from this ancestral pair extant species of tsetse and associated *Wigglesworthia* strains radiated without horizontal transfer events between species. Application of molecular clock methods suggests that this symbiosis is approximately 80 million years old (Chen et al. 1999). *Wigglesworthia*'s genome displays several hallmarks that are comparable

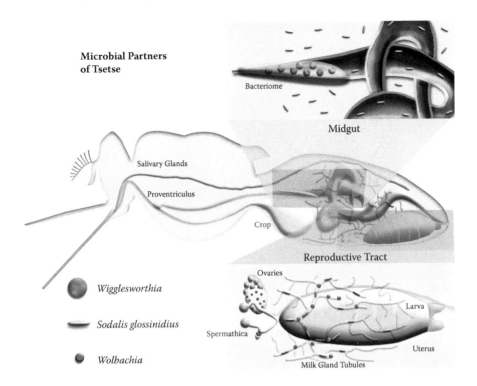

FIGURE 9.1　See color insert. Localization of symbiotic bacteria in tsetse flies. Tsetse harbors three distinct vertically transmitted endosymbionts, *Wigglesworthia glossinidia*, *Sodalis glossinidius*, and members of the genus *Wolbachia*. Two distinct populations of *Wigglesworthia* are found in tsetse. The first is located within specialized bacteriocyte cells that together comprise an organ called the bacteriome. Tsetse's bacteriome is localized in the anterior midgut. The second *Wigglesworthia* population is extracellular in the milk gland lumen of female flies. *Sodalis* has a broad tissue distribution and can be found both intra- and extracellularly in tsetse's midgut, fat body, muscle, hemolymph, milk gland, and salivary glands of certain species. *Wolbachia*, a parasitic bacterium, is also intracellular and can be found within the fly's reproductive tract.

to those found in other intracellular obligates, including an extremely reduced size (approximately 700 kb) and a high AT bias (over 80%; Akman et al. 2002). In addition, *Wigglesworthia*'s genome contains no evidence of transposon or phage-related sequences. This suggests that since its initial captivity, *Wigglesworthia* may have remained secluded in host tissues with little exposure to outside microbial fauna. *Wigglesworthia* resides intracellularly in bacteriocytes, which form the bacteriome organ located in the anterior midgut. In addition to the intracellular population, extracellular *Wigglesworthia* cells were detected in the milk gland lumen and are thought to be essential for transmission to future progeny (discussed in more detail below; Pais et al. 2008, Ma and Denlinger 1974).

The functional role of insect obligate symbionts is difficult to determine because attempts to eliminate these bacteria result in retarded host growth rates and severely compromised fecundity. These obstacles have thus prevented the generation of insect

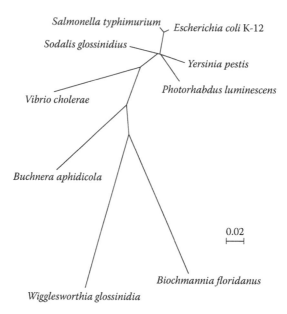

FIGURE 9.2 Phylogenetic tree inferred from 25 concatenated ribosomal protein amino acid sequences from *Sodalis, Wigglesworthia*, and other Enterobacteriaceae, including other obligate insect symbionts. Scale bar represents the length of the branches. Sequences were aligned with ClustalW, and an unrooted tree was generated with NJplot.

lines that lack obligate symbionts. However, *Wigglesworthia*'s putative proteome indicates that this bacterium has retained the pathways required to produce various vitamins, including riboflavin (*ribH*), folate, dihydrofolate (*folA, C, D*, and *K*), thiazole, and thiamine (vitamin B1). In addition, genes involved in the synthesis of lipoic acid (*lipA, B*), and intermediates involved in biotin synthesis, were also detected. The *tktA* gene detected by expression analysis has also been implicated in pyridoxine (vitamin B_6) synthesis (Akman et al., 2002). This information indicates that *Wigglesworthia* likely benefits its host by supplementing essential metabolites that tsetse cannot produce itself or obtain from its vertebrate blood-specific diet.

GENUS *SODALIS* HIGHLIGHTS

Many tsetse colonies, and some natural populations, also harbor a secondary facultative commensal called *Sodalis glossinidius* (Cheng and Aksoy 1999, Rio et al. 2003). Unlike *Wigglesworthia*, *Sodalis* from different tsetse species are closely related, thus indicating this bacterium's recent association with tsetse and the potential for extensive horizontal transfer events. In fact, several other insects, including grain weevils (Heddi et al. 1998), hippoboscid flies (Novakova and Hypsa 2007), slender pigeon lice (Fukatsu et al. 2007), and stinkbugs (Kaiwa et al. 2010), harbor endosymbionts closely allied to *Sodalis*. *Sodalis*'s 4.5 Mb chromosome has not undergone significant size reduction and is more comparable to free-living enteric microbes (Toh et al. 2006). However, *Sodalis*'s genome does contain many pseudogenes, indicating

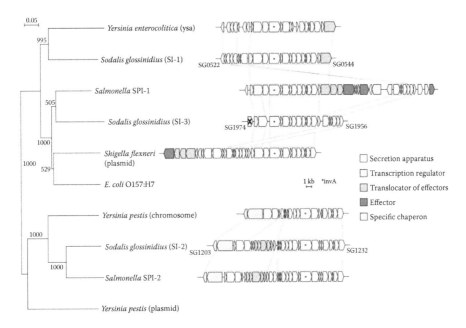

FIGURE 9.3   See color insert. *Sodalis*'s genome encodes type III secretion system (TTSS) structures that are phylogenetically similar to those from pathogenic microbes. The tree was based on amino acid sequence alignments of the *invA* gene product. Scale bars representing branch lengths and bootstrap values are displayed at each internal node. Genes associated with each cluster are depicted as arrows that indicate the direction of transcription. The arrows with a cross denote pseudogenes. The light blue bars between loci indicate the regions of sequence similarity and gene order conservation. The different colors depict the functional roles of their putative products. (Permission to reprint this figure was granted by CSH Press.)

that this bacterium is currently undergoing a transition from free-living to endosymbiotic microbe. Pathways specifically effected by the process of genome reduction in *Sodalis* include those that are likely unnecessary due to tsetse's restricted diet or because of the presence of other symbiotic microbes. For example, *Sodalis* has lost the ability to synthesize thiamine even though it requires this vitamin for growth. Interestingly, this bacterium still retains a thiamine ABC transporter capable of scavenging this vitamin (Snyder et al. 2010). This finding suggests that genetic complementation occurs between *Sodalis* and *Wigglesworthia*, the latter of which produces thiamine as a metabolic supplement to tsetse's vertebrate blood-specific diet (Akman et al. 2002).

Interestingly, *Sodalis*'s chromosome encodes three symbiosis regions (SSR-1, SSR-2, and SSR-3) that share significant homology with type III secretion systems from several pathogenic bacteria, including *Yersinia* and *Salmonella* (Figure 9.3; Toh et al., 2006). *In vitro* assays using insect cell lines previously demonstrated that SSR-1 and SSR-2 have a role in host cell entry and postinvasion processes, respectively (Dale et al. 2001, 2005). Furthermore, *in vivo* expression analysis of SSR-specific *invA* genes indicates that these systems are most likely active during late larval and early pupal stages (Toh et al. 2006). This expression pattern suggests that SSRs assist in maintaining the mutualistic relationship shared between tsetse and *Sodalis*.

Several theories are available to explain *Sodalis*'s function in tsetse. This bacterium has been implicated in tsetse longevity, as flies that had this symbiont selectively eliminated via treatment with the antibiotic streptozotocin exhibited a significantly shorter life span than their wild-type counterparts (Dale and Welburn 2001; Toh et al. 2006). In contrast to this finding, a more recent study demonstrated that *Sodalis*-cured females exhibited no reduction in longevity or fecundity (Weiss et al. 2006). Several studies have indicated that *Sodalis* may also be linked to tsetse's susceptibility to infection with trypanosomes (Welburn et al. 1993; Welburn and Maudlin 1999). This theory is supported by the fact that trypanosome-infected tsetse from natural Liberian populations were found to harbor high *Sodalis* densities (Maudlin et al. 1990). This finding, however, is in contrast to laboratory-based experiments that showed no correlation between *Sodalis* density and trypanosome infection prevalence. In this situation, tsetse treated with ampicillin to remove their *Wigglesworthia* still harbor *Sodalis* at densities comparable to those found in wild-type flies. Despite retaining similar *Sodalis* densities, *Wigglesworthia*-free tsetse were highly susceptible to trypanosome infection, suggesting that *Wigglesworthia*, as opposed to commensal *Sodalis*, may mediate host susceptibility to parasites (Pais et al. 2008). Interestingly, the relationship between *Sodalis* and trypanosome infection in tsetse may be dependent on more than bacterial density. Geiger et al. (2007) found that individuals within a population of *G. palpalis gambiensis* harbor genetically distinct populations of *Sodalis*. Furthermore, these distinct populations of *Sodalis* were linked with the ability of different trypanosomes subspecies to establish an infection in challenged tsetse. These data may suggest that heterogeneities in *Sodalis*'s genotype in wild fly populations could be partially responsible for variations in tsetse vector competence.

TSETSE PARATRANSGENESIS USING *SODALIS*

Genetic transformation in many insects is achieved by microinjection of a transposable element (plasmid or viral vectors) into syntinal embryos (germline transformation). The viviparous reproductive biology of tsetse (see below) prohibits the use of germline transformation to genetically modify this insect. However, *Sodalis*, which lives in close proximity to pathogenic trypanosomes, can be exploited to express foreign gene products that subsequently affect parasite transmission (Beard and Aksoy 1998, Beard et al. 1993). This process, called paratransgenesis, involves genetically modifying *Sodalis in vitro* so that it expresses and secretes an effector molecule that interferes with disease transmission. The recombinant symbionts are then introduced into fertile female flies, where they are passed on to subsequent offspring (Cheng and Aksoy 1999).

Sodalis is well suited for this purpose for several reasons. First, *Sodalis* is highly resistant to many trypanocidal effector molecules (Hu and Aksoy 2005; Haines et al. 2003). Second, *Sodalis* isolates from one tsetse species can be transinfected into different tsetse species to streamline the paratransgenesis process (Weiss et al. 2006). Third, *Sodalis*'s genome is completely sequenced and annotated, and this information will serve as a valuable resource that can be exploited to improve the efficiency of our expression system (Toh et al. 2006). Finally, *Sodalis* has highly restricted metabolic capabilities, and a specific, anchored relationship with its host genera, that would severely hinder (or most likely completely eliminate) survival outside of its normal host (Rio et al. 2003; Toh et al. 2006). Developing an effective and efficient tsetse paratransgenesis strategy is a work in progress, and future studies must be done to identify potent effector molecules and a gene drive mechanism(s) to replace wild-type populations with refractory ones.

SYMBIONT TRANSMISSION BIOLOGY AND GENERATION OF *WIGGLESWORTHIA*-CURED TSETSE

The physiology of reproduction in tsetse is unique among vector arthropods. Tsetse undergoes viviparous reproduction (deposition of live larvae rather than eggs). The mother develops a single oocyte at a time, then carries and nourishes the resulting embryo and larvae in an intrauterine environment for the duration of its immature development. Larval nutrition is provided via a modified accessory gland (milk gland) that connects to the uterus and expands throughout the abdominal cavity of the fly as bifurcating tubules. Females give birth to a fully developed third-instar larva, which pupates within half an hour of deposition. During its life span, a female can birth 8 to 10 progeny. Both *Wigglesworthia* and *Sodalis* are transmitted to tsetse progeny through the mother's milk. While *Wigglesworthia* is free in the lumen, both intracellular and extracellular forms of *Sodalis* can be detected in milk gland tubules (Figure 9.4; Attardo et al. 2008).

Selective elimination of tsetse symbionts allows for the investigation of their individual physiological roles within the host. For example, treatment of pregnant females with the antibiotic ampicillin results in the clearance milk gland-associated extracellular *Wigglesworthia*, while intracellular bacteriome-associated cells are left undisturbed. Ampicillin-treated females thus retain their fecundity but do not transfer *Wigglesworthia* to their progeny. These offspring (Gmm^{Wgm-}) thus lack their obligate symbiont but still harbor *Sodalis* and *Wolbachia* (Pais et al. 2008). Longevity of Gmm^{Wgm-} flies is not significantly different than that of their wild type (WT) counterparts. However, by measuring midgut hemoglobin content, it was determined that Gmm^{Wgm-} individuals exhibit a compromised ability to digest their blood meal. Additionally, Gmm^{Wig-} flies are highly susceptible to parasitism with trypanosomes, indicating that they may have an impaired immune system as adults (Pais et al. 2008).

SYMBIOTIC INFLUENCES ON PARASITISM AND TOLERANCE TO *WIGGLESWORTHIA*

Host tolerance of beneficial fauna is an important concept in the field of symbiosis. Suggested mechanisms of tolerance include symbiont downregulation of host immunity as a form of self-protection, or the absence of molecular signals on symbiont cell surfaces that are typically recognized by insects following exposure to microbes. In

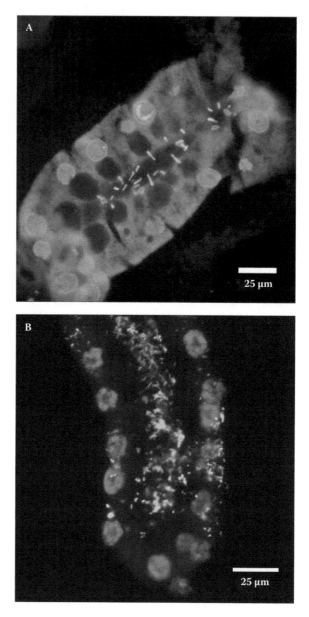

FIGURE 9.4   See color insert. *In situ* staining of *Wigglesworthia* and *Sodalis* in pregnant female milk gland. Tissue sections were stained with DAPI and *Wigglesworthia-* and *Sodalis*-specific DIG-labeled 16S ribosomal RNA probe. (A) The staining pattern shows *Wigglesworthia* extracellularly in secretory cell ducts and in the milk gland lumen. (B) Concentrated *Sodalis* are visible in the extracellular space of the lumen as well as intracellularly within the cytoplasm of the secretory cells.

the case of gram-negative bacteria, microbe-associated molecular patterns (MAMPs) are components of the bacterial surface architecture. MAMPs consist primarily of peptidoglycan (PGN), outer membrane proteins (OMPs), and capsular structures that are often bacteria species specific. These structures are recognized by insect pattern recognition receptors (PRRs), which serve as the initial component of signal transduction pathways that induce innate immunity (Gross et al. 2009). Recent studies have focused on host mechanisms in the context of how *Wigglesworthia* may escape immune recognition, and how the absence of *Wigglesworthia* throughout development can give rise to flies Gmm^{Wgm-} that are immunocompromised and thus transmit parasites at high levels.

Insects have major innate immune pathways that may act alone or synergistically, depending on the stimulus. One pathway in particular, the immune-deficient (IMD) pathway, responds to the presence of gram-negative bacteria via the production of antimicrobial peptides (AMPs). The IMD pathway also plays an important role in resistance to trypanosomes in tsetse (Nayduch and Aksoy 2007). When expression levels of IMD pathway effector AMPs (*attacin* and *cecropin*) and the IMD pathway transcriptional regulator *relish* were reduced by double-stranded RNA (dsRNA)-based RNA interference (RNAi), tsetse had significantly higher midgut parasite infection prevalence (Hu and Aksoy 2006). In addition, the AMPs diptericin and attacin displayed trypanocidal activity both *in vitro* and *in vivo* in tsetse's midgut (Hao et al. 2001; Hu and Aksoy 2005).

A major inducer of the IMD pathway is the pathogen-specific peptidoglycan (PGN). A family of insect PGN recognition proteins (PGRPs) that bind bacterial PGN molecules regulates activation of the IMD pathway. In *Drosophila*, one PGRP (PGRP-LB) also exhibits amidase activity, which can degrade PGN (Royet and Dziarski 2007). In this way, metabolically costly insect immune responses are not activated in response to low-level environmental microbes, but are instead activated when the level of the endogenous PGRP-LB is exhausted following an infection. In tsetse, PGRP-LB is expressed at high levels in the *Wigglesworthia*-containing bacteriome organ (Wang et al. 2009). Furthermore, the level of *pgrp-lb* expression increases as a function of host age and *Wigglesworthia* density (Wang et al. 2009). This positive correlation between *Wigglesworthia* density and host *pgrp-lb* expression levels was also demonstrated in natural field populations (Wang et al. 2009). Furthermore, flies that lack *Wigglesworthia*, such as Gmm^{Wig-}, also expressed significantly less *pgrp-lb* in their bacteriomes (Wang et al. 2009). Thus, the spatial and temporal expression profile of *pgrp-lb* indicates that this molecule is expressed in *Wigglesworthia*-associated tissues as a function of symbiont densities. When *pgrp-lb* levels were reduced through the use of RNAi, wild-type (Gmm^{WT}) flies exhibited high susceptibility to trypanosomes. This phenotype was similar to Gmm^{Wig-} trypanosome infection levels, suggesting that the high susceptibility of Gmm^{Wig-} flies to trypanosome infections may partially arise from low *pgrp-lb* levels in their anterior midgut. Although the antiprotozoal activity of tsetse PGRP-LB has not yet been documented, one PGRP in *Drosophila* and all PGRPs in humans have direct antibacterial properties (Dziarski and Gupta 2006).

In addition to playing a possible antitrypanosomal role, tsetse PGRP-LB also controls *Wigglesworthia* densities. In Gmm^{WT} flies, depletion of *pgrp-lb* through RNAi results in the activation of the IMD pathway, and in turn leads to induced synthesis of AMPs (Wang et al. 2009). This upregulation of the IMD pathway results in a decrease in *Wigglesworthia* numbers. This finding indicates that IMD pathway-modulated host immune responses have the ability to compromise the fitness of *Wigglesworthia* even though this bacterium is essential for host fecundity. Collectively, these results indicate that PGRP-LB has a dual function in tsetse. First, PGRP-LB expressed in response to *Wigglesworthia* protects tsetse's mutualistic symbiosis by negatively modulating the activity of the IMD pathway, which, when induced, can be harmful to this bacterium. Second, PGRP-LB also benefits tsetse by preventing establishment of trypanosome infections, which in turn negatively impact tsetse's reproductive fitness by increasing the larval incubation period (Hu et al. 2008). In the weevil, *Sitophilus zeamais*, which also has mutualistic symbionts, *pgrp-lb* expression was also detected in the bacteriome organ and found to be upregulated in the nymphal phase during a time when the symbionts are released from host cells (Anselme et al. 2006, 2008). Thus, symbiont density regulation through the action of PGRPs may be a general mechanism that insects use to downregulate their immune responses, which could otherwise be damaging to their symbiotic fauna. Interestingly, the same PGRPs can also protect hosts from other microbial infections, such as trypanosome infections in the case of tsetse flies, which can induce host immune pathways that would otherwise damage symbiosis.

TOLERANCE OF TSETSE TO *SODALIS*

Ribosomal protein-based phylogenetic analysis indicates that *Sodalis* is closely related to several bacterial pathogens (Toh et al. 2006; Weiss et al. 2008). These findings raise the question of how *Sodalis* exhibits commensal characteristics in tsetse despite its genotypic similarities to pathogenic bacteria. Several immunogenic components of *Sodalis*'s cell membrane are altered in comparison to *E. coli*. These include a truncated lipopolysaccharide (LPS), missing O-antigen, and modified outer membrane protein A (OmpA; Weiss et al. 2008). OmpA is a major component of the outer membrane in the Enterobacteriaceae, and functions as both an evasin and target of the mammalian immune system (Smith et al. 2007). Furthermore, OMPs from *Leptospira interrogans* (Ristow et al. 2007), *Salmonella enterica* (Singh et al. 2003), and *E. coli* (Prasadarao et al. 1996) exhibit directly virulent phenotypes upon host infection with these bacteria. Examination of *Sodalis*'s genome indicated that several components of its outer cell wall are modified when compared to their *E. coli* homologues (Toh et al. 2006). These modifications included a truncated LPS, missing O-antigen, and altered OmpA (Toh et al. 2006). Because OmpA from other bacteria exhibits direct toxicity, it was presumed that alterations in *Sodalis*'s homologue may be influential in promoting this bacterium's symbiosis with tsetse. A detailed structural analysis of *E. coli* OmpA revealed that the first 171 amino acids form 8 membrane-traversing β-barrels connected by 4 loops that are exposed to the external environment (Pautsch and Schulz 1998). Loops 1 and 2 of pathogenic *E. coli* K1 are composed of amino acid residues that generate highly virulent exposed domains (Prasadarao et al. 1996). An alignment of the putative OmpA transmembrane domains from several pathogenic and symbiotic eubacteria identified a pattern that might explain why infection with *E. coli* is lethal to tsetse while *Sodalis* is not (Figure 9.5A). Membrane-embedded β-barrel domains from both symbionts and pathogens exhibited a high degree of conservation. However, the exposed loops (especially loop 1) had diverged so that symbiont domains contained amino acid insertions and substitutions that were absent from their pathogenic counterparts (Weiss et al. 2008).

In contrast to their well-characterized role in pathogenesis, little is known about how OMPs function to modulate invertebrate host tolerance of symbiotic bacteria. Tsetse is a useful model system to identify bacterial phenotypes that permit host tolerance of symbiosis. To test the role of OMP modifications, Weiss et al. (2008) infected tsetse with a recombinant *Sodalis* genetically modified to produce *E. coli* OmpA. This strain exhibited a virulent phenotype upon inoculation into tsetse so that all flies were dead by 8 days postinfection (Figure 9.5B). As a control, tsetse that were inoculated with WT *Sodalis* were able to survive the 14-day experimental course with little mortality. In addition, the normally nonpathogenic *E. coli* K12 strain rapidly proliferated in tsetse so that flies perished by 10 days postinfection, similar to our findings with rec*Sodalis* (Figure 9.5C). When tsetse were inoculated with an *E. coli* K12 strain lacking the OmpA gene, flies survived the 14-day experimental course with little mortality (Figure 9.5D). Taken together, the OmpA sequence alignment information and above-mentioned functional studies strongly suggest that bacterium-specific polymorphisms in host-exposed OmpA epitopes may be responsible for the differential infection outcomes observed when tsetse were infected with *Sodalis* and *E. coli*. Furthermore, polymorphisms in OmpA may account for one adaptation that many symbiotic bacteria have evolved to achieve homeostasis with their hosts.

Polymorphisms in OmpA represent a bacterium-specific adaptation that may facilitate the establishment of many invertebrate symbioses. In the tsetse-*Sodalis* model system, the functionality of this adaptation appears to be contingent on the interaction between OmpA and the host immune system. To investigate this phenomenon further, flies were infected with bacteria displaying differential OmpA phenotypes and evaluating host immune gene expression profiles. Interestingly, immunity-related gene expression in tsetse was suppressed following infection with virulent bacteria (wild-type *E. coli* and rec*Sodalis* that express *E. coli* OmpA), while infection with avirulent strains (wild-type *Sodalis* and *E. coli* OmpA mutants) activated a robust response (Figure 9.5E). For example, expression of the antimicrobial peptide (AMP) cecropin was significantly lower in flies infected with virulent bacteria. Insufficient *cecropin* expression may account for the lethal effects of virulent strains, as this AMP exhibits powerful bactericidal activity against gram-negative bacteria. On the other hand, avirulent bacterial strains induced considerably more *cecropin* expression. This response likely led to the elimination of mutant *E. coli*. *Sodalis*, on the other hand, exhibits a strong innate resistance to tsetse AMPs and was thus most likely unaffected (Hu and Aksoy 2005). Notably, virulent bacterial strains resulted in increased expression of tsetse *pgrp-lb*. In *Drosophila* PGRP-LB is an amidase that directly inhibits the IMD pathway (Zaidman-Rémy et al. 2006). This observed induction of tsetse *pgrp-lb* may have prevented the activation of humoral immune responses necessary

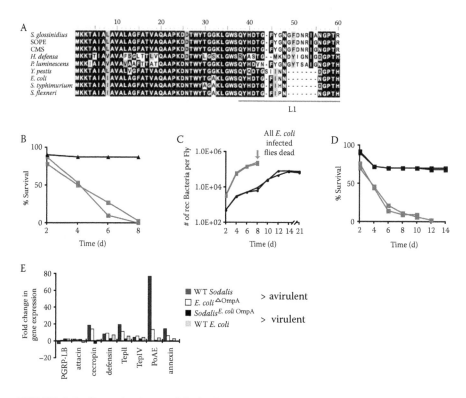

FIGURE 9.5   See color insert. Mechanism of tsetse's tolerance of *Sodalis*. (A) Multiple sequence alignment of the first 60 N-terminal putative amino acids of OmpA from *Sodalis*, SOPE (*Sitophilus oryzae* principal endosymbiont), CMS (*C. melbae* symbiont), *H. defensa*, *P. luminescens*, *Y. pestis*, *E. coli* 536 (UPEC), *S. typhimurium*, and *S. flexneri*. External loop 1 is underlined. Conserved residues are outlined in black, and substitutions in grey. (B) Systemic infection of tsetse with 1×10^3 CFU of *Sodalis* (red squares) genetically modified to express *E. coli* OmpA. Controls (black triangles) received 1×10^3 CFU of WT *Sodalis*. (C) Number of cells per fly over time after injecting 1×10^3 CFU of luciferase-expressing *Sodalis* (black circles) and *E. coli* (red squares) into tsetse's hemocoel. All rec*E. coli*$_{pIL}$-infected flies perished by 10 dpi. (D) Survival of tsetse infected with mutant *E. coli* K12 that do not express OmpA (black squares) and mutant *E. coli* K12 transcomplemented to express their native OmpA (red squares). In B–D, d = days postinfection. (E) Fold change in the expression of tsetse immunity-related genes following infection with avirulent and virulent bacterial strains. All genes were normalized against the constitutively expressed tsetse β-tubulin gene.

to eliminate WT *E. coli* and *Sodalis* that express *E. coli* OmpA. This hypothesis was verified when RNAi-mediated knockdown of host *pgrp-lb* expression reversed the lethal phenotype of WT *E. coli* infection on tsetse (Weiss et al. 2008). Taken together, these findings suggest that polymorphisms in the exposed loop domains of OmpA represent a microbial adaptation that mediates host tolerance of endogenous symbiotic bacteria.

CONCLUSION

Tsetse and its symbionts, obligate *Wigglesworthia* and commensal *Sodalis*, represent intricately associated symbioses. Physiological adaptations by tsetse and its symbiotic fauna have resulted in the evolution of a beneficial symbiotic relationship. In this case, the tsetse fly host provides a safe residential niche for its symbionts in exchange for dietary supplementation and immune system regulation. With the tsetse genome currently undergoing annotation, further detailed investigations into the host-symbiont interactions and their eventual impact on tsetse's physiology will shortly be possible.

REFERENCES

Akman, L., Yamashita, A., Watanabe, H., Oshima, K., Shiba, T., Hattori, M., and Aksoy, S. 2002. Genome sequence of the endocellular obligate symbiont of tsetse flies, *Wigglesworthia glossinidia. Nature Genetics* 32: 402–407.

Aksoy, S. 1995. *Wigglesworthia* gen. nov. and *Wigglesworthia glossinidia* sp. nov., taxa consisting of the mycetocyte-associated, primary endosymbionts of tsetse flies. *International Journal of Systematic Bacteriology* 45: 848–851.

Anselme, C., Perez-Brocal, V., Vallier, A., Vincent-Monegat, C., Charif, D., Latorre, A., Moya, A., and Heddi, A. 2008. Identification of the weevil immune genes and their expression in the bacteriome tissue. *BMC Biology* 6: 43.

Anselme, C., Vallier, A., Balmand, S., Fauvarque, M. O., and Heddi, A. 2006. Host PGRP gene expression and bacterial release in endosymbiosis of the weevil *Sitophilus zeamais. Applied Environmental Microbiology* 72: 6766–6772.

Attardo, G. M., Lohs, C., Heddi, A., Alam, U. H., Yildirim, S., and Aksoy, S. 2008. Analysis of milk gland structure and function in *Glossina morsitans*: milk protein production, symbiont populations and fecundity. *Journal of Insect Physiology* 54: 1236–1246.

Beard, B. C., and Aksoy, S. 1998. Genetic manipulation of insect symbionts. In Crampton, J., Beard, B. C., and Louis, C. (Eds.), *Molecular biology of insect disease vectors.* 555–560. London: Chapman and Hall.

Beard, C. B., O'Neill, S. L., Mason, P., Mandelco, L., Woese, C. R., Tesh, R. B., Richards, F. F., and Aksoy, S. 1993. Genetic transformation and phylogeny of bacterial symbionts from tsetse. *Insect Molecular Biology* 1: 123–131.

Chen, X., Li, S., and Aksoy, S. 1999. Concordant evolution of a symbiont with its host insect species: molecular phylogeny of genus *Glossina* and its bacteriome-associated endosymbiont, *Wigglesworthia glossinidia. Journal of Molecular Evolution* 48: 49–58.

Cheng, Q., and Aksoy, S. 1999. Tissue tropism, transmission and expression of foreign genes *in vivo* in midgut symbionts of tsetse flies. *Insect Molecular Biology* 8: 125–132.

Dale, C. and Maudlin, I. 1999. *Sodalis* gen. nov. and *Sodalis glossinidius* sp. nov., a micro-aerophilic secondary endosymbiont of the tsetse fly *Glossinia morsitans morsians. International Journal of Systematic Bacteriology* 49: 267–275.

Dale, C. and Welburn, S. C. 2001. The endosymbionts of tsetse flies: manipulating host-parasite interactions. *International Journal of Parashology* 31: 628–631.

Dale, C., Jones, T., and Pontes, M. 2005. Degenerative evolution and functional diversification of type-III secretion systems in the insect endosymbiont *Sodalis glossinidius. Molecular Biology and Evolution* 22: 758–766.

Dale, C., Young, S. A., Haydon, D. T., and Welburn, S. C. 2001. The insect endosymbiont *Sodalis glossinidius* utilizes a type III secretion system for cell invasion. *Proceedings of the National Academy of Sciences of the United States of America* 98: 1883–1888.

Dziarski, R., and Gupta, D. 2006. Mammalian PGRPs: novel antibacterial proteins. *Cellular Microbiology* 8: 1059–1069.

Fukatsu, T., Koga, R., Smith, W. A., Tanaka, K., Nokoh, N., Sasaki-Fukatsu, K., Yoshizawa, K., Dale, C., and Clayton, D. 2007. Bacterial endosymbiont of the slender pigeon louse, *Columbicola columbae*, allied to endosymbionts of grain weevils and tsetse flies. *Applied and Environmental Microbiology* 73: 6660–6668.

Geiger, A., Ravel, S., Matielle, T., Janelle, J., Patrel, D., Cuny, G., and Frutos, R. 2007. Vector competence of *Glossina gambiensis* for *Trypanosoma brucei* s.l. and genetic diversity of the symbiont *Sodalis glossinidius*. *Molecular Biology and Evolution* 24: 102–109.

Gross, R., Vavre, F., Heddi, A., Hurst, G. D., Zchori-Fein, E., and Bourtzis, K. 2009. Immunity and symbiosis. *Molecular Microbiology* 73: 751–759.

Haines, L. R., Hancock, R. E., and Pearson, T. W. 2003. Cationic antimicrobial peptide killing of African trypanosomes and *Sodalis glossinidius*, a bacterial symbiont of the insect vector of sleeping sickness. *Vector Borne and Zoonotic Disease* 3: 175–186.

Hao, Z. R., Kasumba, I., Lehane, M. J., Gibson, W. C., Kwon, J., and Aksoy, S. 2001. Tsetse immune responses and trypanosome transmission: implications for the development of tsetse-based strategies to reduce trypanosomiasis. *Proceedings of the National Academy of Sciences of the United States of America* 98: 12648–12653.

Heddi, A., Charles, H., Khatchadourian, C., Bonnot, G., and Nardon, P. 1998. Molecular characterization of the principal symbiotic bacteria of the weevil *Sitophilus oryzae*: a peculiar G+C content of an endosymbiotic DNA. *Journal of Molecular Evolution* 47: 52–61.

Hu, C., and Aksoy, S. 2006. Innate immune responses regulate trypanosome parasite infection of the tsetse fly *Glossina morsitans morsitans*. *Molecular Microbiology* 60: 1194–1204.

Hu, C., Rio, R. V., Medlock, J., Haines, L. R., Nayduch, D., Savage, A. F., Guz, N., Attardo, G. M., Pearson, T. W., Galvani, A. P., and Aksoy, S. 2008. Infections with immunogenic trypanosomes reduce tsetse reproductive fitness: potential impact of different parasite strains on vector population structure. *PLoS Neglected Tropical Diseases* 2: e192.

Hu, Y., and Aksoy, S. 2005. An antimicrobial peptide with trypanocidal activity characterized from *Glossina morsitans morsitans*. *Insect Biochemistry and Molecular Biology* 35: 105–115.

Kaiwa, N., Hosokawa, T., Kikuchi, Y., Nikoh, N., Meng, X. Y., Kimura, N., Ito, M., and Fukatsu, T. 2010. Primary gut symbiont and secondary, *Sodalis*-allied symbiont of the scutellerid stinkbug *Cantao ocellatus*. *Applied and Environmental Microbiology* 76: 3486–3494.

Ma, W. C., and Denlinger, D. L. 1974. Secretory discharge and microflora of milk gland in tsetse flies. *Nature* 247: 301–303.

Maudlin, I., Welburn, S. C., and Mehlitz, D. 1990. The relation between *Rickettsia*-like organisms and trypanosome infections in natural populations of tsetse in Liberia. *Tropical Medicne and Parasitology* 41: 265–267.

Nayduch, D., and Aksoy, S. 2007. Refractoriness in tsetse flies (Diptera: Glossinidae) may be a matter of timing. *Journal of Medical Entomology* 44: 660–665.

Novakova, E., and Hypsa, V. 2007. A new *Sodalis* lineage from the bloodsucking fly *Craterina melbae* (Diptera, Hippoboscoidea) originated independently of the tsetse fly symbiont *Sodalis glossinidius*. *FEMS Microbiology Letters* 269: 131–135.

Pais, R., Lohs, C., Wu, Y., Wang, J., and Aksoy, S. 2008. The obligate mutualist *Wigglesworthia glossinidia* influences reproduction, digestion, and immunity processes of its host, the tsetse fly. *Applied and Environmental Microbiology* 74: 5965–5974.

Pautsch, A., and Schulz, G. E. 1998. Structure of the outer membrane protein A transmembrane domain. *Nature Structural Biology* 5: 1013–1017.

Prasadarao, N. V., Wass, C. A., Weiser, J. N., Stins, M. F., Huang, S. H., and Kim, K. S. 1996. Outer membrane protein A of *Escherichia coli* contributes to invasion of brain microvascular endothelial cells. *Infection and Immunity* 64: 146–153.

Rio, R. V., Lefevre, C., Heddi, A., and Aksoy, S. 2003. Comparative genomics of insect-symbiotic bacteria: influence of host environment on microbial genome composition. *Applied and Environmental Microbiology* 69: 6825–6832.

Ristow, P., Bourhy, P., da Cruz-McBride, F. W., Figueira, C. P., Huerre, M., Ave, P., Girons, I. S., Ko, A. I., and Picardeau, M. 2007. The ompA-like protein Loa22 is essential for Leptospiral virulence. *PLoS Pathogens* 3: 894–903.

Royet, J., and Dziarski, R. 2007. Peptidoglycan recognition proteins: pleiotropic sensors and effectors of antimicrobial defences. *Nature Reviews Microbiology* 5: 264–277.

Singh, S. P., Williams, Y. U., Miller, S., and Nikaido, H. 2003. The C-terminal domain of *Salmonella enterica* serovar typhimurium OmpA is an immunodominant antigen in mice but appears to be only partially exposed on the bacterial cell surface. *Infection Immunity* 71: 3937–3944.

Smith, G. J., Mahon, V., Lambert, M. A., and Fagan, R. P. 2007. A molecular Swiss army knife: OmpA structure, function and expression. *FEMS Microbiology Letters* 273: 111.

Snyder, A. K., Deberry, J. W., Runyen-Janecky, L., and Rio, R. V. 2010. Nutrient provisioning facilitates homeostasis between tsetse fly (Diptera: Glossinidae) symbionts. *Proceedings of the Royal Society B* 277: 2389–2397.

Toh, H., Weiss, B. L., Perkin, S. A., Yamashita, A., Oshima, K., Hattori, M., and Aksoy, S. 2006. Massive genome erosion and functional adaptations provide insights into the symbiotic lifestyle of *Sodalis glossinidius* in the tsetse host. *Genome Research* 16: 149–156.

Wang, J., Wu, Y., Yang, G., and Aksoy, S. 2009. Interactions between mutualist *Wigglesworthia* and tsetse peptidoglycan recognition protein (PGRP-LB) influence trypanosome transmission. *Proceedings of the National Academy of Sciences of the United States of America* 106: 12133–12138.

Weiss, B. L., Mouchotte, R., Rio, R. V., Wu, Y. N., Wu, Z., Heddi, A., and Aksoy, S. 2006. Inter-specific transfer of bacterial endosymbionts between tsetse species: infection establishment and effect on host fitness. *Applied and Environmental Microbiology* 72: 7013–7021.

Weiss, B. L., Wu, Y., Schwank, J. J., Tolwinski, N. S., and Aksoy, S. 2008. An insect symbiosis is influenced by bacterium-specific polymorphisms in outer-membrane protein A. *Proceedings of the National Academy of Sciences of the United States of America* 105: 15088–15093.

Welburn, S. C., Arnold, K., Maudlin, I., and Gooday, G. W. 1993. Rickettsia-like organisms and chitinase production in relation to transmission of trypanosomes by tsetse flies. *Parasitology* 107: 141–145.

Welburn, S. C., and Maudlin, I. 1999. Tsetse-trypanosome interactions: rites of passage. *Parasitology Today,* 15: 399–403.

Zaidman-Rémy, A., Hervé, M., Poidevin, M., Pili-Floury, S., Kim, M. S., Blanot, D., Oh, B. H, Ueda, R., Mengin-Lecreulx, D., and Lemaitre, B. 2006. The *Drosophila* amidase PGRP-LB modulates the immune response to bacterial infection. *Immunity* 24: 463–473.

10 *Rickettsia* Get Around

Yuval Gottlieb, Steve J. Perlman,
Elad Chiel, and Einat Zchori-Fein

CONTENTS

Genus	*Rickettsia*
Family	Rickettsiaceae
Order	Rickettsiales
Description year	1916
Origin of name	After Dr. Howard Taylor Ricketts (1871–1910), an American pathologist who discovered the causative organisms and mode of transmission of Rocky Mountain spotted fever and epidemic typhus, and died from a typhus infection
Description	
Reference	Rocha Lima (1951)

Dr. Howard Taylor Ricketts (1871–1910). (Reproduced from NIAID/RML website. With permission.)

INTRODUCTION

The best-known members of the bacterial genus *Rickettsia* are associates of blood-feeding arthropods and are pathogenic when transmitted to vertebrates. These species include the agents of acute human diseases such as typhus and spotted fever. However, recent surveys of bacteria associated with arthropods and other invertebrates have uncovered many other *Rickettsia* in hosts that have no relationship with vertebrates (Braig et al. 2008). It may therefore be more appropriate to consider *Rickettsia* as symbionts that are transmitted primarily vertically within arthropod hosts and occasionally horizontally between and within vertebrate hosts. Indeed, one of the most striking aspects of *Rickettsia* biology is the diversity of transmission modes. Perhaps even more than most other secondary symbionts, *Rickettsia* are characterized by multiple transitions between strains that are transmitted strictly vertically and those that exhibit mixed (horizontal and vertical) transmission. Moreover, the array of organisms that may harbor *Rickettsia* spans multiple kingdoms, including amoebae, annelids, arthropods, vertebrates, and plants. This unusually broad host range, combined with the ability of *Rickettsia* to actively move between and within living cells, makes this bacterium an excellent model system for studying both the evolution of transmission pathways and transitions between mutualism and pathogenicity (Merhej and Raoult 2010). Indeed, the major insights that have come from studying pathogenic *Rickettsia* will serve to illuminate our understanding of the evolution of infection strategies across this fascinating group of symbionts.

DISCOVERY AND PHYLOGENY

The order Rickettsiales forms a coherent group of obligate intracellular symbionts of eukaryotic cells within the alpha-Proteobacteria, and includes the families Holosporaceae, Anaplasmataceae, and Rickettsiaceae. The latter includes the genera *Orientia*, the agent of scrub typhus, *Wolbachia*, and the subject of this chapter, *Rickettsia*. *Rickettsia* was discovered in 1910 by Howard Taylor Ricketts, who found the bacterium in both ticks and lice and, together with his assistant Russell M. Wilder, suggested that it was the causative agent of both the tick-transmitted Rocky Mountain spotted fever and the louse-borne tabardillo disease, a Mexican form of typhus (Hektoen 1910). In 1910, Ricketts died in Mexico, most likely from the bite of an infected insect (Hektoen 1910). The six bacteria he discovered were described as the genus *Rickettsia* and were named after him in 1916 (Rocha Lima 1951).

Rickettsia were originally known from blood-feeding arthropods, and were traditionally classified into three groups: (1) the typhus group, including the louse-borne *R. prowazekii* (the agent of epidemic typhus) and the flea-borne *R. typhi* (the agent of endemic typhus); (2) the spotted fever group, including the tick-borne pathogens *R. rickettsii* (the agent of Rocky Mountain spotted fever) and *R. conorii* (the agent of Mediterranean spotted fever); and (3) the agent of scrub typhus, transmitted by chigger mites. However, recent molecular surveys and a renewed interest in characterizing the diversity of arthropod-associated microbes have completely changed our understanding of *Rickettsia* evolution and diversity. First, these surveys have unearthed an enormous diversity of *Rickettsia*, the vast majority of which are found in hosts that do not feed on blood. In fact, the *Rickettsia* strains that are primarily associated with disease, found in the spotted fever and typhus groups, are fairly closely related to each other and represent only a small fraction of *Rickettsia* diversity. Also, the causal agent of scrub typhus, transmitted by chigger mites and which until recently was classified as *Rickettsia*, is quite distantly related and is now classified as *Orientia tsutsugamushi*, the sole representative of its genus.

DIVERSITY OF *RICKETTSIA*: HOST DISTRIBUTION AND EVOLUTIONARY PATTERNS

Rickettsia strains have been found to infect a wide range of eukaryotes, primarily arthropods, including hosts in nine insect orders (wasps, flies, beetles, moths, lacewings, booklice, lice, fleas, and true bugs) and five additional arthropod orders (spiders, springtails, ticks, and two mite orders). Infections have also been identified in leeches and a nucleariid amoeba (i.e., amoebae that are characterized by filose pseudopods and discoid cristae mitochondria) that parasitizes fish (Perlman et al. 2006; Weinert et al. 2009a).

A recent study by Weinert and colleagues (2009a) made a major advance in understanding *Rickettsia* evolution and diversity. The authors screened a large sample of arthropods in order to discover novel host-*Rickettsia* associations, and then used three protein-coding genes (citrate syntase—*gltA*, cytochrome C oxidase subunit I—*coxA*, ATP synthase F1 alpha subunit—*atpA*), in addition to 16S rRNA, to

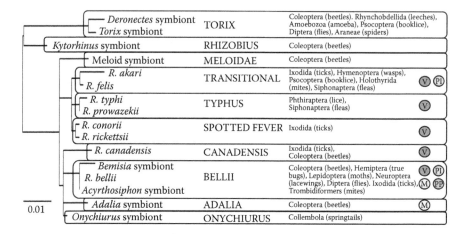

FIGURE 10.1   16S rRNA ML phylogeny of the major groups of *Rickettsia*, as identified by Weinert et al. (2009a). Representative strains from each group were chosen. Listed to the right of the species group name are all the currently known hosts for strains in that group. Listed at the far right are known phenotypes of some strains, including vertebrate pathogens (V), plant pathogens (PP), male killers (M), and parthenogenesis inducers (PI).

construct a comprehensive phylogenetic tree. They identified 10 major lineages of *Rickettsia*, including confirmation of many that had been previously characterized (Figure 10.1). Some of these lineages consist of strains that infect taxonomically related hosts, such as the spotted fever and adalia groups, found in ticks and ladybird beetles, respectively. Other groups appear to be united primarily by the ecology of their hosts, such as typhus group strains, found in blood-feeding lice and fleas, and torix group strains, found primarily in hosts associated with aquatic lifestyles, including leeches, amoeba parasites of fish, and craneflies. Finally, some groups are found in a broad range of taxonomically and ecologically unrelated hosts. For example, the 13 strains in the bellii group that are well characterized (i.e., have multiple genes sequenced) infect hosts from seven arthropod orders, including aphids, ticks, beetles, and springtails.

As had been found in other studies based solely on 16S rRNA, Weinert et al. (2009a) highlight two major patterns in the evolution of *Rickettsia*. First, host shifts are rampant, and closely related strains can be found infecting extremely distantly related hosts. Second, some groups of hosts appear to be hotspots for infection, and harbor closely related strains. Lineages that appear to be commonly infected with certain *Rickettsia* include ladybird beetles, ticks, dytiscid water beetles, and leeches. Both patterns are also found in many facultative maternally transmitted symbionts of arthropods, such as *Wolbachia* and *Cardinium*, where transmission is almost exclusively maternal over ecological timescales, but new hosts are repeatedly colonized over evolutionary timescales (Werren et al. 1995; Zchori-Fein and Perlman 2004). These patterns suggest that although transfers are common, host phylogeny is a major determinant in their success (Russell et al. 2009), an observation that has been confirmed in many experimental symbiont infections, where initial transfer and subsequent inheritance are most likely to succeed in related hosts (e.g., Tinsley and Majerus 2007). How inherited intracellular symbionts colonize novel species is not well understood and remains one of the holy grails in the field. As discussed further, much progress is being made in this area with respect to *Rickettsia*, and we will argue that the way that that bacterium invades cells and tissues may make it much more successful in terms of horizontal transfer than other symbionts.

In addition to phylogenetic evidence for host shifts, a major recent advance in the study of *Rickettsia* evolution has been the discovery that *Rickettsia* commonly harbor conjugation genes, often found on plasmids, as well as transposable elements, that are horizontally transferred between strains and other bacteria species, and may promote wholesale genomic reshuffling and acquisition of novel genes (Baldridge et al. 2010; Weinert et al. 2009b; Ogata et al. 2006; Merhej and Raoult 2010). This is very different from the early view of *Rickettsia* genome evolution, which was biased by focusing on many of the closely related human pathogen strains. For example, the typhus group strains appear to be at the extreme of genome reduction and stasis, and lack conjugation genes and selfish elements (Blanc et al. 2007). Also, while the first few fully sequenced spotted fever strains all lacked plasmids, these are in fact widely distributed in this group (Baldridge et al. 2010). As an example of wholesale genome reshuffling, the survey of Weinert et al. (2009a) uncovered a strain in a beetle that shows evidence of recombination between two distant strains. In addition to transfer between strains, plasmids and transposable elements serve as potential agents of transfer between unrelated symbionts. For example, a recent transfer of a transposable element between *Rickettsia* and *Cardinium* was documented (Felsheim et al. 2009), and it is tantalizing to speculate that these elements facilitate the movement of genes involved in diverse symbiont life history strategies, such as male killing (see below).

PREVALENCE OF *RICKETTSIA* BETWEEN AND WITHIN SPECIES

The limited available data from large-scale surveys of *Rickettsia* offer mixed conclusions on global patterns of infection frequency. Only two studies have undertaken large surveys of *Rickettsia* infection that were unbiased in terms of the host taxa that were targeted. In their screen of 136 arthropod species, Duron and colleagues (2008) found only one species infected with *Rickettsia*; however, the primers they used may have targeted only a subsample of *Rickettsia* diversity. In a survey of 847 individuals, each representing a different species (i.e., n = 1), nine were found to harbor *Rickettsia* (Weinert et al. 2009a). Although both studies found low (~1%) overall infection frequencies, they probably underestimate the actual diversity. Indeed, surveys targeting specific host taxa have found higher infection frequencies, including 8/21 ladybird beetles (Weinert et al. 2009a), 28/122 spiders (Goodacre et al. 2006), 4/6 dytiscid water beetles (Küchler et al. 2009), 4/35 gall wasps and their parasitoids (Weinert et al. 2009a), and 3/4 leeches (Kikuchi and Fukatsu 2005).

Infection frequencies within species also appear to vary greatly. Some infections approach 100% prevalence, including *Rickettsia* that cause parthenogenesis in the wasps *Neochrysocharis formosa* (Hagimori et al. 2006) and *Pnigalio soemius* (Giorgini et al. 2010), as well as infections in *Torix tagoi* leeches (96%; Kikuchi and Fukatsu 2005), *Liposcelis bostrychophila* booklice (100%; Perroti et al. 2006; Behar et al. 2010), *Deronectes platynotus* water beetles (100%; Küchler et al. 2009), *Amblyomma rotundum* ticks (100%; Labruna et al. 2004), and *Onychiurus sinensis* springtails (100%; Frati et al. 2006). Medium frequencies have also been reported, for example, in *Hemiclepsis marginata* leeches (29%; Kikuchi and Fukatsu 2005) and *Deronectes aubei* and *D. semirufus* water beetles (39 and 33%; Küchler et al. 2009). There have been many studies of *Rickettsia* prevalence in blood feeders, and these report a wide range of infection frequencies, including many low-frequency infections (<5%) (e.g., Sabatini et al. 2010; Loftis et al. 2006; Parola 2006; Hornok et al. 2010). These very diverse infection rates are probably influenced by factors such as the effect of *Rickettsia* on their hosts, the mode and efficiency of transmission (see below), and the presence of other symbionts.

A number of detailed surveys on two agricultural insect pests have documented wide population variation in *Rickettsia* infection. For example, frequencies in *Acyrthosiphon pisum* (pea aphids) populations range from 4% in Japan (Tsuchida et al. 2002), to 0 to 10% on aphids collected on alfalfa in France (Simon et al. 2003), to 30 to 37% on aphids collected on peas in France (Simon et al. 2003), to 48% in California (Chen et al. 1996). Infection rates in populations of the sweet potato whitefly, *Bemisia tabaci*, sampled around the world range from 0 to 100% (Himler et al. 2011). Interestingly, some of these patterns might be explained by the presence of other symbionts, as coinfection with *Cardinium* was never observed, while coinfection with other symbionts was common (Gueguen et al. 2010). Moreover, reports on infection with more than one *Rickettsia* species in the same individual are rare (Macaluso et al. 2002; Carmichael and Fuerst 2006; Eremeeva et al. 2008), suggesting that competitive interactions may determine specific prevalence. Indeed, competitive exclusion has been implicated as a major force in the epidemiology of pathogenic *Rickettsia* (Burgdorfer et al. 1981).

EFFECTS OF *RICKETTSIA* ON HOST FITNESS

While the influence of *Rickettsia* on most hosts is not known, a number of interrelated factors can serve to inform studies in this area, including mode and efficiency of transmission, infection frequency, and distribution within host tissues.

It has been previously argued that *Rickettsia* should be viewed primarily as vertically transmitted endosymbionts of arthropods (e.g., Perlman et al. 2006). Indeed, one of the major unifying themes in the study of the ecology and evolution of infectious microorganisms is the major role of transmission in determining the evolution and nature of symbiosis (i.e., its location on the mutualism-pathogenicity continuum), and in shaping infection frequencies (Werren and O'Neill 1997). Many *Rickettsia* appear to be transmitted exclusively maternally, often in the egg cytoplasm, and as such, their fitness is intimately and exclusively tied to that of their hosts. Theory predicts that over the long term, if transmission is less than perfect, such symbionts can only persist in their hosts by either (1) increasing host fitness, (2) manipulating host reproduction in order to increase the fitness or frequency of infected females, or (3) exhibiting increased horizontal transmission (Werren and O'Neill 1997).

Two major types of reproductive manipulation have been described in arthropods: cytoplasmic incompatibility and sex ratio distortion (Stouthamer et al. 1999; Werren et al. 2008). In cytoplasmic incompatibility, symbionts cause mating incompatibilities between infected males and uninfected females; this strategy has only been demonstrated in *Wolbachia* and *Cardinium* (Breeuwer and Werren 1990; O'Neill et al. 1992; Hunter et al. 2003). Sex ratio-distorting symbionts bias host sex ratios in favor of females (the sex that can transmit symbionts), either by transforming male hosts into females or by killing infected males, often early in development. Quite a number of bacterial endosymbionts of insects have independently evolved sex ratio distortion capabilities, including *Wolbachia*, *Cardinium*, *Spiroplasma*, *Flavobacterium*, *Arsenophonus*, *Hamiltonella*, and *Rickettsia*. Two types of sex ratio distortion have been documented in *Rickettsia*. First, strains allied with *R. felis* and *R. belli* induce parthenogenesis in the wasps *Neochrysocharis formosa* (Hagimori et al. 2006) and *Pnigalio soemius* (Giorgini et al. 2010), respectively. Wasps (order Hymenoptera) are haplodiploid (diploid individuals are females, haploid individuals are males), and this mode of sex determination is particularly susceptible to sex ratio-distorting symbionts. In addition, a number of parthenogenetic species, such as the booklouse *L. bostrychophila* (Yusuf et al. 2000; Behar et al. 2010) and the tick *Amblyomma rotundum* (Labruna et al. 2004), are fixed (100% prevalence) for *Rickettsia* infection, although the role of the symbiont is not known.

Male killing has evolved independently in *Rickettsia* at least twice, in the buprestid beetle *Brachys tesselatus* (Werren et al. 1994; Lawson et al. 2001), and in many strains infecting ladybird beetles. Ladybird beetles appear to be a hotspot for male-killing symbionts, including *Wolbachia*, *Spiroplasma*, *Flavobacterium*, *Arsenophonus*, *Hamiltonella*, and *Rickettsia* (Engelstädter and Hurst 2009; Majerus and Majerus 2010). Ladybird beetle eggs are laid in clutches and often suffer severe early mortality due to resource limitation. As a result, sibling cannibalism is common, and it is thought that male killing can provide a fitness benefit to infected lineages when sisters gain crucial resources by consuming their dead brothers (Majerus and Hurst 1997). As such, male killers can be thought to reside on a continuum between parasitism (killing up to 50% of offspring) and mutualism (providing a fitness benefit to infected relatives).

Interestingly, although common in most lineages of inherited symbionts (and predicted by theory; Werren and O'Neill 1997), there is little evidence that strains of *Rickettsia* provide direct fitness benefits to their hosts, and can be thus thought of as mutualistic beneficial symbionts. This may be due to a number of reasons, including a historical focus on the vertebrate pathogenic effects of *Rickettsia*. It is also often difficult to uncover the effects of a symbiont on host fitness, as symbiont infections are often difficult to manipulate in the lab, and fitness itself can be highly labile. *Torix* (but not *Hemiclepsis*) leeches are larger when infected, and infections in the former (but not the latter) approach 100% prevalence (Kikuchi and Fukatsu 2005). *Bemisia tabaci* whiteflies infected with *Rickettsia* produce more offspring, have higher survival rates to adulthood, develop faster, and produce a higher proportion of females. In that system *Rickettsia* functions as both a mutualist and reproductive manipulator, and it probably explains the rapid spread of the symbiont into new host populations (Himler et al. 2011). On the other hand, the bacteria also appear to be associated with increased susceptibility to pesticides (Kontsedalov et al. 2008).

Development of cat fleas also appears to be faster when they are infected with *R. felis* (Wedincamp and Foil 2002). Antibiotic-treated *L. bostrychophila* booklice produce virtually no offspring, and it has been suggested that this may be due to their *Rickettsia* symbiont (Perotti et al. 2006). This interesting finding would constitute the first example of a *Rickettsia* acting as an obligate mutualist. However, other studies, such as those showing that *Rickettsia* infection reduces host fitness in pea aphids (Sakurai et al., 2005) and in some blood feeders (below), suggest that more work is needed, including ruling out the possibly toxic effects of the antibiotics themselves.

There are surprisingly relatively few studies on the effects of *Rickettsia* on the fitness of blood-feeding hosts. Here we might predict that strains that have significant levels of horizontal transmission may reduce host fitness. For example, *R. rickettsii* is often lethal to its tick host *Dermacentor andersonii* (Niebylski et al. 1999). *Rhipicephalus sanguineus* ticks infected with *R. rhipicephali* exhibit cytopathology, including mitochondrial changes, membrane breakage, and general loss of ground substance, although feeding and oviposition are not affected (Hayes and Burgdorfer 1979). Also, artificial infection with *R. prowazekii*, *R. conorii*, and *R. typhi* resulted in the death of louse hosts (Houhamdi et al. 2002, 2003; Houhamdi and Raoult 2006). However, many pathogenic *Rickettsia* exhibit highly efficient vertical transmission (e.g., *R. conorii* (Socolovschi et al. 2009)), and fitness reductions would not be expected in these cases.

DISTRIBUTION OF *RICKETTSIA* WITHIN HOST TISSUES AND ITS RELATIONSHIP WITH SYMBIONT TRANSMISSION

Studies characterizing the distribution of *Rickettsia* within host tissues have revealed two trends: (1) *Rickettsia* are often abundant in tissues associated with vertical transmission, and (2) *Rickettsia* are among the most promiscuous inherited symbionts in terms of the diversity of tissues they can colonize. As we discuss below, this may also be related to the ability of many *Rickettsia* to move between cells.

Many studies have demonstrated the presence of *Rickettsia* in oocytes, using microscopy or molecular biology techniques (e.g., Yusuf and Turner 2004; Frati et al. 2006); this is, of course, a hallmark of transovarial transmission. Only a few studies, however, have followed the route by which *Rickettsia* penetrates the germline in order to be transmitted to the next generation. In the booklouse *L. bostrychophila*, *Rickettsia* was observed moving into the cytoplasm of early oocytes from the surrounding nurse cells, while the follicular epithelium cells did not contain the symbiont (Perotti et al. 2006). In contrast, in the ladybird beetle *Adalia bipunctata*, *Rickettsia* penetrate into the oocytes through the cytoplasm of follicular epithelium cells (Sokolova et al. 2002).

The *Rickettsia* infecting *B. tabaci* has a particularly interesting distribution (Gottlieb et al. 2006). *Bemisia tabaci* is a species complex composed of about 20 biotypes, and contains up to 7 symbionts (Gueguen et al. 2010). Symbiont content is generally biotype dependent, and some symbiont species are virtually never found together in the same individual. *Rickettsia* is one of the symbionts that is not biotype dependent, and has a unique localization. While most secondary symbionts are primarily confined within the bacteriocyte together with the primary symbiont *Portiera aleyridodarum*, or have limited hemolymph dispersal, *Rickettsia* are scattered throughout the whitefly body, excluding the bacteriome, with higher density in the follicle cells and along the gut. The only case in which *Rickettsia* are seen in the bacteriocyte is in young eggs, and it is assumed that inhabiting the bacteriocyte is a useful means for transovarial transmission (Gottlieb et al. 2006). In addition, a second *Rickettsia* phenotype that is strictly confined within the bacteriocytes has also been found in some populations (Gottlieb et al. 2008). These two phenotypes may exist in the same population. *Rickettsia* from both the confined and scattered phenotypes are transmitted to the oocyte within a bacteriocyte. Interestingly, the scattered phenotype *Rickettsia* appear to use the bacteriocyte only as a vector to hitchhike into the oocyte, and once there, the bacterium leaves the bacteriocyte and starts multiplying in the ooplasm (Gottlieb et al. 2006, 2008). It thus seems that although the routes taken by *Rickettsia* to ensure their vertical transmission vary from host to host, their ability to actively move between cells is a crucial component in all systems. Similarly, a number of other studies have documented *Rickettsia* scattered throughout the host hemolymph. In addition to oocytes (Yusuf and Turner 2004), *Rickettsia* in the booklouse *L. bostrychophila* infects Malpighian tubules, fat bodies, epithelial cells, and ganglia (Chapman 2003; Perotti et al. 2006). Massive infections were also found in the dytiscid beetle *Deronectes platynotus,* which is fixed for *Rickettsia* infection. All tissues that are in contact with the hemolymph are affected, with lower numbers in muscles. All developmental stages carry the symbiont, and females have a higher abundance of *Rickettsia* than males. Interestingly, *Rickettsia* infections occur at lower frequencies (~40%) in three related dytiscids species where bacterial densities within individual beetles also appear lower (Küchler et al. 2009). In the springtail *Onychiurus sinensis*, *Rickettsia* is found to massively infect the gonads of males and females, and as a result is suspected of interfering with reproduction (Frati et al. 2006).

Localization of *Rickettsia* symbionts in diverse host tissues may provide clues as to their potential role in host biology. For example, we might predict that reproductive manipulators are abundant in gonads, strains that protect their host against natural enemies occur in hemolymph, and symbionts involved in detoxification occur in Malpighian tubules.

RICKETTSIA MOVEMENT

Although *Rickettsia* are obligate intracellular bacteria, they frequently move inside and between host cells and tissues, as well as between host individuals. Within-cell *Rickettsia* movement was first observed by Schaechter et al. (1957). Using transmission electron microscopy, Munderloh et al. (1998) were able to show that in cell culture, *Rickettsia* infected new cells by inducing the formation of a phagocytic vacuole. Soon after invading a new tick cell, the bacteria were stripped (or escaped) from the host cell membrane and were then found in direct contact with host cell cytoplasm. The authors reported that the endoplasmic reticulum membrane was damaged during entry, and during exit the host membrane that was stretched around the bacteria was lost before reentry to the new cell (Munderloh et al. 1998).

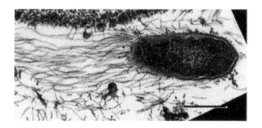

FIGURE 10.2 Transmission electron micrograph of myosin S1-decorated actin tails of *Rickettsia conorii* in epithelial-like human cell line (HEp2). Bar, 0.5 μm. (Reproduced with permission from Gouin et al., *Journal of Cell Science*, 112, 1697–1708, 1999.)

Once inside the new host cell, *Rickettsia* induce actin tail formation by manipulating host cytoskeletal proteins (summarized in Goldberg 2001; Gouin et al. 2005). Under normal circumstances, actin polymerization and thus cell motility are tightly regulated via signal transduction cascades, and the bacteria not only use the actin cytoskeleton to move around the host cell, but also have developed mechanisms to overcome and challenge the regulating factors of this process. The force generated by actin assembly propels the bacteria through the cytoplasm and into neighboring cells, promoting cell-to-cell spread (e.g., Teysseire et al. 1992; Heinzen et al. 1993). A detailed study of *R. conorii* showed that its intracellular cycle consists of engulfment, formation of a membrane-bound vacuole, actin polymerization with formation of an actin cloud, rearrangement of F-actin in tails, formation of a protrusion, and generation of a two-membrane vacuole, which is then lysed to allow release of the bacteria (Gouin et al. 1999; Figure 10.2). The organization of actin filaments in *Rickettsia* tails is unique compared to other motile intracellular pathogens, suggesting that *Rickettsia* utilize a different mechanism of actin tail assembly. The *Rickettsia* protein, RickA, was reported to play an essential role in actin tail formation and intracellular movement of pathogenic rickettsiae, including *R. conorii* and virulent and avirulent strains of *R. rickettsii, R. montanensis, R. parkeri, R. australis,* and *R. monacensis* (Teysseire et al. 1992; Heinzen et al. 1993; Gouin et al. 2004; Baldridge et al. 2005). However, the presence and *in vivo* expression of RickA was not enough to promote actin polymerization in the nonmotile *R. raoultii* (Balraj et al. 2008). Recently, an additional bacterial actin assembly gene, surface cell antigen 2 (*sca2*), was found to functionally mimic eukaryotic formin proteins to enable host-independent motility (Haglund et al. 2010). Compared with their virulent relatives, the genomes of avirulent strains of *Rickettsia* lack functional genes related to actin-based movement, suggesting that this ability determines virulence (Ellison et al. 2008, Felsheim et al. 2009). In addition, plasmids and repetitive insertion sequences may alter virulence (Felsheim et al. 2009). The ability of *Rickettsia* to move within arthropod tissues and reach salivary glands may also be dependent on actin-based movement, and it seems that this ability can be acquired or lost under specific, yet unknown, conditions that determine whether *Rickettsia* are symbionts or pathogens.

HORIZONTAL TRANSMISSION

Despite the progress made in the last decade, the mechanisms and routes of *Rickettsia* movement have been studied almost exclusively in cell lines, and there is very little information on actual movement within and between whole organisms. The possible bias of *in vitro* systems can be demonstrated by the report that the shape, length, and formation time of *R. conorii* actin tails vary depending on which specific cell line is used (Gouin et al. 1999). Obviously, the challenges facing an intracellular bacterium when it moves between hosts as far away as different kingdoms require sophisticated mechanisms for all processes involved. Empirical demonstration of natural horizontal transmission among different organisms is very scarce (Werren et al. 1986; Heath et al. 1999; Huigens et al. 2000; Varaldi et al. 2005), and the possible mechanisms behind it are still largely unknown. This supports the notion that horizontal transmission, especially interspecific, is a rare event, although cofeeding by several blood-feeder individuals on the same host is an accepted mechanism (e.g., Gern and Rais 1996). Successful inter- and intraspecific horizontal transmission of *Rickettsia* by microinjection has been reported by Chen and Purcell (1997). Hemolymph from *Rickettsia*-carrying *Acyrthosiphon pisum* pea aphids was injected to *Rickettsia*-free conspecifics as well as to *Rickettsia*-free blue alfalfa aphids, *Acyrthosiphon kondoi*. The newly injected *Rickettsia* were successfully established in ~75% of *A. pisum* and in 25% of *A. kondoi* recipient mothers, and were maintained for multiple generations by efficient maternal transmission. In contrast to the artificial method of microinjection, attempts to transfer *Rickettsia* from infected to uninfected aphids via a shared host plant were unsuccessful (Chen and Purcell 1997; Chen et al. 2000). It is not known why aphid *Rickettsia* is not transmitted via the host plant; this could be explained by either the failure of *Rickettsia* to colonize the aphids' salivary glands or the plant cells or its unsuccessful transmission out from the gut lumen to other tissues. Horizontal transmission of *Rickettsia* via a shared host plant has so far only been inferred from a study of a strain closely related to *R. bellii* that has been implicated in papaya bunchy top disease (Davis et al. 1998). In this study, *R. bellii*-like bacteria were detected in diseased papaya plants and in the leafhopper vector, but not in healthy plants. This is highly suggestive that a *Rickettsia* is indeed the causative agent of this disease, and that it is transmitted by the leafhopper. However, further experimentation is required to prove that this *Rickettsia* is transmitted from infected leafhoppers to healthy plants, followed by transmission from the newly infected plants to uninfected leafhoppers (ecological horizontal transmission), and then from those newly infected leafhoppers to their progeny (evolutionary horizontal transmission) (Davis et al. 1998). Some secondary symbionts may also be transmitted during mating (Moran and Dunbar 2006), but to the best of our knowledge, this mode of transmission has not been demonstrated in *Rickettsia*.

Another possible route for horizontal transmission is via predation or parasitism. The intimate interaction between hosts and their endoparasitoids would seem to provide opportunities for horizontal transmission of symbionts, as parasitoid larvae consume nothing but symbiont-contaminated food throughout their development. Yet, to our knowledge, there is no experimental evidence of permanent acquisition of arthropods' symbionts by their natural enemies. A recent study showed that when a *Rickettsia*-infected *B. tabaci* is parasitized by wasps of the genus *Eretmocerus*, *Rickettsia* invade the parasitoid larvae, persist in adults and in females, and reach the ovaries (Chiel et al. 2009). However, *Rickettsia* do not penetrate wasp oocytes, but instead are localized only in the follicular epithelial cells, and consequently are not transmitted vertically. Interestingly, in parasitoids of the genus *Encarsia*, *Rickettsia* were merely transient in the digestive tract and were excreted with the meconia before wasp pupation (Chiel et al. 2009). Additionally, adults of both parasitoid genera frequently acquired *Rickettsia* by host feeding, but the bacteria were excreted within a few days and did not become established. These findings suggest that symbionts may potentially acquire new hosts via host-parasitoid transmission. This study also shows that predators and parasitoids could be wrongly diagnosed by PCR as being hosts for *Rickettsia* and other bacteria, while in fact the symbionts are simply present in the gut along with the prey or host material.

Rickettsia is emerging as a bacterial genus that is capable of both colonizing diverse tissues and hosts and inducing a wide array of phenotypes. However, while its fascinating repertoire is gradually being revealed, focus should also remain on the many species that have pathogenic relationships with vertebrates. In fact, these two very different roles, along with the wide range of transmission modes, make *Rickettsia* especially intriguing, as they offer an unusual opportunity to answer questions about the origins, mechanisms, and transitions involved in manipulation of both arthropods and vertebrates.

ACKNOWLEDGMENTS

The authors thank Lucy Weinert and Frank Jiggins for help in producing Figure 10.1. This research was partially supported by the Israel Science Foundation (grants 262/09 to E.Z.-F. and 456/10 to Y.G.). S.P. acknowledges support from the Natural Sciences and Engineering Research Council of Canada (NSERC) and the Canadian Institute for Advanced Research (CIFAR).

REFERENCES

Baldridge G.D., Burkhardt N., Herron M.J., Kurtti T.J., Munderloh U.G. (2005). Analysis of fluorescent protein expression in transformants of *Rickettsia monacensis*, an obligate intracellular tick symbiont. *Appl. Environ. Microbiol.* 71:2095–2105.

Baldridge G.D., Burkhardt N.Y., Oliva A.S., Kurtti T.J., Munderloh U.G. (2010). Rickettsial *ompB* promoter regulated expression of GFPuv in transformed *Rickettsia montanensis*. *PLoS ONE* 5:e8965.

Balraj P., Karkouri K.E., Vestris G., Espinosa L., Raoult D., Renesto P. (2008). RickA expression is not sufficient to promote actin-based motility of *Rickettsia raoultii*. *PLoS ONE* 3:e2582.

Behar A., McCormick L.J., Perlman S.J. (2010). *Rickettsia felis* infection in a common household insect pest, *Liposcelis bostrychophila* (Psocoptera: Liposcelidae). *Appl. Environ. Microbiol.* 76:2280–2285.

Blanc G., Ogata H., Robert C., Audic S., Suhre K., Vestris G., Claverie J.-M., Raoult D. (2007). Reductive genome evolution from the mother of *Rickettsia*. *PLoS Genet* 3:e14.

Braig H.R., Turner B.D., Perotti M.A. (2008). Symbiotic *Rickettsia*. In K. Bourtzis, T.A. Miller (Eds.), *Insect symbiosis 3*. CRC Press, Boca Raton, FL, 221–249.

Breeuwer J.A.J., Werren J.H. (1990). Microorganisms associated with chromosome destruction and reproductive isolation between two insect species. *Nature* 346:558–560.

Burgdorfer W., Hayes, S.F., Mavros, A.J. (1981). Nonpathogenic rickettsiae in *Dermacentor andersoni*: a limiting factor for the distribution of *Rickettsia rickettsii*. In W. Burgdorfer, Anacker, R.L. (Eds.), *Rickettsiae and rickettsial diseases*. Academic Press, New York, 585–594.

Carmichael J.R., Fuerst P.A. (2006). A rickettsial mixed infection in a *Dermacentor variabilis* tick from Ohio. *Ann. NY Acad. Sci.* 1078:334–337.

Chapman G.B. (2003). Pharynx, esophagus, and associated structures in the booklouse, *Liposcelis divinatorius*. *Invert. Biol.* 122:52–60.

Chen D.-Q., Campbell B.C., Purcell A.H. (1996). A new *Rickettsia* from a herbivorous insect, the pea aphid *Acyrthosiphon pisum* (Harris). *Curr. Microbiol.* 33:123–128.

Chen D.-Q., Montllor C.B., Purcell A.H. (2000). Fitness effects of two facultative endosymbiotic bacteria on the pea aphid, *Acyrthosiphon pisum*, and the blue alfalfa aphid, *A. kondoi*. *Entomol. Exp. Appl.* 95:315–323.

Chen D.-Q., Purcell A.H. (1997). Occurrence and transmission of facultative endosymbionts in aphids. *Curr. Microbiol.* 34:220–225.

Chiel E., Zchori-Fein E., Inbar M., Gottlieb Y., Adachi-Hagimori T., Kelly S.E., Asplen M.K., Hunter M.S. (2009). Almost there: transmission routes of bacterial symbionts between trophic levels. *PLoS ONE* 4:e4767.

Davis M.J., Ying Z., Brunner B.R., Pantoja A., Ferwerda F.H. (1998). Rickettsial relative associated with papaya bunchy top disease. *Curr. Microbiol.* 36:80–84.

Duron O., Bouchon D., Boutin S., Bellamy L., Zhou L., Engelstadter J., Hurst G. (2008). The diversity of reproductive parasites among arthropods: *Wolbachia* do not walk alone. *BMC Biol.* 6:27.

Ellison D.W., Clark T.R., Sturdevant D.E., Virtaneva K., Porcella S.F., Hackstadt T. (2008). Genomic comparison of virulent *Rickettsia rickettsii* Sheila Smith and avirulent *Rickettsia rickettsii* Iowa. *Infect. Immun.* 76:542–550.

Engelstädter J., Hurst G.D.D. (2009). The ecology and evolution of microbes that manipulate host reproduction. *Ann. Rev. Ecol. Evol. Syst.* 40:127–149.

Eremeeva M.E., Warashina W.R., Sturgeon M.M., Buchholz A.E., Olmsted G.K., Park S.Y., Effler P.V., Karpathy S.E. (2008). *Rickettsia typhi* and *R. felis* in rat fleas (*Xenopsylla cheopis*), Oahu, Hawaii. *Emerg. Infect. Dis.* 14:1613–1615.

Felsheim R.F., Kurtti T.J., Munderloh U.G. (2009). Genome sequence of the endosymbiont *Rickettsia peacockii* and comparison with virulent *Rickettsia rickettsii*: identification of virulence factors. *PLoS ONE* 4:e8361.

Frati F., Negri I., Fanciulli P.P., Pellecchia M., Dallai R. (2006). Ultrastructural and molecular identification of a new *Rickettsia* endosymbiont in the springtail *Onychiurus sinensis* (Hexapoda, Collembola). *J. Invert. Pathol.* 93:150–156.

Gern L., Rais O. (1996). Efficient transmission of *Borrelia burgdorferi* between cofeeding *Ixodes ricinus* ticks (Acari: Ixodidae). *J. Med. Entomol.* 33:189–192.

Giorgini M., Bernardo U., Monti M.M., Nappo A.G., Gebiola M. (2010). *Rickettsia* symbionts cause parthenogenetic reproduction in the parasitoid wasp *Pnigalio soemius* (Hymenoptera: Eulophidae). *Appl. Environ. Microbiol.* 76:2589–2599.

Goldberg M.B. (2001). Actin-based motility of intracellular microbial pathogens. *Microbiol. Mol. Biol. Rev.* 65:595–626.

Goodacre S.L., Martin O.Y., Thomas C.F.G., Hewitt G.M. (2006). *Wolbachia* and other endosymbiont infections in spiders. *Mol. Ecol.* 15:517–527.

Gottlieb Y., Ghanim M., Chiel E., Gerling D., Portnoy V., Steinberg S., Tzuri G., Horowitz A.R., Belausov E., Mozes-Daube N., Kontsedalov S., Gershon M., Gal S., Katzir N., Zchori-Fein E. (2006). Identification and localization of a *Rickettsia* sp. in *Bemisia tabaci* (Homoptera: Aleyrodidae). *Appl. Environ. Microbiol.* 72:3646–3652.

Gottlieb Y., Ghanim M., Gueguen G., Kontsedalov S., Vavre F., Fleury F., Zchori-Fein E. (2008). Inherited intracellular ecosystem: symbiotic bacteria share bacteriocytes in whiteflies. *FASEB J* 22:2591–2599.

Gouin E., Egile C., Dehoux P., Villiers V., Adams J., Gertler F., Li R., Cossart P. (2004). The RickA protein of *Rickettsia conorii* activates the Arp2/3 complex. *Nature* 427:457–461.

Gouin E., Gantelet H., Egile C., Lasa I., Ohayon H., Villiers V., Gounon P., Sansonetti P., Cossart P. (1999). A comparative study of the actin-based motilities of the pathogenic bacteria *Listeria monocytogenes*, *Shigella flexneri* and *Rickettsia conorii*. *J. Cell Sci.* 112:1697–1708.

Gouin E., Welch M.D., Cossart P. (2005). Actin-based motility of intracellular pathogens. *Curr. Opin. Microbiol.* 8:35–45.

Gueguen G., Vavre F., Gnankine O., Peterschmitt M., Charif D., Chiel E., Gottlieb Y., Ghanim M., Zchori-Fein E., Fleury F. (2010). Endosymbiont metacommunities, mtDNA diversity and the evolution of the *Bemisia tabaci* (Hemiptera: Aleyrodidae) species complex. *Mol. Ecol.* 19:4365–4376.

Hagimori T., Abe Y., Date S., Miura K. (2006). The first finding of a *Rickettsia* bacterium associated with parthenogenesis induction among insects. *Curr. Microbiol.* 52:97–101.

Haglund C.M., Choe J.E., Skau C.L, Kovar D.R., Welch M.D. (2010). *Rickettsia* Sca2 is a bacterial formin-like mediator of actin-based motility. *Nature Cell Biol.* 12:1057–1063.

Hayes S.F., Burgdorfer W. (1979). Ultrastructure of *Rickettsia rhipicephali*, a new member of the spotted fever group rickettsiae in tissues of the host vector *Rhipicephalus sanguineus*. *J. Bacteriol.* 137:605–613.

Heath B.D., Butcher R.D.J., Whitfield W.G.F., Hubbard S.F. (1999). Horizontal transfer of *Wolbachia* between phylogenetically distant insect species by a naturally occurring mechanism. *Curr. Biol.* 9:313–316.

Heinzen R.A., Hayes S.F., Peacock M.G., Hackstadt T. (1993). Directional actin polymerization associated with spotted fever group *Rickettsia* infection of Vero cells. *Infect. Immun.* 61:1926–1935.

Hektoen L. (1910). Howard Taylor Ricketts. *Science* 32:585–587.

Himler A.G., Adachi-Hagimori T., Bergen J.E., Kozuch A., Kelly S.E., Tabashnick B.G., Chiel E., Duckworth V.E., Dennehy T.J., Zchori-Fein E., Hunter M.S. (2011). Rapid spread of a bacterial symbiont in an invasive whitefly is driven by fitness benefits and female bias. *Science* 332:254–256.

Hornok S., Hofmann-Lehmann R., Fernández de Mera I.G., Meli M.L., Elek V., Hajtós I., Répási A., Gönczi E., Tánczos B., Farkas R., Lutz H., de la Fuente J. (2010). Survey on blood-sucking lice (Phthiraptera: Anoplura) of ruminants and pigs with molecular detection of *Anaplasma* and *Rickettsia* spp. *Vet. Parasitol.* 174:355–358.

Houhamdi L., Fournier P.-E., Fang R., Lepidi H., Raoult D. (2002). An experimental model of human body louse infection with *Rickettsia prowazekii*. *J. Infect. Dis.* 186:1639–1646.

Houhamdi L., Fournier P.-E., Fang R., Raoult D. (2003). An experimental model of human body louse infection with *Rickettsia typhi*. *Ann. NY Acad. Sci.* 990:617–627.

Houhamdi L., Raoult D. (2006). Experimentally infected human body lice (*Pediculus humanus humanus*) as vectors of *Rickettsia rickettsii* and *Rickettsia conorii* in a rabbit model. *Am. J. Trop. Med. Hyg.* 74:521–525.

Huigens M.E., Luck R.F., Klaassen R.H.G., Maas M.F.P.M., Timmermans M.J.T.N., Stouthamer R. (2000). Infectious parthenogenesis. *Nature* 405:178–179.

Hunter M.S., Perlman S.J., Kelly S.E. (2003). A bacterial symbiont in the Bacteroidetes induces cytoplasmic incompatibility in the parasitoid wasp *Encarsia pergandiella*. *Proc. Royal Soc. London B* 270:2185–2190.

Kikuchi Y., Fukatsu T. (2005). *Rickettsia* infection in natural leech populations. *Micro. Ecol.* 49:265–271.

Kontsedalov S., Zchori-Fein E., Chiel E., Gottlieb Y., Inbar M., Ghanim M. (2008). The presence of *Rickettsia* is associated with increased susceptibility of *Bemisia tabaci* (Homoptera: Aleyrodidae) to insecticides. *Pest Manag. Sci.* 64:789–792.

Küchler S.M., Kehl S., Dettner K. (2009). Characterization and localization of *Rickettsia* sp. in water beetles of genus *Deronectes* (Coleoptera: Dytiscidae). *FEMS Microbiol. Ecol.* 68:201–211.

Labruna M.B., Whitworth T., Bouyer D.H., McBride J., Camargo L.M.A., Camargo E.P., Popov V., Walker D.H. (2004). *Rickettsia bellii* and *Rickettsia amblyommii* in *Amblyomma* ticks from the state of Rondonia, Western Amazon, Brazil. *J. Med. Entomol.* 41:1073–1081.

Lawson E.T., Mousseau T.A., Klaper R., Hunter M.D., Werren J.H. (2001). *Rickettsia* associated with male-killing in a buprestid beetle. *Heredity* 86:497–505.

Loftis A.D., Reeves W.K., Szumlas D.E., Abbassy M.M., Helmy I.M., Moriarity J.R., Dasch G.A. (2006). Surveillance of Egyptian fleas for agents of public health significance: *Anaplasma, Bartonella, Coxiella, Ehrlichia, Rickettsia,* and *Yersinia pestis. Am. J. Trop. Med. Hyg.* 75:41–48.

Macaluso K.R., Sonenshine D.E., Ceraul S.M., Azad A.F. (2002). Rickettsial infection in *Dermacentor variabilis* (Acari: Ixodidae) inhibits transovarial transmission of a second *Rickettsia. J. Med. Entomol.* 39:809–813.

Majerus M., Hurst G. (1997). Ladybirds as a model system for the study of male-killing symbionts. *BioControl* 42:13–20.

Majerus T.M.O., Majerus M.E.N. (2010). Intergenomic arms races: detection of a nuclear rescue gene of male-killing in a ladybird. *PLoS Pathog.* 6:e1000987.

Merhej V., Raoult D. (2010). Rickettsial evolution in the light of comparative genomics. *Biol. Rev.* August 17.

Moran N.A., Dunbar H.E. (2006). Sexual acquisition of beneficial symbionts in aphids. *Proc. Natl. Acad. Sci. USA* 103:12803–12806.

Munderloh U.G., Hayes S.F., Cummings J., Kurtti T.J. (1998). Microscopy of spotted fever *Rickettsia* movement through tick cells. *Micros. Microanal.* 4:115–121.

Niebylski M.L., Peacock M.G., Schwan T.G. (1999). Lethal effect of *Rickettsia rickettsii* on its tick vector (*Dermacentor andersoni*). *Appl. Environ. Microbiol.* 65:773–778.

O'Neill S.L., Giordano R., Colbert A.M., Karr T.L., Robertson H.M. (1992). 16S rRNA phylogenetic analysis of the bacterial endosymbionts associated with cytoplasmic incompatibility in insects. *Proc. Natl. Acad. Sci. USA* 89:2699–2702.

Ogata H., La Scola B., Audic S., Renesto P., Blanc G., Robert C., Fournier P.-E., Claverie J.-M., Raoult D. (2006). Genome sequence of *Rickettsia bellii* illuminates the role of amoebae in gene exchanges between intracellular pathogens. *PLoS Genet* 2:e76.

Parola P. (2006). Rickettsioses in sub-Saharan Africa. *Ann. NY Acad. Sci.* 1078:42–47.

Perlman S.J., Hunter M.S., Zchori-Fein E. (2006). The emerging diversity of *Rickettsia. Proc. Royal Soc. B* 273:2097–2106.

Perotti M.A., Clarke H.K., Turner B.D., Braig H.R. (2006). *Rickettsia* as obligate and mycetomic bacteria. *FASEB J.* 20:2372–2374.

Rocha Lima H. (1951). *Rickettsia prowazeki*; its discovery and characterization constituting a new group of microorganisms. *Rev. Bras. Med.* 8:311–20.

Russell J.A., Goldman-Huertas B., Moreau C.S., Baldo L., Stahlhut J.K., Werren J.H., Pierce N.E. (2009). Specialization and geographic isolation among *Wolbachia* symbionts from ants and lycaenid butterflies. *Evolution* 63:624–640.

Sabatini G.S., Pinter A., Nieri-bastos F.A., Marcili A., Labruna M.B. (2010). Survey of ticks (Acari: Ixodidae) and their *Rickettsia* in an Atlantic rain forest reserve in the state of Sao Paulo, Brazil. *J. Med. Entomol.* 47:913–916.

Sakurai M., Koga R., Tsuchida T., Meng X.-Y., Fukatsu T. (2005). *Rickettsia* symbiont in the pea aphid *Acyrthosiphon pisum*: novel cellular tropism, effect on host fitness, and interaction with the essential symbiont *Buchnera. Appl. Environ. Microbiol.* 71:4069–4075.

Schaechter M., Bozeman F.M., Smadel J.E. (1957). Study on the growth of rickettsiae. II. Morphologic observations of living rickettsiae in tissue culture cells. *Virology* 3:160–172.

Simon J.-C., Carré S., Boutin M., Prunier-Leterme N., Sabater-Muñoz B., Latorre A., Bournoville R. (2003). Host-based divergence in populations of the pea aphid: insights from nuclear markers and the prevalence of facultative symbionts. *Proc. R. Soc. London B* 270:1703–1712.

Socolovschi C., Bitam I., Raoult D., Parola P. (2009). Transmission of *Rickettsia conorii conorii* in naturally infected *Rhipicephalus sanguineus*. *Clin. Microbiol. Infect.* 15:319–321.

Sokolova M.I., Zinkevich N.S., Zakharov I.A. (2002). Bacteria in ovarioles of females from maleless families of ladybird beetles *Adalia bipunctata* L. (Coleoptera: Coccinellidae) naturally infected with *Rickettsia*, *Wolbachia*, and *Spiroplasma*. *J. Invert. Pathol.* 79:72–79.

Stouthamer R., Breeuwer J.A.J., Hurst G.D.D. (1999). *Wolbachia pipientis*: microbial manipulator of arthropod reproduction. *Ann. Rev. Microbiol.* 53:71–102.

Teysseire N., Chiche-Portiche C., Raoult D. (1992). Intracellular movements of *Rickettsia conorii* and *R. typhi* based on actin polymerization. *Res. Microbiol.* 143:821–829.

Tinsley M., Majerus M. (2007). Small steps or giant leaps for male-killers? Phylogenetic constraints to male-killer host shifts. *BMC Evol. Biol.* 7:238.

Tsuchida T., Koga R., Shibao H., Matsumoto T., Fukatsu T. (2002). Diversity and geographic distribution of secondary endosymbiotic bacteria in natural populations of the pea aphid, *Acyrthosiphon pisum*. *Mol. Ecol.* 11:2123–2135.

Varaldi J., Fouillet P., Boulétreau M., Fleury F. (2005). Superparasitism acceptance and patch-leaving mechanisms in parasitoids: a comparison between two sympatric wasps. *Animal Behav.* 69:1227–1234.

Wedincamp J.J., Foil L.D. (2002). Vertical transmission of *Rickettsia felis* in the cat flea (*Ctenocephalides felis* Bouche). *J. Vector Ecol.* 27:96–101.

Weinert L., Welch J.J., Jiggins F.M. (2009b). Conjugation genes are common throughout the genus *Rickettsia* and are transmitted horizontally. *Proc. R. Soc. London B* 276:3619–3627.

Weinert L., Werren J., Aebi A., Stone G., Jiggins F.M. (2009a). Evolution and diversity of *Rickettsia* bacteria. *BMC Biol.* 7:6.

Werren J.H., Baldo L., Clark M.E. (2008). *Wolbachia*: master manipulators of invertebrate biology. *Nature Rev. Microbiol.* 6:741–751.

Werren J.H., Hurst G.D., Zhang W., Breeuwer J.A., Stouthamer R., Majerus M.E. (1994). Rickettsial relative associated with male killing in the ladybird beetle (*Adalia bipunctata*). *J. Bacteriol.* 176:388–394.

Werren J.H., O'Neill S.L. (1997). The evolution of heritible symbiont. In S.L. O'Neill, A.A. Hoffmann, J.H. Werren (Eds.), *Influential passengers: inherited microorganisms and arthropod reproduction.* pp. 1–41. Oxford University Press, New York.

Werren J.H., Skinner S.W., Huger A.M. (1986). Male-killing bacteria in a parasitic wasp. *Science* 231:990–992.

Werren J.H., Windsor D., Guo L. (1995). Distribution of *Wolbachia* among neotropical arthropods. *Proc. R. Soc. London B* 262:197–204.

Yusuf M., Turner B. (2004). Characterisation of *Wolbachia*-like bacteria isolated from the parthenogenetic stored-product pest psocid *Liposcelis bostrychophila* (Badonnel) (Psocoptera). *J. Stored Products Res.* 40:207–225.

Yusuf M., Turner B., Whitfield P., Miles R., Pacey J. (2000). Electron microscopical evidence of a vertically transmitted *Wolbachia*-like parasite in the parthenogenetic, stored-product pest *Liposcelis bostrychophila* Badonnel (Psocoptera). *J. Stored Products Res.* 36:169–175.

Zchori-Fein E., Perlman S.J. (2004). Distribution of the bacterial symbiont *Cardinium* in arthropods. *Mol. Ecol.* 13:2009–2016.

11 Cardinium
The Next Addition to the Family of Reproductive Parasites

Hans Breeuwer, Vera I.D. Ros, and Tom V.M. Groot

CONTENTS

Genus	*Cardinium*
Species	*hertigii*
Family	Flexibacteraceae
Order	Sphingobacteriales
Phylum	Bacteroidetes
Description year	2004
Origin of name	Species name is after Marshall Hertig, an American pathologist. He and Dr. Wolbach discovered in 1924, intracellular bacteria in *Culex pipiens* and described them as *Rickettsia*-like organisms. In 1936 Hertig named the genus *Wolbachia*.
Description	
Reference	Zchori-Fein et al. (2004)
Anecdote	The bacterium genus comes from the brush-like parallel structures observed within the bacteria under electron microscope. These structures reminded one of the authors of the columns that flank the main axis of a Roman town.

INTRODUCTION

Cardinium bacteria are a recently discovered group that form a new monophyletic bacterial taxon within the phylum Bacteroidetes. They have adopted a symbiotic lifestyle and reside in the cytoplasm of their eukaryotic host. Many of them have developed the fascinating capability to manipulate host reproduction in various ways.

Much research on reproductive parasites over the last 25 years has focused on *Wolbachia*. The reason for this focus was that *Wolbachia* is widespread among arthropods, and *Wolbachia*'s capability to manipulate host reproduction in a variety of ways to increase the production or survival of infected female hosts. Specific primers for *Wolbachia* added to its popularity. It was long believed that *Wolbachia* was the only major reproductive parasite, but recent findings offset this presumption. The phylogenetic diversity of male-killing symbionts was a first sign that there might be other reproductive manipulators out there (Majerus 2006). It is now well established that certain groups of bacteria, such as the *Rickettsia* (Weinert et al. 2009a), *Spiroplasma* (Clark 1982), and *Arsenophonus* (Nováková et al. 2009), are also widespread in arthropods. The phenotypic effects of these bacteria on their arthropod hosts are largely unexplored, with some exceptions (male-killing *Rickettsia* and *Arsenophonus*, *Rickettsia* inducing parthenogenesis (Giorgini et al. 2010); and *Anaplasma* bacteria that enhance cold resistance of ticks (Neelakanta et al. 2010)). Nevertheless, they clearly suggest that there are more microorganisms that can manipulate arthropods and other hosts. The discovery of *Cardinium* bacteria is very exciting because it is unrelated to *Wolbachia*, but shows remarkable congruence in phenotype. Already, *Cardinium* has been found to induce three of the reproductive manipulations that are known for *Wolbachia* in a variety of arthropod hosts (see Weeks and Breeuwer 2002). This chapter is an update of our current knowledge of *Cardinium* endosymbionts.

INCIDENCE OF *CARDINIUM*

Since its discovery, *Cardinium* has been found in several hexapod orders (Table 11.1), Hymenoptera, Hemiptera (including the Homoptera), and Diptera, and a number of Chelicerate orders, including Araneae (spiders), Ixodida (ticks), Mesostigmata (e.g., predatory mites), and Prostigmata (e.g., spider mites) (Chang et al. 2010).

The incidence of inherited bacteria is routinely determined by polymerase chain reaction (PCR) screens using more or less specific primer pairs. The 16S rDNA gene is the most commonly used in bacterial studies. For *Cardinium* a few primer sets are routinely used (Table 11.2). The most commonly used for the detection of *Cardinium* are the CLO and Ch primer sets.

PCR-based screening for *Cardinium* or other inherited bacteria is prone to a number of pitfalls. Specificity of primers is a mixed blessing. It avoids amplification of contaminating bacteria that could result in false positives. On the other hand, infection with *Cardinium* or close relatives that have slightly different sequences than are targeted by primers will result in false negatives because the target amplicon fails to amplify. Variation in amplicon concentrations due to variation in bacterial density and composition among host individuals or variation in PCR conditions within and between laboratories creates additional uncertainty about the infection status of host

TABLE 11.1
Incidence of *Cardinium*

Taxon	Reference
Coleoptera	Zchori-Fein et al. 2004
Diptera	Zchori-Fein et al. 2004
	Nakamura et al. 2009
Hemiptera/Homoptera	Zchori-Fein et al. 2004
	Gruwell et al. 2009
	Nakamura et al. 2009
Hymenoptera	Zchori-Fein et al. 2004
	Jeong et al. 2009
Acari	Zchori-Fein et al. 2004
	Liu et al. 2006
	Martin and Goodacre 2009
	Nakamura et al. 2009
	De Luna et al. 2009
	Ros and Breeuwer 2009
Araneae	Martin and Goodacre 2009
	Duron et al. 2008b
	Perlman et al. 2010
Scorpiones	Martin and Goodacre 2009
Opiliiones	Martin and Goodacre 2009
	Duron et al. 2008b
	Chang et al. 2010
Nematoda	Noel and Atibalentja 2006

TABLE 11.2

Primer Sets for *Cardinium*

	Sequence	Size	Reference
16S rDNA			
CLO-F	GCG GTG TAA AATGAG CGT G	500	Weeks et al. 2003
			Zchori-Fein et al. 2004
CLO-R	ACC TMT TCT TAA CTC AAG CCT		
Ch-F	TAC TGT AAG AAT AAG CAC CGG C	500	Zchori-Fein and
			Perlman 2004
Ch-R	GTG GAT AC TTA ACG CTT TCG		
Car-sp-F	CGG CTT ATT AAG TCA GTT GTG AAA TCC TAG	544	Nakamura et al. 2009
Car-2p-R	TCC TTC CTC CCG CTT ACA CG		
CLO-f1	GGA ACC TTA CCT GGG CTA GAA TGT ATT	468	Gotoh et al. 2007a
CLO-r1	GCC ACT GTC TTC AAG CTC TAC CAA C		
Gyrase B			
F	GTT ACC GTA TAC CGA AAT GG	700	Groot and Breeuwer
			2006
R	TGC TTT CCG RGC MGC TTG		
Car gyrB2F	GGK GTY TCB TGT GTA AAT GC	1267	Nakamura et al. 2009
Car gyrB2R	TAS TGY TCT TCT TTR TCT CG		
Car gyrB1F	CAA AGA YAC CTA TAA RAT TTC TG	1389	Nakamura et al. 2009
Car gyrB1R	GTA ACG TTG TAC ARA KAC RGC AT		

samples. One way to circumvent this problem is to use different primer sets for the same gene. Nakamura et al. (2009) reported that different primer pairs gave different positive and negative results in the same samples. They showed, for example, that bacteria from biting midges had substitutions at the CLO-forward primer site. Consequently, estimates of infection frequency and incidence may be lower than in reality, and novel bacteria remain undetected if one relies on specific primer sets (e.g., Weinert et al. 2007). More systematic screening for *Cardinium* in arthropod species is required for the development of appropriate primer sets.

It is also clear that many hosts house more than one kind of bacterium, and that the composition of symbionts may change in time and space (e.g., Chang and Musgrave 1972; Chiel et al. 2007, 2009; Skaljac et al. 2010; Gottlieb et al. 2008; Sacchi et al. 2008). These bacteria can be closely related and represent different haplotypes, as is frequently observed in *Wolbachia* (e.g., Bordenstein and Werren 2007) and possibly in *Cardinium* (Groot 2006). They could also be distantly related, as, for example, *Wolbachia* and *Cardinium* double infections have been observed in *Bryobia* mites and *Encarsia* wasps (Ros and Breeuwer 2009; White et al. 2009; Perlman et al. 2006). This may lead to the wrong conclusion that symbionts are not associated with the host traits under study. Therefore, screens for endosymbionts should not focus on single groups of bacteria and should make use of combinations of primers targeted at various known symbiont groups, and we should keep our eyes open for the possibility of yet unknown bacteria.

Another problem is introduced by the sampling strategy. Infection frequencies may not be fixed, and low sampling intensity combined with rarity of the infection will also result in false negatives (Duron et al. 2008a, 2008b; Weinert et al. 2007; Jiggins et al. 2001). In particular, nonessential symbionts such as those that cause sex ratio distortion or cytoplasmic incompatibility may have more variable prevalence at geographic and temporal scales of their host (Weeks et al. 2002; Jiggins et al. 2001; Weinert et al. 2007).

Nevertheless, a number of interesting patterns are emerging in PCR screens of arthropods for *Wolbachia*, *Cardinium*, and other known reproductive manipulators: *Cardinium* incidence is estimated between 4 and 7% (Zchori-Fein and Perlman 2004; Weeks et al. 2003; Duron et al. 2008b); the incidence of *Cardinium* in arachnids ranges between 22 and 33% and is much higher than in insects (Duron et al. 2008b; Martin and Goodacre 2009; Chang et al. 2010; Perlman et al. 2010; Weeks et al. 2003); and double infections of *Cardinium* with *Wolbachia* and other bacteria have been documented (Ros and Breeuwer 2009; Duron et al. 2008; Nakamura et al. 2009; Perlman et al. 2010; White et al. 2009). In insects sampled so far, *Cardinium* infection seems to be restricted to a limited number of insect orders with a more restricted host range than *Wolbachia*. It should be pointed out that sampling has very much focused on insects and arachnids. Sampling of other arthropods, such as crustaceans, or outside the arthropods has been very limited. Recently, sequence analysis established that bacteria of a number of plant pathogenic nematodes are probably *Cardinium* or at least closely related (Atibalentja and Noel 2008; Nakamura et al. 2009).

It is now well established that *Cardinium* is widespread in arthropods and appears more abundant in arachnids than in insects. The next step is to determine the phenotypic effects and evolutionary consequences of *Cardinium* on its hosts.

PHENOTYPE AND MECHANISM OF MANIPULATION

In the majority of host species reported to be infected with *Cardinium*, phenotypic effects have not been established. Typically, correlations between host sex ratio or mode of reproduction and *Cardinium* incidence are mentioned, which may suggest that all female populations or parthenogenetic reproduction may be caused by the *Cardinium* infection. Other possible phenotypes, such as cytoplasmic incompatibility, male killing, or host fitness effects are less obvious and remain unknown in studies that focus on the incidence and phylogeny of *Cardinium*. At present, the full suite of phenotypic effects of *Cardinium* is largely unexplored.

Antibiotic studies comparing infected and cured hosts show that *Cardinium* has the same suite of phenotypic effects on its host as *Wolbachia*: parthenogenesis in the parasitoid *Encarsia pergandiella* (Zchori-Fein et al. 2001), feminization in *E. hispida* (Giorgini et al. 2009), false spider mite species of the genus *Brevipalpus* (Weeks et al. 2001; Groot and Breeuwer 2006), and cytoplasmic incompatibility (CI) in several mite species, *Eotetranychus suginamensis* (Gotoh et al. 2007a), *Bryobia sarothamni* (Ros and Breeuwer 2009), and a sexual population of *E. pergandiella* (Hunter et al. 2003; Perlman et al. 2008).

Fitness effects such as enhanced fecundity have been determined in the predatory mite *Metaseiulus occidentalis* (Weeks and Stouthamer 2004). Harris et al. (2010) suggest that the CI-inducing *Cardinium* in *E. pergandiella* may have a cryptic fitness benefit since CI-inducing *Cardinium* were able to spread in population cage experiments even when their initial frequency was below the predicted invasion threshold frequency. Finally, in a sample of the two-spotted spider mite *Tetranychus urticae* that was doubly infected with *Cardinium* and *Wolbachia*, neither bacterium seemed to have a phenotype; cross compatibility, egg production, and hatchability were not affected (Gotoh et al. 2007b).

Surprisingly, the phenotypic effect of *Cardinium* in the first host it was discovered in, the tick *Ixodus scapularis* (Kurtti et al. 1996), and its incidence in the field are still unknown.

Possibly, *Cardinium* plays a role in the nutrition of the tick to compensate for the nutritional imbalance of blood (see Douglas 2009) or enhance host fitness under certain environmental conditions (see Neelakanta et al. 2010). On the other hand, *Cardinium* could be parasitic; alpha-proteobacterial symbionts have been described in the tick *I. ricinus* that invade and destroy mitochondria in ovarian cells (Sassera et al. 2007).

In a number of cases, hosts have been found that are infected with both *Cardinium* and *Wolbachia* (Weeks et al. 2003; Zchori-Fein and Perlman 2004; Gotoh et al. 2007b; Enigl and Schausberger 2007; Duron et al. 2008; White et al. 2009; Ros and Breeuwer 2009). These systems provide opportunities to study interactions among symbionts. Differential curing will enable us to determine the effects of either symbiont. The parasitoid wasp *Encarsia inaron* is naturally infected with *Cardinium* and *Wolbachia* (White et al. 2009), and doubly infected wasps show the CI phenotype (Perlman et al. 2006). Differentially curing *E. inaron* of its endosymbionts followed by crossing hosts with different infection status showed that *Wolbachia* was responsible for the CI phenotype, whereas *Cardinium* was not and its role remains unknown (White et al. 2009).

The crossing patterns in the doubly infected spider mite *Bryobia sarothamni* were quite different (Ros and Breeuwer 2009). Cytoplasmic incompatibility was absent in doubly infected mites. However, in singly infected mites *Cardinium* induced strong CI, reducing egg hatch rate by 60% in the incompatible cross, whereas *Wolbachia* did not affect hatchability. This suggests that *Wolbachia* interferes with the expression of the CI phenotype of *Cardinium*. Interactions between *Wolbachia* and *Cardinium* were further investigated in a series of crosses involving doubly infected mites, and crosses between differentially infected mites (Ros and Breeuwer 2009). All incompatible crosses involved males that were singly infected with *Cardinium* and females infected with *Wolbachia*, together or not with *Cardinium*, or uninfected females. The latter combination is the typical incompatible cross. Clearly, *Wolbachia* influences the expression of CI phenotype of *Cardinium*. In doubly infected males *Cardinium* no longer induces CI, and in doubly infected females *Cardinium* is no longer capable of restoring compatibility with sperm from males singly infected with *Cardinium*.

The current proposed model for CI induction of *Wolbachia* is the modification-rescue model (Hoffmann and Turelli 1997; Werren 1997). Sperm is modified in the testes by the male symbiont, and as a result, the paternal chromosomes behave abnormally during early mitotic divisions in the fertilized egg. This modification is "rescued" if the appropriate symbiont strain is present in the egg. Only when males and females are infected with the same symbiont strains with the same modification-rescue system will fertilization be successful. Presuming that CI induction by *Cardinium* follows the same modification-rescue model, Ros and Breeuwer (2009) suggest that *Wolbachia* prevents both the *Cardinium* modification of sperm in males and the rescuing effect of *Cardinium* in doubly infected mite eggs. One mechanistic explanation they offer is that *Cardinium* densities are reduced in cells of doubly infected hosts due to competition and fall below a threshold that is needed for successful modification or rescue.

It may be difficult to unravel the phenotypic effects of symbionts in hosts that house multiple infections. In particular, if the host depends upon symbionts that are essential to its survival and reproduction, such as in the Hemiptera, standard antibiotic treatment of infected hosts and subsequent observations on changes in host phenotype will not be possible, since it will also affect the primary symbionts and their function in the host. In addition, differential curing may not render singly infected hosts, and it will be impossible to obtain uninfected hosts.

Cytological studies using fluorescent *in situ* hybridization (FISH) can provide important insights in the role of *Cardinium* when antibiotic treatments fail (Chiel et al. 2009b). In *Wolbachia* this approach has yielded insights into the transmission during oogenesis, its distribution in spermatogenesis, and its association with cellular structures during development, in particular of the mitotic and meiotic apparatus (Serbus et al. 2008). Also, additional sampling in the field may yield populations with different or no infections that can be used for comparison of reproductive mode, sex ratio, and crossing experiments (e.g., Ros and Breeuwer 2009; Weinert et al. 2007).

The phenotypic effects of *Cardinium* show remarkable similarities with those of *Wolbachia*. They show the same diversity in phenotypic effects; close relatives within each group can have entirely different phenotypic effects on their host, and phenotypic effects of infection in closely related hosts can also be very different. This may suggest that *Cardinium* and *Wolbachia* may have similar genes that cause these effects in their hosts, and that they may target the same system in their host.

MORPHOLOGY AND LOCALIZATION WITHIN THE HOST

Cardinium has a curious and distinctive ultrastructure that is characterized by a two-layer envelope and an array of microfilament structures attached to the inner membrane (Figure 11.1), the function of which is unknown (Chang and Musgrave 1972; Shepherd et al. 1973; Endo 1979; Hess and Hoy 1982; Costa et al. 1995; Kurtti et al. 1996; Zchori-Fein et al. 2001; Bigliardi et al. 2006; Kitajima et al. 2007). This structure was first observed in bacteria in cells of a leafhopper *Helochara communis* (Hemiptera: Cicadellidae) and plant nematode species, and later in the predatory mite *Phytoseiulus occidentalis* and in two biotypes of the whitefly *Bemisia tabaci*. It was this unusual structure that led to the discovery by Zchori-Fein et al. (2001) that ticks and wasps harbor the same bacteria, and subsequent confirmation of their phylogenetic relatedness based on 16S sequence information that led to the proposal of a new genus and species: *"Candidatus* Cardinium hertigii" (Zchori-Fein et al.

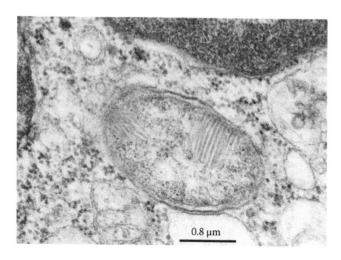

FIGURE 11.1   EM photograph of *Cardinium* bacteria in *Scaphoideus titanus* showing the microtubule-like structures and double membrane. (Photo kindly provided by Luciano Sacchi, University of Pavia.)

2004). Sequence information later confirmed the link between ultrastructural observations and *Cardinium* in the same species in the case of *B. tabaci*, *M. occidentalis*, or other species in the same group; the cicadellid leafhopper *Scaphoideus titanus* (Marzorati et al. 2006, 2008); and plant-parasitic nematodes (Shepherd et al. 1973; Noel and Atibalentja 2006). This distinctive microfilament structure seems to be absent in the closest known relative of *Cardinium*, *Amoebophilus* endosymbionts that infect *Acanthamoebe* sp. (Horn et al. 2001; Schmitz-Esser et al. 2008). Instead, *Amoebophilus* are surrounded by a host-derived membrane that is packed with ribosomes (Horn et al. 2001; Schmitz-Esser et al. 2008). Thus, this microfilament structure seems to be diagnostic for *Cardinium* at the ultrastructural level.

Another feature of *Cardinium* is that it is surrounded by two membranes and appears to be immersed directly in the cytoplasm of infected cells (Bigliardi et al. 2006; Costa et al. 1995; Kitajima et al. 2007; Sacchi et al. 2008). This is different from *Wolbachia*, which is located within a vacuole, separated from the cytoplasm by a membrane (Louis and Nigro 1989). *Wolbachia* cells are typically spherical, surrounded by a membrane-bound vacuole, and can be observed as individual organisms or in small groups of bacteria. The significance of this difference is unclear, but may play a role in transmission and localization in host tissue and infection dynamics.

Cytological studies show that *Cardinium* is predominantly localized in the reproductive tissues and mainly inside the follicle and nurse cells (Marzorati et al. 2006, 2008; Matalon et al. 2007; Zchori-Fein et al. 2001, 2004). In *Brevipalpus* mites *Cardinium* not only is present in reproductive tissue, but also is observed in epidermal and nerve tissue (Kitajima et al. 2007). In Hemiptera, *Cardinium* is present in the bacteriocytes of the whitefly *B. tabaci* (Costa et al. 1995; Skaljac et al. 2010) or fat cells of the leafhopper *S. titanus* (Sacchi et al. 2008). The recently described bacterium *Paenicardinium endonii*, the closest known relative of *Cardinium* and a symbiont of the plant pathogenic nematode *Heterodera glycines*, is observed in

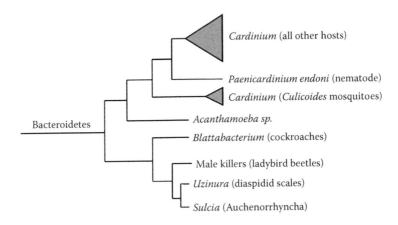

FIGURE 11.2 Evolutionary relationships of heritable symbionts. The phylogeny is based on widely supported findings from studies listed in the citations.

tissues such as the pseudocoelom and intestine, in addition to reproductive tissues of the nematode (Noel and Atibalentja, 2006). The localization of bacteria in host tissues that play a role in digestion and nutrition is often an indication that the symbionts have mutual beneficial relationships with their host (Wernegreen 2002; Wu et al. 2006). *Cardinium* is also reported from salivary glands and fat cells in *S. titanus* (Marzorati et al. 2006).

In the false spider mite, the distribution of *Cardinium* is entirely different; *Cardinium* occurs in every tissue of the *Brevipalpus*, including epidermis, muscle, eye, fat body, digestive system epithelium, nervous cells (ganglia and innervations), tracheal cells, prosomal gland, ovarium, and developing eggs, and also in body parts such as the legs and palps (Kitajima et al. 2007). Whether this is a peculiarity of *Brevipalpus* mites or a general characteristic of mites or the *Cardinium* in mites or arachnids is not clear, since similar studies in other *Cardinium*-Acari symbioses have not been done yet.

PHYLOGENY

Cardinium belongs to the phylum Bacteroidetes, which harbors a number of clades of arthropod endosymbionts (Figure 11.2), including two clades of primary endosymbionts from sap-sucking Hemiptera. One clade contains the endosymbionts of auchenorrhynchan Hemiptera genus *Sulcia*, which are closely related to the *Blattabacteria*, symbionts of cockroaches, and male killers of ladybird beetles (Moran et al. 2005; Bressan et al. 2009). A second clade includes the endosymbionts of two scale insect families of the hemipteran suborder Sternorrhyncha, the Diaspididae (armored scales) and Monophlebidae (giant and cushion scales) (Gruwell et al. 2007, 2009; Matsuura et al. 2009). *Cardinium* endosymbionts form the third emerging clade and include a broad host range of arthropods and plant pathogenic nematodes. Whereas the two aforementioned clades contain mainly primary endosymbionts and mutualistic interactions with their host have been shown in a number of cases (Wu et al.

2006; Wren and Cochran, 1987), the *Cardinium* clade contains bacteria that manipulate host reproduction (e.g., Weeks et al. 2001; Zchori-Fein et al. 2001; Hunter et al. 2003, but see Weeks and Stouthamer 2004). The *Cardinium* clade is only distantly related to the other two clades (Gruwell et al. 2007, 2009). Sisters to the *Cardinium* clade are the *Amoebophilus* symbionts of *Acanthamoeba* (Horn et al. 2001; Schmitz-Esser et al. 2008).

Relationships within *Cardinium* are still difficult to determine because only part of the 16S rDNA, and *gyr*B in a limited number of cases, has been sequenced. The first signs of phylogenetic structuring within *Cardinium* are emerging. Nakamura et al. (2009) suggest three monophyletic groups: the A group contains *Cardinium* from arthropods except the biting midges, the B group contains *Cardinium* (i.e., "*Candidatus* Paenicardinium endonii") (Noel and Atibalentja 2006; Atibalentja and Noel 2008) from the plant-parasitic nematodes, and the C group is from the biting midges. The emerging group of *Cardinium* symbionts of nematodes has also been observed for filarial nematodes infected with *Wolbachia* (Taylor et al. 2005). Monophyletic *Cardinium* clades are also found in *Cybeaeus* spiders (Perlman et al. 2010) and Opiliones daddy long-leg spiders (Chang et al. 2010), based on *gyr*B and 16S rRNA sequences, respectively. This suggests that some host clades may have acquired *Cardinium* only once in the evolutionary past.

However, confusion over the taxonomic revision is already arising as Gruwell et al. (2009) and Nakamura et al. (2009) distinguish monophyletic groups within the *Cardinium* clade, but with different criteria. The phylogeny of Perlman et al. (2010) suggests that the groups distinguished by Gruwell et al. (2009) are at a higher level and fall within the A group of Nakamura et al. (2009).

Hence, it is still early to detect the presence of groups within *Cardinium* and assign them systematic names. It is clear that we need more sequence information from bacteria from a greater number of arthropod species to construct a robust phylogeny that will allow for the detection of natural clades and avoid systematic confusion. Phylogenies should also include the same representative species that have been used previously, because taxon sampling can affect the resulting phylogeny. Preferably such phylogenies should be based on multilocus sequences, since recombination between relatives and lateral transfer of genetic material can blur phylogenetic relationships that are based on single genes (Baldo et al. 2006; Weinert et al. 2009b; Nováková et al. 2010; Thomas and Greub 2010). For example, *Cardinium* has probably donated transposons by lateral transfer to *Rickettsia*, causing extensive reorganizations in the genome of the endosymbionts *R. peacockii* (Felsheim et al. 2009; see also Thomas and Greub 2010). Also, the streamlining of microbial genomes once they have adopted an intracellular lifestyle by losing genes made redundant by the host environment (Akman et al. 2002; Wernegreen 2002; Wernegreen et al. 2003; Toh et al. 2006; Tamas et al. 2008) will differentially affect evolutionary rates of different genes, and consequently phylogenies. The construction of a database based on the multilocus sequence typing approach that has been taken in *Wolbachia* and more recently in *Rickettsia* (Baldo et al. 2006; Weinert et al. 2009) can be used as a template for *Cardinium*.

TRANSMISSION AND LATERAL TRANSFER

As with other intracellular microorganisms, *Cardinium* is thought to be primarily vertically transmitted via the cytoplasm of the eggs. This is based on the fact that *Cardinium* resides in the cytoplasm of reproductive organs. In that respect, *Cardinium* is not different from other intracellular symbionts such as *Wolbachia* or mycetome symbionts. The comparison of host and symbiont phylogenies is a popular method to determine the mode of transmission (see Moran and Baumann 1994). Such comparison may also provide insight into the questions of how and when *Cardinium* spread to such a variety of arthropod hosts. Phylogenetic analyses so far suggest that there is some congruence between host and symbiont phylogenies (Nakamura et al. 2009; Chang et al. 2010; Perlman et al. 2010; Provencher et al. 2005), and that there may be some degree of specialization of *Cardinium* types on certain host species. However, in other parts of the *Cardinium* phylogeny no congruence with host phylogeny is observed, suggesting extensive horizontal transfer between host species.

Groot (2006) studied the mode of transmission of *Cardinium* in three closely related *Brevipalpus* mites by comparing the phylogeny of the symbiont based on *gyr*B and the host based on the cytochrome oxidase I mitochondrial gene (Figure 11.3). Although there is general congruence between the two phylogenies, incongruences were observed at various levels, suggesting that horizontal transmission has occurred within and between the three mite species. Especially clade B symbionts that are associated with *B. phoenicis* appear to be quite mobile. For example, symbiont haplotype G10 was also found in some clonal lineages of the other two species, *B. obovatus* and *B. californicus*. Additionally, double and possibly triple *Cardinium* infections were observed in some host lineages, e.g., *B. obovatus* line C01. This indicates that *Cardinium* is horizontally transmitted, although the timescale at which it occurs is unclear.

Cardinium phylogeny does not follow the phylogeny of its *Cybeaeus* spider host. Instead, Perlman et al. (2010) found that *Cardinium* haplotypes clustered within geographically close spider species suggest horizontal transfer. Finally, Gruwell et al. (2007) found that the 16S rRNA phylogeny of *Cardinium* from armored scale insects was not monophyletic and highly diverse. Moreover, the phylogeny of *Cardinium* haplotypes from scale insects was intertwined with the *Cardinium* haplotypes of their parasitoid wasps. The association between armored scale insects and their parasitoids may facilitate horizontal transfer of *Cardinium*. Alternatively, some groups of mites are predators or scavengers of armored scale insects, and these behaviors provide routes for horizontal transmission (Gruwell et al. 2009). The presence of *Cardinium* in somatic tissues, particularly the salivary glands and digestive system (Marzorati et al. 2006; Kitajima et al. 2007), is of interest since it may facilitate horizontal transmission.

CONCLUSION

The number of documented *Cardinium* infections has grown rapidly over the last decade and makes *Cardinium* a major player in the evolution of arthropods and nematodes. It is also clear that *Cardinium* and *Wolbachia* share many fascinating

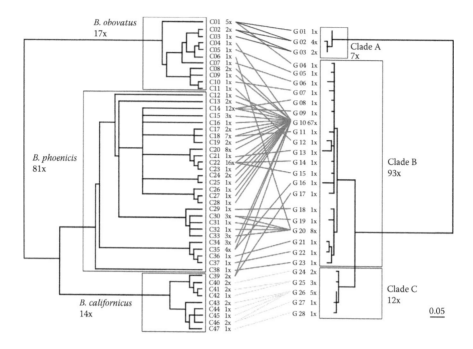

FIGURE 11.3   See color insert. Mite *COI* tree (cladogram) facing symbiont *gyr*B tree (phylogram). The names of the mite species and symbiont clades are followed by the number of samples that comprise that species or clade. Haplotypes are numbered; haplotype numbers are preceded by a C for *COI*, and a G for *gyr*B. Each haplotype name is followed by the number of times the haplotype was encountered. Both trees are midpoint rooted. The colored lines connect individual mites with their symbionts. When several lines connect a single mite haplotype to various symbionts, this means that several mites with the same haplotypes were found to contain different symbionts. The bar with the *gyr*B phylogram indicates the branch length that represents 5% maximum likelihood distance.

characteristics: they are widespread among arthropods; manipulate host reproduction in a number of ways that increase the frequency of the preferred sex, female hosts; are predominantly vertically transmitted; and seem to have given up a free-living lifestyle. They may not be unique among bacteria. Genomics and phylogenetics generated renewed interest in groups such as *Rickettsia* and *Spiroplasma*, or showed that previously odd bacteria, such as male-killing *Arsenophonus* bacteria, are also members of a monophyletic group of *Arsenophonus* bacteria that are widespread among arthropods. The diversity of phenotypic effects of these groups on their hosts needs to be established.

We need to document in a systematic way the incidence of *Cardinium*, not only in insects and arachnids, but in other arthropods and outside arthropods as well. Such surveys should simultaneously investigate the presence of other bacteria. The next challenge is then to establish the phenotypic effects and host-symbiont interactions. These have only been established in a limited number of cases for *Cardinium* (e.g. Kenyon and Hunter 2007). This will involve ecological experimentation and careful monitoring.

Comparative genomics will play a crucial role in revealing interesting evolutionary patterns that impact the evolutionary trajectory of the symbiosis. Already endosymbiont genomics has provided exciting observations. Bacterial genomes are stripped of genes that are made redundant when they enter their niche within the host environment (Akman et al. 2002; Wernegreen et al. 2003; Toh et al. 2006; Tamas et al. 2008). It makes the bacteria dependent on their host, and this probably explains why it is difficult to grow them outside their host. In cases where the host depends on their microbes and horizontal transmission of the symbiont is lost, the bacteria also commonly lose the genetic systems that encode recombination (Dale and Moran 2006). On the other hand, genetic exchange between endosymbionts and other bacteria within the host can take place, probably mediated by transposons, plasmids, or Pox viruses (Felsheim et al. 2009; Weinert et al. 2009a, 2009b; Werren et al. 2010; Schmitz-Esser et al. 2010). Endosymbionts may also exchange genetic material with their hosts. In the genomes of *Nasonia* parasitoid wasps several genes have been found that are of *Wolbachia* origin (Werren et al. 2010). The genome of endobacteria-free filarial nematodes contains *Wolbachia* DNA, indicating past horizontal genetic transfer, and may explain why they can reproduce without endosymbionts (McNulty et al. 2010). Transfer of genes that cause the feminizing trait from the symbiont to the host may also explain why in some *B. obovatus* lineages *Cardinium* is not detected in PCR screens, and yet they are asexual (Groot and Breeuwer 2006).

The congruence between *Cardinium* and *Wolbachia* is very exciting and generates many interesting questions. As mentioned before, do they have the same suite of genes that cause these effects in their hosts? And if so, how then did these genes end up in unrelated bacteria? Alternatively, one could argue that, for example, the reproductive system of their hosts is conserved in that *Cardinium* and *Wolbachia* have evolved different ways to manipulate the host reproductive machinery, and thus have a different suite of "manipulating" genes. If so, this would indicate that reproductive manipulation traits evolved multiple times and relatively easily. Again, comparative genomics of these systems is crucial for understanding the mechanism and evolutionary dynamics of these symbiotic interactions.

REFERENCES

Akman, L., A. Yamashita, H. Watanabe, et al. 2002. Genome sequence of the endocellular obligate symbiont of tsetse flies, *Wigglesworthia glossinidia*. *Nature Genetics* 32: 402–407.

Atibalentja, N., and G.R. Noel. 2008. Bacterial endosymbionts of plant-parasitic nematodes. *Symbiosis* 46: 87–93.

Baldo, L., J.C. Dunning Hotopp, K.A. Jolley, et al. 2006. Multilocus sequence typing system for the endosymbiont *Wolbachia pipientis*. *Applied and Environmental Microbiology*. 72: 7098–110.

Bigliardi, E., L. Sacchi, M. Genchi, et al. 2006. Ultrastructure of a novel *Cardinium* sp. symbiont in *Scaphoideus titanus* (Hemiptera: Cicadellidae). *Tissue and Cell* 38: 257–261.

Bordenstein, S.R., and J.H. Werren. 2007. Bidirectional incompatibility among divergent *Wolbachia* and incompatibility level differences among closely related *Wolbachia* in *Nasonia*. *Heredity* 99: 278–287.

Bressan, A., J. Arneodo, M. Simonato, W.P. Haines, and E. Boudon-Padieu. 2009. Characterization and evolution of two bacteriome-inhabiting symbionts in cixiid planthoppers (Hemiptera: Fulgoromorpha: Pentastirini). *Environmental Microbiology* 11: 3265–3279.

Chang, J., A. Masters, A. Avery, and J.H. Werren. 2010. A divergent *Cardinium* found in daddy long-legs (Arachnida: Opiliones). *Journal of Invertebrate Pathology*. 105: 220–227.

Chang, K.P., and A.J. Musgrave. 1972. Multiple symbiosis in a leafhopper, *Helochara communis* Fitch (Cicadellidae: Homoptera): envelopes, nucleoids and inclusions of the symbiotes. *Journal of Cell Science* 11: 275-293.

Chiel, E., Y. Gottlieb, E. Zchori-Fein et al. 2007. Biotype-dependent secondary symbiont communities in sympatric populations of *Bemisia tabaci. Bulletin of Entomological Research* 97: 407–413.

Chiel, E., M. Inbar, N. Mozes-Daube, J.A. White, M.S. Hunter, and E. Zchori-Fein. 2009a. Assessments of fitness effects by the facultative symbiont *Rickettsia* in the sweetpotato whitefly (Hemiptera: Aleyrodidae). *Annals of the Entomological Society of America* 102: 413–418.

Chiel, E., E. Zchori-Fein, I.M. Inbar, et al. 2009b. Almost there: transmission routes of bacterial symbionts between trophic levels. *PLoS One* 4: e4767. DOI: 10.1371/journal.pone.0004767.

Clark, T.B. 1982. Spiroplasmas: diversity of arthropod reservoirs and host-parasite relationships. *Science* 217: 57–59.

Costa, H.S., D.M. Wescot, D.E. Ullman, R. Rosell, J.K. Brown, and M.W. Johnson. 1995. Morphological variation in *Bemesia* endosymbionts. *Protoplasma* 189: 194–202.

Dale, C., and N.A. Moran. 2006. Molecular interactions between bacterial symbionts and their hosts. *Cell* 126: 453–465.

De Luna, C.J., C.V. Moro, J.H. Guy, L. Zenner, and O.A.E. Sparagano. 2009. Endosymbiotic bacteria living inside the poultry red mite (*Dermanyssus gallinae*). *Experimental and Applied Acarology* 48: 105–113.

Douglas, A.E. 2009. The microbial dimension in insect nutritional ecology. *Functional Ecology* 23: 38–47.

Duron, O., D. Bouchon, S. Boutin, L. Bellamy, L.Q. Zhou, J. Engelstadter, and G.D.D. Hurst. 2008a. The diversity of reproductive parasites among arthropods: *Wolbachia* do not walk alone. *BMC Biology* 6.

Duron, O., G.D.D. Hurst, E.A. Hornett, J.A. Josling, and J. Engelstadter. 2008b. High incidence of the maternally inherited bacterium *Cardinium* in spiders. *Molecular Ecology* 17: 1427–1437.

Endo, B.Y. 1979. The ultrastructure and distribution of an intracellular bacterium-like microorganism in tissues of larvae of the soybean cyst nematode, *Heterodera glycines. Journal of Ultrastructure Research* 67: 1–14.

Enigl, M., and P. Schausberger. 2007. Incidence of the endosymbionts *Wolbachia, Cardinium* and *Spiroplasma* in phytoseiid mites and associated prey. *Experimental and Applied Acarology* 42: 75–85.

Felsheim, R.F., T.J. Kurtti, and U.G. Munderloh. 2009. Genome sequence of the endosymbiont *Rickettsia peacockii* and comparison with virulent *Rickettsia rickettsii*: identification of virulence factors. *PLoS One* 4:e8361.

Giorgini, M., U. Bernardo, M.M. Monti, A.G. Nappo, and M. Gebiola. 2010. *Rickettsia* symbionts cause parthenogenetic reproduction in the parasitoid wasp *Pnigalio soemius* (Hymenoptera: Eulophidae). *Applied and Environmental Microbiology* 76: 2589–2599.

Giorgini, M., M.M. Monti, E. Caprio, R. Stouthamer, and M.S. Hunter. 2009. Feminization and the collapse of haplodiploidy in an asexual parasitoid wasp harboring the bacterial symbiont *Cardinium. Heredity* 102: 365–371.

Gotoh, T., H. Noda, and S. Ito. 2007a. *Cardinium* symbionts cause cytoplasmic incompatibility in spider mites. *Heredity* 98: 13–20.

Gotoh, T., J. Sugasawa, H. Noda, and Y. Kitashima. 2007b. *Wolbachia*-induced cytoplasmic incompatibility in Japanese populations of *Tetranychus urticae* (Acari: Tetranychidae). *Experimental and Applied Acarology* 42: 1–16.

Gottlieb, Y., M. Ghanim, G. Gueguen, et al. 2008. Inherited intracellular ecosystem: symbiotic bacteria share bacteriocytes in whiteflies. *FASEB Journal* 22: 2591–2599.

Groot, T.V.M. 2006. The effects of symbiont induced haploid thelytoky on the evolution of *Brevipalpus* mites. PhD thesis, University of Amsterdam, Amsterdam, the Netherlands. http://dare.uva.nl/document/33053.

Groot, T.V.M., and J.A.J. Breeuwer. 2006. *Cardinium* symbionts induce haploid thelytoky in most clones of three closely related *Brevipalpus* species. *Experimental and Applied Acarology* 39: 257–271.

Gruwell, M.E., G.E. Morse, and B.B. Normark. 2007. Phylogenetic congruence of armored scale insects (Hemiptera: Diaspididae) and their primary endosymbionts from the phylum Bacteroidetes. *Molecular Phylogenetics and Evolution* 44: 267–280.

Gruwell, M.E., J. Wu, and B.B. Normark. 2009. Diversity and phylogeny of *Cardinium* (Bacteroidetes) in armored scale insects (Hemiptera: Diaspididae*). Annals of the Entomological Society of America* 102: 1050–1061.

Harris, L.R., S.E. Kelly, M.S. Hunter, and S.J. Perlman. 2010. Population dynamics and rapid spread of *Cardinium*, a bacterial endosymbiont causing cytoplasmic incompatibility in *Encarsia pergandiella* (Hymenoptera: Aphelinidae). *Heredity* 104: 239–246.

Hess, R.T., and M.A. Hoy. 1982. Microorganisms associated with the spider mite predator *Metaseiulus* (= *Typhlodromus*) *occidentalis* (Acrina, Phytoseiidae)—electron-microscope observations. *Journal of Invertebrate Pathology* 401: 98–106.

Hoffmann, A.A., and M. Turelli. 1997. Cytoplasmic incompatibility in insects. In *Influential passengers*, ed. S.L. O'Neill, A.A. Hoffmann, and J.H. Werren, 42–80. Oxford: Oxford University Press

Horn, M., M.D. Harzenetter, T. Linner, et al. 2001. Members of the Cytophaga-Flavobacterium-Bacteroides phylum as intracellular bacteria of *Acanthamoebae*: proposal of 'Candidatus Amoebophilus asiaticus.' *Environmental Microbiology* 37: 440–449.

Hunter, M.S., S.J. Perlman, et al. 2003. A bacterial symbiont in the Bacteroidetes induces cytoplasmic incompatibility in the parasitoid wasp *Encarsia pergandiella. Proceedings of the Royal Society of London Series B—Biological Sciences* 270: 2185–2190.

Jeong, G., K. Lee, J. Choi, et al. 2009. Incidence of *Wolbachia* and *Cardinium* endosymbionts in the Osmia community in Korea. *Journal of Microbiology* 47: 28–32.

Jiggins, F.M., J.K. Bentley, M.E.N. Majerus, and G.D.D. Hurst. 2001. How many species are infected with *Wolbachia*? Cryptic sex ratio distorters revealed to be common by intensive sampling. *Proceedings of the Royal Society B—Biological Sciences* 268: 1123–1126.

Kenyon, S.G., and M.S. Hunter. 2007. Manipulation of oviposition choice of the parasitoid wasp, *Encarsia pergandiella*, by the endosymbiotic bacterium *Cardinium. Journal of Evolutionary Biology* 20: 707–716.

Kitajima, E.W., T.V.M. Groot, V.M. Novelli, J. Freitas-Astua, G. Alberti, and G.J. de Moraes. 2007. *In situ* observation of the *Cardinium* symbionts of *Brevipalpus* (Acari: Tenuipalpidae) by electron microscopy. *Experimental and Applied Acarology* 42: 263–271.

Kurtti, T.J., U.G. Munderloh, et al. 1996. Tick cell culture isolation of an intracellular prokaryote from the tick *Ixodes scapularis*. *Journal of Invertebrate Pathology* 67: 318–321.

Liu, Y., H. Miao, and X.Y. Hong. 2006. Distribution of the endosymbiotic bacterium *Cardinium* in Chinese populations of the carmine spider mite *Tetranychus cinnabarinus* (Acari: Tetranychidae). *Journal of Applied Entomology* 130: 523–529.

Louis, C., and L. Nigro. 1989. Ultrastructural evidence of *Wolbachia*-Rickettsiales in *Drosophila simulans* and their relationships with unidirectional cross-incompatibility. *Journal of Invertebrate Pathology* 54: 39–44.

Martin, O.Y., and Goodacre, S.L. 2009. Widespread infections by the bacterial endosymbiont *Cardinium* in arachnids. *Journal of Arachnology* 37: 106–108.

Marzorati, M., A. Alma, L. Sacchi, et al. 2006. A novel bacteroidetes symbiont is localized in *Scaphoideus titanus*, the insect vector of flavescence doree in *Vitis vinifera*. *Applied and Environmental Microbiology* 72: 1467–1475.

Marzorati, M., M. Pajoro, M. Clementi, et al. 2008. Characterization of the microflora associated to *Scaphoideus titanus*, the insect vector of the "flavescence doree." *Bulletin of Insectology* 61: 215–216.

Matalon, Y., N. Katzir, Y. Gottlieb, V. Portnoy, and E. Zchori-Fein. 2007. *Cardinium* in *Plagiomerus diaspidis* (Hymenoptera: Encyrtidae). *Journal of Invertebrate Pathology* 96: 106–108.

McNulty, S.N., J.M. Foster, M. Mitreva et al. 2010. Endosymbiont DNA in endobacteria-free filarial nematodes indicates ancient horizontal genetic transfer. *PLoS One* 5: 1–9, e11029.

Moran, N., and P. Baumann. 1994. Phylogenetics of cytoplasmically inherited microorganisms of arthropods. *Trends in Ecology and Evolution* 9: 15–20.

Moran, N.A., P. Tran, and N.M. Gerardo. 2005. Symbiosis and insect diversification: an ancient symbiont of sap-feeding insects from the bacterial phylum Bacteroidetes. *Applied and Environmental Microbiology* 71: 8802–8810.

Nakamura, Y., S. Kawai, F. Yukuhiro, et al. 2009. Prevalence of *Cardinium* bacteria in planthoppers and spider mites and taxonomic revision of "*Candidatus* Cardinium hertigii" based on detection of a new *Cardinium* group from biting midges. *Applied and Environmental Microbiology* 75: 6757–6763.

Neelakanta, G., H. Sultana, F. Durland, et al. 2010. *Anaplasma phagocytophilum* induces *Ixodes scapularis* ticks to express an antifreeze glycoprotein gene that enhances their survival in the cold. *Clinical Investigation* 120: 3179–3190.

Noel, G.R., and N. Atibalentja. 2006. '*Candidatus* Paencardinium endonii', an endosymbiont of the plant-parasitic nematode *Heterodera glycines* (Nemata: Tylenchida), affiliated to the phylum Bacteroidetes. *International Journal of Systematic and Evolutionary Microbiology* 56: 1697–1702.

Nováková, E., V. Hypsa, and N.A. Moran. 2009. *Arsenophonus*, an emerging clade of intracellular symbionts with a broad host distribution. *BMC Microbiology* 9.

Perlman, S.J., S.E. Kelly, and M.S. Hunter. 2008. Population biology of cytoplasmic incompatibility: maintenance and spread of *Cardinium* symbionts in a parasitic wasp. *Genetics* 178: 1003–1011.

Perlman, S.J., S.E. Kelly, E. Zchori-Fein, and M.S. Hunter. 2006. Cytoplasmic incompatibility and multiple symbiont infection in the ash whitefly parasitoid, *Encarsia inaron*. *Biological Control* 39: 474–480.

Perlman, S.J., S.A. Magnus, and C.R. Copley. 2010. Pervasive associations between *Cybaeus* spiders and the bacterial symbiont *Cardinium*. *Journal of Invertebrate Pathology* 103: 150–155.

Provencher, L.M., G.E. Morse, A.R. Weeks, and B.B. Normark. 2005. Parthenogenesis in the *Aspidiotus nerii* complex (Hemiptera: Diaspididae): a single origin of a worldwide, polyphagous lineage associated with *Cardinium* bacteria. *Annals of the Entomological Society of America* 98: 629–635.

Ros, V.I.D., and J.A.J. Breeuwer. 2009. The effects of, and interactions between, *Cardinium* and *Wolbachia* in the doubly infected spider mite *Bryobia sarothamni*. *Heredity* 102: 413–422.

Sacchi, L., M. Genchi, E. Clementi, et al. 2008. Multiple symbiosis in the leafhopper *Scaphoideus titanus* (Hemiptera: Cicadellidae): details of transovarial transmission of *Cardinium* sp. and yeast-like endosymbionts. *Tissue and Cell* 40: 231–242.

Sassera, D., L. Lo, E.A.P. Bouman, S. Epis, M. Mortarino, and C. Bandi. 2007. "*Candidatus* Midichloria" endosymbionts bloom after the blood meal of the host, the hard tick *Ixodes ricinus*. *Heredity* 98: 13–20.

Schmitz-Esser, S., P. Tischler, R. Arnold, et al. 2010. The genome of the Amoeba symbiont "Candidatus *Amoebophilus asiaticus*" reveals common mechanisms for host cell interaction among amoeba-associated bacteria. *Journal of Bacteriology* 192: 1045–1057.

Schmitz-Esser, S., E.R. Toenshoff, S. Haider, et al. 2008. Diversity of bacterial endosymbionts of environmental *Acanthamoeba* isolates. *Applied and Environmental Microbiology* 74: 5822–5831.

Serbus, L.R., C. Casper-Lindley, F. Landmann, and W. Sullivan. 2008. The genetics and cell biology of *Wolbachia*-host interactions. *Annual Review of Genetics* 42: 683–707.

Shepherd, A.M., S.A. Clark, and A. Kempton. 1973. An intracellular micro-organism associated with tissues of *Heterodera* spp. *Nematologica* 19: 31–34.

Skaljac, M., K. Zanic, S.G. Ban, S. Kontsedalov, and M. Ghanim. 2010. Co-infection and localization of secondary symbionts in two whitefly species. *BMC Microbiology* 10.

Tamas, I., J.J. Wernegreen, B. Nystedt, et al. 2008. Endosymbiont gene functions impaired and rescued by polymerase infidelity at poly(A) tracts. *Proceedings of the National Academy of Sciences of the United States of America* 105: 14934–14939.

Taylor, M.J., C. Bandi, and A. Hoerauf. 2005. A *Wolbachia* bacterial endosymbiont of filarial nematodes. *Advances in Parasitology* 60: 245–284.

Thomas, V., and G. Greub. 2010. Amoeba/amoebal symbiont genetic transfers: lessons from giant virus neighbours. *Intervirology* 53: 254–267.

Toh, H., B.L. Weiss, S.A.H. Perkin, A. Yamashita, K. Oshima, M. Hattori, and S. Aksoy. 2006. Massive genome erosion and functional adaptations provide insights into the symbiotic lifestyle of *Sodalis glossinidius* in the tsetse host. *Genome Research* 16: 149–156.

Weeks, A.R., and J.A.J. Breeuwer. 2002. A new bacterium from the Cytophaga-Flavobacterium-Bacteroides phylum that causes sex-ratio distortion. In *Insect symbiosis*, ed. K. Bourtzis and T.A. Miller, 165–176. Boca Raton, FL: CRC Press.

Weeks, A.R., F. Marec, and J.A.J. Breeuwer. 2001. A mite species that consists entirely of haploid females. *Science* 292: 2479–2482.

Weeks, A.R., K.T. Reynolds, and A.A. Hoffmann. 2002. *Wolbachia* dynamics and host effects: what has (and has not) been demonstrated? *Trends in Ecology and Evolution* 17: 257–262.

Weeks, A.R., and R. Stouthamer 2004. Increased fecundity associated with infection by a Cytophaga-like intracellular bacterium in the predatory mite, *Metaseiulus occidentalis*. *Proceedings of the Royal Society of London Series B—Biological Sciences* 271: S193–S195.

Weeks, A.R., R. Velten, and R. Stouthamer. 2003. Incidence of a new sex-ratio-distorting endosymbiotic bacterium among arthropods. *Proceedings of the Royal Society of London Series B—Biological Sciences* 270: 1857–1865.

Weinert, L.A., M.C. Tinsley, et al. 2007. Are we underestimating the diversity and incidence of insect bacterial symbionts? A case study in ladybird beetles. *Biology Letters* 3: 678–681.

Weinert, L.A., J.J. Welch, and F.M. Jiggins. 2009b. Conjugation genes are common throughout the genus *Rickettsia* and are transmitted horizontally. *Proceedings of the Royal Society of London Series B—Biological Sciences* 276: 3619–3627.

Weinert, L.A., J.H. Werren, A. Aebi, G.N. Stone, and F.M. Jiggins. 2009a. Evolution and diversity of *Rickettsia* bacteria. *BMC Biology* 7.

Wernegreen, J.J. 2002. Genome evolution in bacterial endosymbionts of insects. *Nature Reviews Genetics* 3: 850–861.

Wernegreen, J.J., P.H. Degnan, A.B. Lazarus, C. Palacios, and S.R. Bordenstein. 2003. Genome evolution in an insect cell: distinct features of an ant-bacterial partnership. *Biological Bulletin* 204: 221–231.

Werren, J.H. 1997. Biology of *Wolbachia*. *Annual Review of Entomology* 42: 587–609.

Werren, J.H., S. Richards, C.A. Desjardins, et al. 2010. Functional and evolutionary insights from the genomes of three parasitoid *Nasonia* species. *Science* 327: 343–348.

White, J.A., S.E. Kelly, S.J. Perlman, and M.S. Hunter. 2009. Cytoplasmic incompatibility in the parasitic wasp *Encarsia inaron*: disentangling the roles of *Cardinium* and *Wolbachia* symbionts. *Heredity* 102: 483–489.

Wren, H.N., and D.G. Cochran 1987. Xanthine dehydrogenase-activity in the cockroach symbiont *Blattabacterium cuenoti* (Mercier 1906) Hollande and Favre 1931 and in the cockroach fat-body. *Comparative Biochemistry and Physiology B—Biochemistry and Molecular Biology* 88: 1023–1026.

Wu, D., S.C. Daugherti, S.E. van Aken, et al. 2006. Metabolic complementarity and genomics of the dual bacterial symbiosis of sharpshooters. *PloS Biology* 4: 1079–1092.

Zchori-Fein, E., Y. Gottlieb, S.E. Kelly, et al. 2001. A newly discovered bacterium associated with parthenogenesis and a change in host selection behavior in parasitoid wasps. *Proceedings of the National Academy of Sciences of the United States of America* 98: 12555–12560.

Zchori-Fein, E., and S.J. Perlman. 2004. Distribution of the bacterial symbiont *Cardinium* in arthropods. *Molecular Ecology* 13: 2009–2016.

Zchori-Fein, E., S.J. Perlman, S.E. Kelly, N. Katzir, and M.S. Hunter. 2004. Characterization of a 'Bacteroidetes' symbiont in *Encarsia* wasps (Hymenoptera: Aphelinidae): proposal of '*Candidatus* Cardinium hertigii.' *International Journal of Systematic and Evolutionary Microbiology* 54: 961–968.

12 The Genus *Arsenophonus*

*Timothy E. Wilkes, Olivier Duron,
Alistair C. Darby, Václav Hypša,
Eva Nováková, and Gregory D. D. Hurst*

CONTENTS

Genus	*Arsenophonus*
Type species	*nasoniae*
Family	Gamma-Proteobacteria
Order	Enterobacteriacae
Description year	1991
Origin of name	*Arsen* = male, *phonus* = slayer
Description	
Reference	Gherna et al. (1991)

INTRODUCTION TO THE GENUS ARSENOPHONUS

In 1985, individual lines of the wasp *Nasonia vitripennis* were observed to produce strongly female-biased broods, containing just 0–5% sons (Skinner 1985). The trait was maternally inherited, and the sex ratio bias found to be associated with the death of male embryos. Pertinent to this book, the trait was additionally found to be infectiously transmitted between wasp lines following superparasitism. This latter observation implied the presence of an infectious agent. Detailed microbiological study led to the isolation of a new bacterium in liquid culture, for which Koch's postulates were then fulfilled (Werren et al. 1986). The bacterium was formally described as *Arsenophonus nasoniae* sp. nov. gen. nov., the type species of the genus *Arsenophonus* (Gherna et al. 1991). The formal status of the genus was established by Werren in 2005 (Werren 2005).

Since its initial discovery, membership of the genus *Arsenophonus* has expanded, and it is now known to be both widespread and biologically diverse. In terms of incidence it is one of the "big four" inherited symbionts of arthropods, being present in ca. 5% of species. Infections have been described in a diverse range of arthropod hosts, including arachnids, ticks, cockroaches, hemipterans, hymenopterans, lice, flies, and coleopterans (Duron et al. 2008). In terms of symbiotic diversity, several different stages of symbiosis can be found within the genus, and interactions include facultative and obligate, parasitic and beneficial. The type species *A. nasoniae* is a parasite with a large genome with substantial metabolic capability (Darby et al. 2010), and can be grown in supplemented cell-free media (Werren et al. 1986). Other strains have reduced genomes, can be grown only in cell culture, and are probably beneficial for host fitness. *Riesia*, the primary symbiont of lice, can be considered a biologically highly derived species of *Arsenophonus*. It has a very diminished genome size (Kirkness et al. 2010), is unculturable, and is required by the host. This makes the *Arsenophonus* clade one where all stages of symbiosis can be identified, and in which we can potentially reconstruct the evolutionary processes that underlie changes in symbiotic relationships.

One other point of note is the placement of the genus compared to other microbes. The genus *Arsenophonus* falls in the gamma subdivision of the Proteobacteria (Figure 12.1). As such, it benefits from possessing a plethora of well-studied "model" organism comparators, including *E. coli*, human pathogens such as *Yersinia pestis*

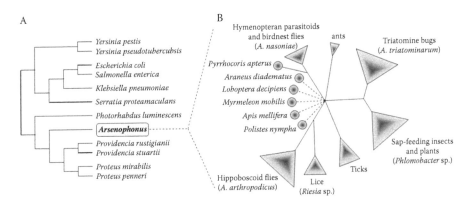

FIGURE 12.1   Phylogeny of *Arsenophonus*. (A) The phylogenetic relationships between *Arsenophonus* and representative members of the Enterobacteriaceae family. (B) An unrooted phylogenetic tree of the main subdivisions of *Arsenophonus*. Dotted lines represent *Arsenophonus* strains for which relationships with the other strains are uncertain. Triangles represent described diversity within each lineage based on a single *Arsenophonus* strain. Circles represent single strains.

and *Salmonella typhimurium*, and insect pathogens such as *Photorhabdus luminescens*. These each provide good hypotheses for gene function and capabilities in the *Arsenophonus* genome, where functional work has previously been lacking. There are also a variety of well-understood insect symbionts to which it can be compared, including facultative symbionts such as *Hamiltonella defensa* and *Sodalis*, and primary symbionts like *Buchnera* and *Wigglesworthia*. Its position within the gamma-Proteobacteria is now well resolved from multigene data, with *Proteus* and *Providencia* clearly the closest allied genera (Darby et al. 2010).

Arsenophonus therefore represents a symbiotic clade of great potential interest. In this chapter, we first describe the known microbial biodiversity within the genus *Arsenophonus* as it is currently recognized, and debates about membership of it. We then summarize the current state of knowledge of the five *Arsenophonus*-host interactions that are best characterized, and review recent findings from draft genome sequences of two members of the clade, *A. nasoniae* and *Riesia pediculicola*. Finally, we point to four areas of future research for workers on this clade of symbionts.

BIODIVERSITY WITHIN THE GENUS *ARSENOPHONUS*

Within the genus *Arsenophonus*, there are currently two species formally described beyond *A. nasoniae*, the type. These are Cand. *Arsenophonus triatominarum*, isolated from triatomine bugs, and Cand. *Arsenophonus arthropodicus*, isolated from the pigeon louse fly, *Pseudolynchia canariensis*. There has been a recent debate about inclusion of other bacteria, notably Cand. *Phlomobacter frageriae* and Cand. *Riesia pediculicola*.

Cand. *Phlomobacter fragariae*, the causative agent of marginal chlorosis in strawberry plants and endosymbiont of the planthopper *Cixius wagneri*, was described by Danet et al. (2003). In 2005, Werren (2005) suggested *Phlomobacter* fell within the clade *Arsenophonus*. Subsequent phylogenetic analysis using the sequence of the 16S rRNA gene has indeed indicated that this bacterium falls within the genus *Arsenophonus*, and being junior to it in description, should be incorporated within it, and it is now regarded as a member of the clade *Arsenophonus* (Bressan et al. 2009). Cand. *Riesia pediculicola*, a genus of symbionts found in lice, is a very different bacterium from described *Arsenophonus* species in terms of symbiosis, microbiology, and genomic content. Notwithstanding the clear biological differentiation of the microbe, *Riesia* falls phylogenetically within the genus *Arsenophonus* rather than as a distinct clade. Without prejudging this debate, we include *Riesia* in the chapter by virtue of the interestingly different nature of this related bacterium.

Beyond these described species, *Arsenophonus* strains are generally known only from the 16S rRNA gene sequence recovered from insect specimens using general eubacterial primers (see Table 12.1). In some cases, such as *Arsenophonus* in *Bemisia*, fluorescence *in situ* hybridization (FISH) reveals the infection to be present within bacteriocytes inside the insect (Gottlieb et al. 2008). In most other cases, presence is indicated solely by a polymerase chain reaction (PCR) positive assay using template material derived from whole arthropods. While there is no direct confirmation of symbiosis in these cases, it is highly likely that these represent symbiotic partners of the insects in question.

The incidence of *Arsenophonus* is best estimated from a "blind" survey. Duron et al. (2008) used a PCR assay for *Arsenophonus* to investigate how commonly it occurred in field-collected insects. They reported infection in 6 of 136 species, representing an estimated incidence of 5%. The survey concluded there was no obvious bias in the taxa infected, and it is notable the same survey did conclude that *Cardinium*, *Wolbachia*, and *Spiroplasma ixodetis* were more common in spiders than insects, a pattern not found for *Arsenophonus*. The survey data provide a slightly different picture from individual records, which suggests *Arsenophonus* infections may be most common among species of Hemiptera.

Study of the biodiversity of *Arsenophonus* strains has traditionally derived solely from the 16S rRNA gene sequence. However, this marker may perform poorly in resolving patterns of relatedness in the genus, and the interpretation of past investigations is now subject to caution since several recent studies have revealed major failures in the framework for studies of *Arsenophonus* phylogenetics (Novakova et al. 2009; Sorfova et al. 2008). Although 16S rRNA gene sequence marker identifies *Arsenophonus* as a robust monophyletic clade, Novakova et al. (2009) demonstrated that inner topology is unstable, mainly because of methodological artifacts due to insufficient sequence acquisition, preventing clear understanding of the evolutionary relationships between *Arsenophonus* strains. In parallel, Sorfova et al. (2008) showed that several *Arsenophonus* strains carry a high number of rRNA gene copies in their genomes, a characteristic also reported in other Enterobacteriaceae genomes (Moran et al. 2008). Variability between individual rRNA gene copies within a bacterium has serious consequences for phylogenetic inferences, as it depicts inexact evolutionary trajectories for *Arsenophonus* infections.

TABLE 12.1
Details of Currently Known Members of the *Arsenophonus* Clade

Host Organism	Strain Nomenclature	Phenotype	Reference
Arachnida			
Ixodida			
Amblyomma americanum (Ixodidae)	*Arsenophonus* sp.	Not known	Clay et al. 2008
Dermacentor variabilis (Ixodidae)	*Arsenophonus* sp.	Not known	Grindle et al. 2003
Dermacentor andersoni	*Arsenophonus* sp.	Not known	Dergousoff and Chilton 2010
Aranea			
Araneus diadematus (Araneidae)	*Arsenophonus* sp.	Not known	Duron et al. 2008
Insecta			
Blattaria			
Loboptera decipiens (Blatellidae)	*Arsenophonus* sp.	Not known	Duron et al. 2008
Diptera			
Protocalliphora azurea (Calliphoridae)	*Arsenophonus* sp.	Not known	Duron et al. 2008
Various hippoboscid species (Hippoboscidae)[a]	*A. arthropodicus*, *Arsenophonus* sp.	Not known	Dale et al. 2006 Duron et al. 2008 Novakova et al. 2009
Various streblid species (Streblidae)[a]	*Arsenophonus* sp.	Not known	Novakova et al. 2009 Trowbridge et al. 2006
Hemiptera			
Pyrrhocoris apterus (Pyrrhocoridae)	*Arsenophonus* sp.	Not known	Duron et al. 2008
Various triatomine species (Reduviidae)[a]	*A. triatominarum*, *Arsenophonus* sp.	Not known	Hypsa and Dale 1997 Sorfova et al. 2008
Various planthopper species (Cixiidae)[a,b]	*Phlomobacter* sp.	Phytopathogen	Bressan et al. 2008 Danet et al. 2003 Semetey et al. 2007
Nilaparvata lugens (Dephacidae)	*Arsenophonus* sp.	Not known	Wang et al. 2010
Various whiteflies species (Aleyrodidae)[a]	*Arsenophonus* sp.	Bacteriocytes associated	Gottlieb et al. 2008 Thao and Baumann 2004 Zchori-Fein and Brown 2002
Various aphid species (Aphididae)[a]	*Arsenophonus* sp.	Not known	Najar-Rodriguez et al. 2009 Russell et al. 2003
Various psyllid species (Psylloidea)[a]	*Arsenophonus* sp.	Not known	Hansen et al. 2007 Spaulding and von Dohlen 2001 Subandiyah et al. 2000 Thao et al., 2000a 2000b

(continued)

TABLE 12.1 (continued)
Details of Currently Known Members of the *Arsenophonus* Clade

Host Organism	Strain Nomenclature	Phenotype	Reference
Hymenoptera			
Nasonia vitripennis (Pteromelidae)	*A. nasoniae*	Son killer	Gherna et al. 1991 Werren et al. 1986 Skinner 1985 Huger et al. 1985
Polistes nympha (Vespidae)	*Arsenophonus* sp.	Not known	Duron et al. 2008
Apis mellifera (Apidae)	*Arsenophonus* sp.	Not known	Babendreier et al. 2007
Two ant species (Formicidae)	*Arsenophonus* sp.	Not known	Novakova et al. 2009
Neuroptera			
Myrmeleon mobilis (Myrmeleontidae)	*Arsenophonus* sp.	Not known	Dunn and Stabb 2005
Phthiraptera			
Various lice species (Pediculidea)[a]	*Riesia* sp.	Bacteriocytes associated	Allen et al. 2007 Sasaki-Fukatsu et al. 2006 Perotti et al. 2007
Plants			
Beta vulgaris (sugar beet)[b]	*Phlomobacter* sp., SBR bacterium	Phytopathogen	Bressan et al. 2008 Semetey et al. 2007
Fragaria sp. (strawberry)[b]	*Phlomobacter fragariae*	Phytopathogen	Danet et al. 2003 Zreik et al. 1998

[a] Most of the host species of these families are infected and are not detailed exhaustively.
[b] The same *Arsenophonus* (*Phlomobacter*) strains are found in Cixidae hosts and plants.

In light of this, more recent studies of biodiversity have utilized a multilocus system for approaching phylogeny in the clade based on three housekeeping genes. These conformed to the basic properties of a multilocus strain typing (MLST) system—single copy, no internal recombination, with the majority of strains (save *Riesia*) amplifiable with conserved primers (Table 12.2). The sequences of these genes have been used to examine patterns of transmission of *Arsenophonus* between species, particularly *A. nasoniae* in the guild of filth fly parasites (Duron et al. 2010).

HOSTS INFECTED AND NATURE OF THE SYMBIOSES FOUND

Information about the nature of these symbioses is known for a few cases that we detail below. Basic comparative information about these symbionts is given in Table 12.3. The details of these symbioses reveal varying importance of horizontal

TABLE 12.2
Oligonucleotide Primer Sequences, PCR Annealing Temperature and Expected Amplicon Size for a Variety of Genetic Markers in *Arsenophonus* Used in Reconstructing the Relatedness of Strains

Gene	Hypothetical Product	Primers (5′–3′)		Tm	Fragment Size
16S rRNA	Small ribosomal subunit	ArsF	GGGTTGTAAAGTACTTTCAGTCGT	52°C	804 bp
		ArsR2	GTAGCCCTRCTCGTAAGGGCC		
fbaA	Fructose-bisphosphate aldolase class II	fbaAf	GCYGCYAAAGTTCRTTCTCC	52°C	659 bp
		fbaAr	CCWGAACCDCCRTGGAAAACAAAA		
ftsK	Cell division protein (DNA translocase)	ftsKf	GTTGTYATGGTYGATGAATTTGC	52°C	445 bp
		ftsKr	GCTCTTCATCACYATCAWAACC		
yaeT	Outer membrane protein assembly factor	yaeTf	GCATACGGTTCAGACGGGTTTG	52°C	473 bp
		yaeTr	GCCGAAACGCCTTCAGAAAAG		
zapA	Zinc-dependent metalloprotease	zapAf	GGGTCACATACCTATTTT	50°C	594 bp
		zapAr	GTAGTCGCCTGGGTGGG		
aprA	Zinc-dependent metalloprotease	aprAf	CATTTAATTCCAAGAAC	50°C	513 bp
		aprAr	GAAAGTCTGCTTGTCCATCTCC		

Note: zapA and aprA products only obtained in *A. nasoniae*.

and vertical transmission of the symbiont, the presence of sex ratio distorting activity in one strain (but not others), potential beneficial effects in some, and confirmed beneficial effects associated with integration into host anatomy in others.

Arsenophonus Nasoniae/Pteromalid Wasps

Microbial Facts

Rod-like bacterium, 6.9–10.0 μm in length and 0.40–0.57 μm in diameter (Gherna et al. 1991). Possesses a genome of c. 3.5 MB with at least one small and one large plasmid (Darby et al. 2010). Culturable on cell-free GC media with addition of Kellogg's supplement, where it produces colonies in 4–5 days (Werren et al. 1986). *In vitro* growth generally conducted at 25–30°C. May be cryopreserved in glycerol stocks.

Interaction with Host

Widely disseminated infection found throughout host tissues (Huger et al. 1985). Infection induces a female-biased sex ratio in its host, associated with mortality of 80% of male offspring (Skinner 1985).

TABLE 12.3

Summary of Information on the Five Most Studied Members of the Genus *Arsenophonus*

	Microbial Morphology (Length, Diameter)	Grows in Cell-Free Culture?	Maintained in Insect Cell Culture?	Genome Size	Genome Structure	Known Phenotypes
A. nasoniae	Rod 7–10 × 0.4–0.6 μm	Y	Y	~3.5 MB	Chromosome + small plasmid + 1 or more large plasmids	Son killing
C. A. triatominarum	Rod 15 × 1.5 μm	N	Y	~2.2 MB	NA	NA
C. A. arthropodicus	Rod 2–5 × 0.3 μm	Y	Y	~3.5 MB	Chromosome + small plasmid + 1 or more large plasmids	NA
C. Phlomobacter fragariae	Rod 2–2.5 × 0.3 μm	NA	NA	NA	NA	Phytopathogen
C. Riesia pediculicola	Rod Up to 25 μm long	N	N	0.57 MB	Linear chromosome + small plasmid	Obligate mutualist, B vitamin provision

Note: For references, see text. NA, not ascertained.

Around 5% of *Nasonia vitripennis* females are naturally infected with *A. nasoniae*, an infection otherwise known as son killer. The label *son killer* describes the phenotype of the infection. As discussed in the opening paragraph, *Arsenophonus* distorts the sex ratio of its wasp host *Nasonia*, increasing the proportion of daughters by killing up to 90% of sons (Skinner 1985; Werren et al. 1986).

The son-killer trait raises two questions: How does selective killing of sons occur, and why? The mechanistic question has been resolved in terms of the changes that occur to the host. As in other Hymenoptera, the wasp has haplodiploid sex determination: unfertilized (haploid) eggs develop into males, and fertilized (diploid) eggs develop into females. Indeed, *Arsenophonus nasoniae* appears to act by preventing the development of unfertilized eggs. Recently, *A. nasoniae* was shown to kill sons by preventing the formation of maternal centrosomes in unfertilized eggs, which resulted in early developmental arrest (Ferree et al. 2008). What is not clear mechanistically is the nature of *A. nasoniae* effectors that produce this change.

The advantage to the bacterium of son killing is less clear in this system. The general rationale for son killing is given in terms of benefit to a maternally inherited element that cannot pass through sons. Male killing may either directly release resources to female sibling hosts (e.g., cannibalism in ladybirds), release resources indirectly (through reduced resource competition following brood reduction), or reduce the probability of deleterious inbreeding (Hurst and Majerus 1993). *Nasonia* is a gregarious parasite that lays many eggs within a pupa. Death of males may therefore reduce resource competition suffered by female siblings within the pupa. *Nasonia* is also a species that routinely inbreeds, and this inbreeding avoidance has been suggested to provide a benefit to male killing in this system (Werren 1987). Experimental analysis of these factors has not been completed. With respect to the former hypothesis, field data indicated no difference in body size between infected and uninfected wasps (Balas et al. 1996). With respect to the latter, inbreeding occurs regularly in the species. However, inbreeding depression is modest (Luna and Hawkins 2004), and death of males may also result on occasion in females remaining virgin, which prevents the transmission of a maternally inherited element.

There are reasons to believe *A. nasoniae* will be biologically and microbiologically very different from other microbes that display reproductive parasitism. While *Arsenophonus* is a maternally inherited bacterium like *Wolbachia*, it is also horizontally transmitted at high frequency among *Nasonia* wasps developing within the same fly host. The initial observation of this predates the description of the bacterium, with Skinner (1985) noting transfer of the son-killer trait following superparasitism, and Huger et al. (1985) observing bacterial passage across gut epithelia. *Arsenophonus nasoniae* is thus inoculated into the fly host and then ingested by the developing wasp offspring. This peroral transmission of the bacterium to the next generation of wasps is unique, in contrast to the pure cytoplasmic mode of transmission typical of other sex ratio-distorting microorganisms. Unlike other reproductive parasites, *A. nasoniae* can be grown in cell-free media (Werren et al. 1986), allowing greater chance of survival outside the host that could enhance the likelihood of successful transfer between arthropods.

Horizontal transmission of heritable bacteria via ecological interactions has been proposed but has only rarely been demonstrated. Following its initial description in *N. vitripennis*, later studies also indicated *A. nasoniae* presence in *N. longicornis* (Balas et al. 1996). A wider survey of the filth fly community revealed *A. nasoniae* infection in two further parasite species, and in the bird nest fly *Protocalliphora azurea* (Diptera: Calliphoridae) itself (Duron et al. 2010). The infections were, in all cases, MLST identical to the type strain, suggesting the infection jumped recently between the different species in the filth fly community. Laboratory experiments were then used to test the hypothesis that *A. nasoniae* infection would transmit between species following multiparasitism (sharing of a pupal host), in the same way that intraspecific transmission of infection was observed. Host sharing was observed to lead to transfer of infection between species commonly, making *A. nasoniae* unusual among inherited symbionts in the degree to which it passes freely from species to species, certainly at higher rates than is regarded as normal for inherited bacteria (Duron et al. 2010).

Arsenophonus Arthropodicus/Hippoboscid Flies

Microbial Facts

Rod-shaped cells, measuring 2–5 μm in length and 0.3 μm in diameter. Possesses a genome of c. 3.5 MB with one small plasmid of 9.9 kb, and one or more larger plasmids. Culturable on cell-free liquid and solid MM media in a high CO_2/low O_2 atmosphere, where it grows slowly, producing visible colonies in 5 days, and colonies 2–3 mm diameter by 10 days. *In vitro* growth conducted at 25°C in the presence of polymyxin to prevent growth of other bacteria. Can be transformed with a broad host range plasmid (Dale et al. 2006).

Interaction with Host

Widely disseminated infection found throughout host tissues, and is found both intra- and intercellularly. Does not distort host sex ratio.

The blood-feeding flies of Hippoboscidae and Streblidae families harbor a high diversity of *Arsenophonus* strains (Dale et al. 2006; Duron et al. 2008; Novakova et al. 2009; Trowbridge et al. 2006). Two phylogenetically distinct strains are found in these families, but only one of these, Cand. *A. arthropodicus*, has been formally named. *Arsenophonus arthropodicus* is found in the pigeon louse fly, *Pseudolynchia canariensis* (Dale et al. 2006).

Very little is known about the evolutionary ecology of these *Arsenophonus* infections. Because *Arsenophonus* depend on maternal transmission for spread within arthropod populations, persistence of infection in a host species over long periods of time should result in diversification of symbiont alongside the host, the process of cocladogenesis. In accordance with this hypothesis, phylogenetic studies based on 16S rRNA gene sequences have suggested the codiversification of *Arsenophonus* with their hosts: monophyletic groups of *Arsenophonus* have been reported in flies of the families Hippoboscidae and Streblidae (Trowbridge et al. 2006). These results suggested that *Arsenophonus* acquisition could be ancient in these dipteran families, and followed by vertical transmission that tracks host cladogenesis. However, this interpretation is based on 16S rRNA data and awaits confirmation from analysis of other genetic markers.

The phenotype of infections remains to be characterized. The *Arsenophonus* strains isolated from two different Hippoboscidae species show no evidence of sex ratio distortion activity (Dale et al. 2006; Duron et al. 2008): infection is equally present in males and female hosts, in contrast with the behavior of *A. nasoniae* in wasps. Obligate blood-sucking insects, such as the hippoboscids and streblids, live on a diet depauperate in vitamins, and other obligate blood feeders (e.g., lice, tsetse flies) obtain these vitamins from their symbionts. This represents a tempting area of study.

Arsenophonus Triatominarum/Triatomine Bugs

Microbial Facts

Highly filamentous rods, >15 μm in length, 1–1.5 μm in diameter. Cultivated in *Ae. albopictus* cell line C6/36 in MM medium supplemented with fetal calf serum at 25°C, where growth is visible in 72 h. Cryopreservation not yet successfully achieved, and culture on GC medium for *A. nasoniae* failed (Hypsa and Dale 1997). Preliminary work indicates a genome c. 2.2 MB in size. This modestly reduced genome indicates that cell-free culture will be challenging.

Interaction with Host

Infection widely disseminated in host. Bacteria reside intracellularly and extracellularly. Does not distort the sex ratio (Hypsa 1993).

As long ago as 1986, light and electron microscopy surveys of the salivary glands of two triatomine species, *Triatoma infestans* and *Panstrongylus megistus*, revealed the presence of intracellular infections (Louis et al. 1986). Later, in a more detailed study, the bacterium was found to possess a strict tissue tropism and to infect several tissues of the triatominae host. Most typically, and like *A. nasoniae*, infection is concentrated in neural ganglia and in large nests below the neurilemma. Visceral muscles, dorsal vessels, tracheal systems, gonadal sheaths, and hemocytes represent other tissues often invaded by the symbiont. Some tissues, such as somatic muscles or adipocytes in the fat body, were never found to contain the symbiont. Interestingly, developing ovarioles seem to be symbiont-free. However, the presence of symbionts in the embryonal gut indicates they enter the egg at some later stage of oogenesis/embryogenesis (Hypsa 1993).

Phylogenetic analysis of the 16S rRNA gene placed this bacterium as a sister group of the then phylogenetically isolated symbiotic bacterium, *Arsenophonus nasoniae*, which led to its description as a new *Arsenophonus* species, Cand. *A. triatominarum* (Hypsa and Dale 1997). Subsequently, *A. triatominarum* has been reported from other triatomine bugs and is currently known from 17 species (Sorfova et al. 2008). It thus represents the most numerous set of *Arsenophonus* lineages obtained from closely related hosts, making it an ideal system for the study of various aspects of the insect-*Arsenophonus* coevolutionary process, such as the age of the association or the mode of host-symbiont cospeciation. In contrast to some other *Arsenophonus* lineages, *A. triatominarum* seems to have extremely high prevalence within the host population. To date, symbiont infection was detected in all individuals sampled over 17 investigated species of the tribe Triatomini. Its distribution covers the whole phylogenetic span of the tribe; it was detected in basal taxa, such as *Triatoma rubrofasciata*, in the most derived species, e.g., *T. infestans*, and also in several other genera (*Mepraia, Eratyrus, Panstrongylus, Meccus* (Sorfova et al. 2008)). On the other hand, it was not present in any of four tested species of the tribe Rhodniini (unpublished results). This pattern suggests a pronounced host specificity of this *Arsenophonus* lineage.

The coevolutionary history of host and symbiont is as yet unclear. Phylogenetic reconstruction established the strains of *A. triatominarum* as a monophyletic group. However, the phylogenetic arrangement of the *A. triatominarum* lineages does not correspond to the phylogeny of the host species (Sorfova et al. 2008). Since the reciprocal monophyly of host and symbiont clades is difficult to explain without invoking an ad hoc hypothesis (e.g., physiological constraints preventing infection of other host taxa), a coevolutionary history seems to be the most plausible explanation. Sorfova et al. (2008) postulated that the incongruence of host and symbiont phylogenies could be a phylogenetic artifact arising from intragenomic variability of the 16S rRNA gene sequence. Comparing two different copies of the 16S rRNA gene from *Arsenophonus* lineages isolated from four triatominae species, they showed that at this phylogenetic level, intragenome variability is capable of masking true phylogenetic relationships. The sequence of other genetic markers will be needed to resolve the issue.

There are no data available on the nature of the symbiosis between *A. triatominarum* and its host. Aposymbiotic bugs derived by antibiotic treatment remain viable and capable of reproduction (unpublished results). Patterns of molecular evolution and genome degeneration of *A. triatominarum* corresponds to those typical of S-symbionts rather than mutualistic P-symbionts. Compared to the presumably mutualistic long-branched lineages of *Arsenophonus* (i.e., *Riesia* and *Trichobius* symbiont), *A. triatominarum* displays only the standard baseline rate of 16S rRNA gene sequence evolution, and thus forms typical short-branched offshoots within the *Arsenophonus* phylogeny (Novakova et al. 2009). Other traits that make *A. triatominarum* distinct from mutualistic P-symbionts include only modest AT compositional shifts in housekeeping genes and the ability to maintain infection in cell culture.

The absence of any apparent effect of *A. triatominarum* on host fitness or reproduction raises the question of the factors maintaining this bacterium in the host population. Apart from beneficial effects typical for mutualistic symbionts, the most efficient way for a symbiont to spread through a host population is by manipulating host reproduction. However, no evidence of reproductive abnormalities has been observed in our laboratory colony of *T. infestans* (Hypša, personal observation). Furthermore, data obtained from the best studied model of reproductive manipulation, the genus *Wolbachia*, indicates that such manipulation only allows for a transient fixation of the symbiont in the host population, which is in contrast to the radiation of *A. triatominarum* in triatomine bugs.

RIESIA PEDICULICOLA/LICE

Microbial Facts

Filamentous rods, whose size varies through host development, with cells up to 25 μm in length (Perotti 2007). Possesses a genome of 0.574 MB with a single linear chromosome, and one small plasmid of 7.6 kb. Uncultured to date, a highly degenerate genome lacking an intact ATP synthesis pathway suggests microbe will be refractory to culture.

Interaction with Host

Found intracellularly within various mycetome structures, with location varying with host age. Extracellular migration between these structures, and to the ovariole (Perotti 2007). Effect on sex ratio unknown, although lice do produce female-biased broods (Perotti et al. 2004). Infection likely to provide host with B vitamins.

Human body and head lice (*Pediculus humanus*, *P. capitis*), human pubic lice (*Phthirus pubis*), and chimpanzee lice (*Pediculus schaeffi*) are all obligate blood-sucking parasitic insects that carry a required P-symbiont, located classically in stomach discs. Recently, the infection in each of these louse species was identified as belonging to a monophyletic bacterial group, called *Riesia*, which was allied to the genus *Arsenophonus*, with the infection of the three *Pediculus* species being monophyletic with respect to the infection found in pubic lice (Sasaki-Fukatsu et al. 2006; Allen et al. 2007). *Riesia* demonstrates cocladogenesis with its lice hosts, indicating that this is an obligate P-symbiont that has coevolved with these lice over their recent evolutionary past (Allen et al. 2007). All *Riesia* infections are required for host function, putatively providing B vitamins classically lacking in the diet of obligate blood feeders.

The clade is considered to have diverged in the last 13–25 million years, with the 16S rRNA gene in the clade *Riesia* showing the most rapid evolution of that found for this gene in any eubacteria, estimated at 19–34%/50 million years (Allen et al. 2009). The latter diversification contrasts with that of *Buchnera*, estimated at 1%/50 million years (Moran et al. 1995). The fast evolutionary rate is also reflected in the degeneration of the *Riesia* genome (see section below).

Despite the louse mycetome being the first ever discovered (observed by Hook in 1664 and Swammerdam in 1669), the biology of its mycetome symbiont *Riesia* has only recently been investigated fully (Perotti et al. 2007). Infection resides in two mycetomes that have prolonged existence across the louse life history, and move compartments in two transitory mycetomes, with two migration events where the bacteria exited the mycetome to move to either another mycetome or ovarian tissue. When within cells, the bacteria reside inside vacuoles. Outside cells, they are faced by a hostile cellular immune response, with hemocytes described as chasing the bacteria as they move from stomach disc to penetrate the oviduct, and phagocytosing those they encounter. The strength of the pursuit varies, occurring instantaneously on *Riesia* exiting cells in the case of head lice, but being delayed by 2 h in the case of body lice. It is conjectured that the structure of the tunica of the oviduct is an adaptation to allow ready penetration of the bacteria, and persistence of the symbiont despite an active cellular immune response.

PHLOMOBACTER FRAGARIAE/PLANTHOPPERS/PLANTS

Microbial Features

Rod-shaped microbe, 2–2.5 μm in length and 0.3 μm diameter. No culture or genomic information is currently available.

Interaction with Host

Microbe is largely associated with phloem of plants in which it causes disease, and is vectored by planthoppers. Infection in planthoppers is disseminated, including gut, salivary glands, and ovaries. Transmission electron microscopy (TEM) studies of infected insects indicate intracellular infection (Bressan et al. 2009). Effects on insect are unknown.

Arsenophonus strains (known as Phlomobacter spp.) have been found in planthoppers (Cixiidae) and also in the phloem of diverse plant species on which these insects feed, where they have been identified as the causative agents of phytopathologies: the syndrome "basses richesses" of sugar beet and the marginal chlorosis of strawberry (Bressan et al. 2008, 2009; Danet et al. 2003; Semetey et al. 2007; Zreik et al. 1998). Notably, Arsenophonus reduced the biomass and sugar content of sugar beet plants. In contrast, TEM study suggests infection of planthoppers with Arsenophonus is not associated with significant cytopathology.

The Phlomobacter-planthopper-plant interaction has obvious similarities to other phytopathogenic bacteria, where hemipteran hosts vector the bacteria between plants. This pattern is seen for Spiroplasma species (e.g., S. kunkelli, S. citri), as well as certain Rickettsia (e.g., the etelogic agent of papaya bunchy top disease (Davis et al. 1998)). For Phlomobacter, vertical transmission through the insect host occurs very inefficiently (30% of F1 progeny of infected mothers inherit infection (Bressan 2009)). Reflecting the two modes of transmission, qPCR indicates the two most heavily infected tissues of adult female planthoppers to be salivary glands and reproductive organs. Horizontal transmission through plants appears to be an epidemiologically dominant factor, with vertical transmission (and any potential benefit of infection) playing a secondary role in maintenance of the infection in the population (although it may be crucial in maintaining infection over winter in the absence of plants).

Aside from the cixiid bugs infected with Arsenophonus above, hemipterans in general appear to be a hotspot for infection. At least three other families of sap-feeding bugs host Arsenophonus: Aleyrodidae (whiteflies), Aphididae (aphids), and Psylloidea (psyllids) (Table 12.1). In the light of the Phlomobacter study, the role of plants in movement of these other Arsenophonus infections between their arthropod hosts is worthy of investigation. On the one hand, it is likely that some emergent phytopathologies will be linked to Arsenophonus infection, and a thorough examination of infection in hemipterans and their associated plants is now required. On the other hand, not all Arsenophonus infections in insects will have a biology like Phlomobacter. Indeed, recent investigations in whiteflies have found Arsenophonus as obligatory symbionts in bacteriocytes, i.e., inside host cells specifically modified to house bacteria (Gottlieb 2008). This location typifies infections that could provide an advantage to their hosts, although *Arsenophonus* could just benefit from the protection offered by the bacteriocyte.

ARSENOPHONUS GENOMES

Draft genome sequences have recently been completed for *A. nasoniae* (Darby et al. 2010) and *Riesia pediculicola* (Kirkness et al. 2010). Sequencing projects are under way for *A. triatominarum* and *A. arthropodicus*.

THE GENOME OF *A. NASONIAE*

As well as a reduced size relative to free-living relatives, the *A. nasoniae* genome also displays an AT bias. This conforms to the patterns seen in other vertically transmitted bacteria, despite *A. nasoniae*'s high levels of horizontal spread and its ability to infect across the gut epithelium. In terms of gene content, *A. nasoniae* shows no retention of metabolic pathways that obviously function in the provisioning of nutritional supplements to the host, as typify obligate symbiont genomes (Ruby 2008 from Darby et al. 2010). As might be expected of a bacterium that can be grown on supplemented cell-free media, *A. nasoniae* has abundant active metabolic pathways when compared to obligate symbionts such as *Buchnera* or *Wigglesworthia*, but shows a paucity of active biochemical pathways compared to free-living *E. coli* or *P. luminescens*. Analysis of the genome indicates that chitin may represent an important substrate for growth, which would accord with the saprophytic stage this bacterium has in the fly cadaver.

One feature of particular interest in the *A. nasoniae* genome is evidence of lateral transfer of genes from other symbionts. The similarity of phage elements in *Arsenophonus*, *Sodalis*, and *Hamiltonella* is one example of this. However, the presence of an open reading frame (ORF) encoding an outer membrane protein of *Wolbachia* in *A. nasoniae* indicates lateral transfer in the symbiome beyond the gamma-Proteobacteria (Darby et al. 2010). This suggests that mechanisms of symbiosis might readily move between quite unrelated symbionts.

A. nasoniae does, however, display an extensive and varied array of virulence and symbiosis factors (Wilkes et al. 2010). Notable among the virulence factors are four apoptosis-inducing protein (Aip)-like ORFs and several ORFs with an unusual chimeric structure, with an N terminus carrying multiple leucine-rich repeats (known to interact with ligands), and a C terminus carrying RTX and toxB domains. Rather than being used as host-lethal toxins, these ORFs are believed to play a role in host immune avoidance in *A. nasoniae*, probably secreted as a defense to phagocytosis. Other than this, the genome contains two type III secretion systems (TTSSs) and numerous ORFs with similarity to TTSS effectors of *Salmonella* sp. and *Yersinia* sp., as well as various other transport, adhesion, and flagellar apparatus. It is highly likely that these effectors are involved with disabling host innate immunity, and possibly in invasion through the gut. As a whole, it is clear that symbiosis and pathogenesis utilize similar mechanisms—aside from larger genome features, it is hard to discriminate between pathogen and symbiont in terms of the virulence genome.

The presence of regions with similarity to insecticidal genes of various gamma-Proteobacteria, but displaying significant pseudogenization, suggests *A. nasoniae* may have evolved from a generalist insect pathogen. The pseudogenized regions display similarity to the toxin complex (TC) genes of *Photorhabdus*, the repeats in toxin (RTX) genes of *Yersinia* and *E. coli*, and a pathogenicity island with similarity to virulence factors of *Xenorhabdus*, *Clostridium*, and *Ricketsiella* sp..

Despite a large toxin arsenal, the genetic basis of the male-killing phenotype of *A. nasoniae* remains elusive. Work by Ferree et al. (2008) identified male death to be the result of a lack of maternal centrosome formation in developing male (unfertilized) eggs. Candidate effectors are therefore either TTSS-injected or small, membrane-diffusible molecules. The presence of a polyketide synthase system (PKS) represents a possible solution, but further investigation is needed.

Genomic data place *A. nasoniae* at an intermediate stage in the evolution from free-living, insect-associated bacteria to obligate endosymbiont. Future work on the comparative genomics between *A. nasoniae* and other, non-male-killing *Arsenophonus* species will help elucidate the symbiosis set of genes used by all *Arsenophonus* species, and may further identify those factors involved in reproductive parasitism.

THE GENOME OF *R. PEDICULICOLA*

The genome of *Riesia* contains just 557 ORFs, which reflects a typical highly reduced primary endosymbiont genome (Kirkness et al. 2010). Like other primary endosymbionts, it lacks mobile elements. This level of genome degradation is thought to have occurred very rapidly, the origin of *Riesia* in lice lying within the last 13–15 million years (Allen et al. 2009). The linear chromosome is notable, and is unique to date among endosymbionts. This chromosome possesses subtelomeric repeat elements that presumably provide stability against chromosomal erosion.

Comparative genomic analysis found *Riesia* to possess 24 ORFs not present in other obligate symbionts (but found in other bacteria). The majority of these genes are thought to be associated with transport and binding, and enzymes involved in lipopolysaccharide biosynthesis. This latter feature is conjectured as being important in the extracellular phase of *Riesia* life history, where it migrates from mycetome to ovariole. *Riesia* lacks exonuclease genes required for conjugation, and also many genes for ATP synthesis. The conjectured function of *Riesia*, provision of B vitamins to a host who lives on a depauperate blood diet, is reflected in an intact pathway for synthesis of vitamin B5 (pantothenic acid) split between chromosome and plasmid.

PROSPECTS

Members of the *Arsenophonus* genus have been observed to be common and likely to have a variety of interactions with their hosts. As noted above, while the type species *A. nasoniae* is a reproductive parasite, other members of the genus are likely secondary symbionts that could be conditionally beneficial, or primary symbionts, or insect-vectored phytopathogens. This mix of effects makes evolution in the clade *Arsenophonus* rather interesting, as there clearly are transitions between reproductive parasitism and potentially beneficial symbiosis. The variety of pheno-types, once recapitulated to genome sequences, should provide an excellent place in which to generate hypotheses with respect to genes that function in different symbiotic capacities.

Aside from variation in the symbiotic interaction, *Arsenophonus* also provides a more tractable model for functional genetic research than most symbionts. It is culturable and clonable. It has been transformed with broad host range plasmids. It is likely that *Arsenophonus* will be a system in which the standard microbial toolkit can be adapted to produce GFP (green fluorescent protein)-marked strains, GFP-tagged proteins, as well as targeted knockouts. These can then be used for examining symbiont properties within hosts, for examining the targeting of proteins, and for direct testing of gene function.

We can identify four main future objectives for research on this symbiont:

1. Use of the recently developed multilocus typing system to understand how *Arsenophonus*-host interactions have evolved with respect to cospeciation and horizontal transfer, and any patterns in this.
2. Gaining a deeper understanding of the consequences of infection for the host. Currently, this is known only for *A. nasoniae* son killing and for *Riesia* anabolic roles. The effect of infection on other hosts is not known.
3. Further to this, the development of the existing systems above through genetic manipulation, in order to test the role of particular genes and path-ways in establishing symbiosis and in symbiont phenotype.
4. The development of other *Arsenophonus*-host interactions for study. The five interactions detailed to date are likely a subset of the full range.

REFERENCES

Allen, J. M., Light, J. E., Perotti, M. A., Braig, H. R., and Reed, D. L. 2009. Mutational melt-down in primary endosymbionts: selection limits Muller's ratchet. *PLoS ONE*, 4, e4969.

Allen, J. M., Reed, D. L., Perotti, M. A., and Braig, H. R. 2007. Evolutionary relationships of "*Candidatus* Riesia spp.," endosymbiotic Enterobacteriaceae living within hematopha-gous primate lice. *Appl. Environ. Microbiol.*, 73, 1659–1664.

Babendreier, D., Joller, D., Romeis, J., Bigler, F., and Widmer, F. 2007. Bacterial community structures in honeybee intestines and their response to two insecticidal proteins. *FEMS Microbiol. Ecol.*, 59, 600–610.

Balas, M. T., Lee, M. H., and Werren, J. H. 1996. Distribution and fitness effects of the son-killer bacterium in *Nasonia*. *Evol. Ecol.*, 10, 593–607.

Bressan, A., Semetey, O., Arneodo, J., Lherminier, J., and Boudon-Padieu, E. 2009. Vector transmission of a plant-pathogenic bacterium in the *Arsenophonus* clade shar-ing ecological traits with facultative insect endosymbionts. *Phytopathology*, 99, 1289–1296.

Bressan, A., Semetey, O., Nusillard, B., Clair, D., and Boudon-Padieu, E. 2008. Insect vec-tors (Hemiptera: Cixiidae) and pathogens associated with the disease syndrome "basses richesses" of sugar beet in France. *Plant Dis.*, 92, 113–119.

Clay, K., Klyachko, O., Grindle, N., Civitello, D., Oleske, D., and Fuqua, C. 2008. Microbial communities and interactions in the lone star tick, *Amblyomma americanum*. *Mol. Ecol.*, 17, 4371–4381.

Dale, C., Beeton, M., Harbison, C., Jones, T., and Pontes, M. 2006. Isolation, pure culture, and characterization of "*Candidatus* Arsenophonus arthropodicus," an intracellular secondary endosymbiont from the hippoboscid louse fly *Pseudolynchia canariensis*. *Appl. Environ. Microbiol.*, 72, 2997–3004.

Danet, J. L., Foissac, X., Zreik, L., Salar, P., Verdin, E., Nourrisseau, J. G., and Garnier, M. 2003. "*Candidatus* Phlomobacter fragariae" is the prevalent agent of marginal chlorosis of strawberry in French production fields and is transmitted by the planthopper *Cixius wagneri* (China). *Phytopathology*, 93, 644–649.

Darby, A. C., Choi, J. H., Wilkes, T., Hughes, M. A., Werren, J. H., Hurst, G. D. D., and Colbourne, J. K. 2010. Characteristics of the genome of *Arsenophonus nasoniae*, son-killer bacterium of the wasp *Nasonia*. *Insect Mol. Biol.*, 19, 75–89.

Davis, M. J., Ying, Z., Brunner, B. R., Pantoja, A., and Ferwerda, F. H. 1998. Rickettsial relative associated with papaya bunchy top disease. *Curr. Microbiol.*, 36, 80–84.

Dergousoff, S. J., and Chilton, N. B. 2010. Detection of a new *Arsenophonus*-type bacterium in Canadian populations of the Rocky Mountain wood tick, *Dermacentor andersoni*. *Exp. Appl. Acarol.*, 52, 85–91.

Dunn, A. K., and Stabb, E. V. 2005. Culture-independent characterization of the microbiota of the ant lion *Myrmeleon mobilis* (Neuroptera: Myrmeleontidae). *Appl. Environ. Microbiol.*, 71, 8784–8794.

Duron, O., Bouchon, D., Boutin, S., Bellamy, L., Zhou, L., Engelstädter, J., and Hurst, G. D. 2008. The diversity of reproductive parasites among arthropods: *Wolbachia* do not walk alone. *BMC Biol.*, 6, 27.

Duron, O., Wilkes, T. E., and Hurst, G. D. D. 2010. Interspecific transmission of a male-killing bacterium on an ecological timescale. *Ecol. Lett.*, 13, 1139–1148.

Ferree, P. M., Avery, A., Azpurua, J., Wilkes, T., and Werren, J. H. 2008. A bacterium targets maternally inherited centrosomes to kill males in *Nasonia*. *Curr. Biol.*, 18, 1409–1414.

Gherna, R. L., Werren, J. H., Weisburg, W., Cote, R., Woese, C. R., Mandelco, L., and Brenner, D. J. 1991. *Arsenophonus nasoniae* gen.-nov., sp.-nov., the causative agent of the son killer trait in the parasitic wasp *Nasonia vitripennis*. *Int. J. Syst. Bact.*, 41, 563–565.

Gottlieb, Y., Ghanim, M., Gueguen, G., Kontsedalov, S., Vavre, F., Fleury, F., and Zchori-Fein, E. 2008. Inherited intracellular ecosystem: symbiotic bacteria share bacteriocytes in whiteflies. *FASEB J.*, 22, 2591–2599.

Grindle, N., Tyner, J. J., Clay, K., and Fuqua, C. 2003. Identification of *Arsenophonus*-type bacteria from the dog tick *Dermacentor variabilis*. *J. Invertebr. Pathol.*, 83, 264–266.

Hansen, A. K., Jeong, G., Paine, T. D., and Stouthamer, R. 2007. Frequency of secondary symbiont infection in an invasive psyllid relates to parasitism pressure on a geographic scale in California. *Appl. Environ. Microbiol.*, 73, 7531–7535.

Huger, A. M., Skinner, S. W., and Werren, J. H. 1985. Bacterial infections associated with the son-killer trait in the parasitoid wasp *Nasonia* (=Mormeniella) *vitripennis* (Hymenoptera Pteronalidae). *J. Invertebr. Pathol.*, 46, 272–280.

Hurst, G. D. D., and Majerus, M. E. N. 1993. Why do maternally inherited microorganisms kill males? *Heredity*, 71, 81–95.

Hypsa, V. 1993. Endocytobionts of *Triatoma infestans*: distribution and transmission. *J. Invertebr. Pathol.*, 61, 32–38.

Hypsa, V., and Dale, C. 1997. *In vitro* culture and phylogenetic analysis of "Candidatus *Arsenophonus triatominarum*," an intracellular bacterium from the triatomine bug, *Triatoma infestans*. *Int. J. Syst. Bact.*, 47, 1140–1144.

Kirkness, E. F., Haas, B. J., Sun, W. L., Braig, H. R., Perotti, M. A., Clark, J. M., Lee, S. H., Robertson, H. M., Kennedy, R. C., Elhaik, E., Gerlach, D., Kriventseva, E. V., Elsik, C. G., Graur, D., Hill, C. A., Veenstra, J. A., Walenz, B., Tubio, J. M. C., Ribeiro, J. M. C., Rozas, J., Johnston, J. S., Reese, J. T., Popadic, A., Tojo, M., Raoult, D., Reed, D. L., Tomoyasu, Y., Krause, E., Mittapalli, O., Margam, V. M., Li, H. M., Meyer, J. M.,

Johnson, R. M., Romero-Severson, J., Vanzee, J. P., Alvarez-Ponce, D., Vieira, F. G., Aguade, M., Guirao-Rico, S., Anzola, J. M., Yoon, K. S., Strycharz, J. P., Unger, M. F., Christley, S., Lobo, N. F., Seufferheld, M. J., Wang, N. K., Dasch, G. A., Struchiner, C. J., Madey, G., Hannick, L. I., Bidwell, S., Joardar, V., Caler, E., Shao, R. F., Barker, S. C., Cameron, S., Bruggner, R. V., Regier, A., Johnson, J., Viswanathan, L., Utterback, T. R., Sutton, G. G., Lawson, D., Waterhouse, R. M., Venter, J. C., Strausberg, R. L., Berenbaum, M. R., Collins, F. H., Zdobnov, E. M., and Pittendrigh, B. R. 2010. Genome sequences of the human body louse and its primary endosymbiont provide insights into the permanent parasitic lifestyle. *Proc. Natl. Acad. Sci. USA*, 107, 12168–12173.

Louis, C., Drif, L., and Vago, C. 1986. Evidence and ultrastructural-study of *Rickettsia*-like prokaryotes in salivary-glands of Heteroptera-Triatomidae. *Ann. Soc. Entomol. France*, 22, 153–162.

Luna, M. G., and Hawkins, B. A. 2004. Effects of inbreeding versus outbreeding in *Nasonia vitripennis* (Hymenoptera: Pteromalidae). *Environ. Entomol.*, 33, 765–775.

Moran, N. A., Mccutcheon, J. P., and Nakabachi, A. 2008. Genomics and evolution of heritable bacterial symbionts. *Annu. Rev. Genet.*, 42, 165–190.

Moran, N. A., Von Dohlen, C. D., and Baumann, P. 1995. Faster evolutionary rates in endosymbiotic bacteria than in cospeciating insect hosts. *J. Mol. Evol.*, 41, 727–731.

Najar-Rodriguez, A. L., Mcgraw, E. A., Mensah, R. K., Pittman, G. W., and Walter, G. H. 2009. The microbial flora of *Aphis gossypii*: patterns across host plants and geographical space. *J. Invertebr. Pathol.*, 100, 123–126.

Novakova, E., Hypsa, V., and Moran, N. A. 2009. *Arsenophonus*, an emerging clade of intracellular symbionts with a broad host distribution. *BMC Microbiol.*, 9.

Perotti, M. A., Allen, J. M., Reed, D. L., and Braig, H. R. 2007. Host-symbiont interactions of the primary endosymbiont of human head and body lice. *FASEB J.*, 21, 1058–1066.

Perotti, M. A., Catala, S. S., Ormeno, A. D., Zelazowska, M., Bilinski, S. M., and Braig, H. R. 2004. The sex ratio distortion in the human head louse is conserved over time. *BMC Genet.*, 5.

Russell, J. A., Latorre, A., Sabater-Munoz, B., Moya, A., and Moran, N. A. 2003. Side-stepping secondary symbionts: widespread horizontal transfer across and beyond the Aphidoidea. *Mol. Ecol.*, 12, 1061–1075.

Sasaki-Fukatsu, K., Koga, R., Nikoh, N., Yoshizawa, K., Kasai, S., Mihara, M., Kobayashi, M., Tomita, T., and Fukatsu, T. 2006. Symbiotic bacteria associated with stomach discs of human lice. *Appl. Environ. Microbiol.*, 72, 7349–7352.

Semetey, O., Gatineau, F., Bressan, A., and Boudon-Padieu, E. 2007. Characterization of a gamma-3 Proteobacteria responsible for the syndrome "basses richesses" of sugar beet transmitted by *Pentastiridius* sp. (Hemiptera, Cixiidae). *Phytopathology*, 97, 72–78.

Skinner, S. W. 1985. Son-killer: a third extrachromosomal factor affecting sex ratios in the parasitoid wasp *Nasonia vitripennis*. *Genetics*, 109, 745–754.

Sorfova, P., Skerikova, A., and Hypsa, V. 2008. An effect of 16S rRNA intercistronic variability on coevolutionary analysis in symbiotic bacteria: molecular phylogeny of *Arsenophonus triatominarum*. *Syst. Appl. Microbiol.*, 31, 88–100.

Spaulding, A. W., and Von Dohlen, C. D. 2001. Psyllid endosymbionts exhibit patterns of co-speciation with hosts and destabilizing mutation in ribosomal RNA. *Ins. Mol. Biol.*, 10, 57–67.

Subandiyah, S., Nikoh, N., Tsuyumu, S., Somowiyarjo, S., and Fukatsu, T. 2000. Complex endosymbiotic microbiota of the citrus psyllid *Diaphorina citri* (Homoptera: Psylloidea). *Zool. Sci.*, 17, 983–989.

Thao, M. L., Clark, M. A., Baumann, L., Brennan, E. B., Moran, N. A., and Baumann, P. 2000a. Secondary endosymbionts of psyllids have been acquired multiple times. *Curr. Microbiol.*, 41, 300–304.

Thao, M. L., Moran, N. A., Abbot, P., Brennan, E. B., Burckhardt, D. H., and Baumann, P. 2000b. Cospeciation of psyllids and their primary prokaryotic endosymbionts. *Appl. Environ. Microbiol.*, 66, 2898–2905.

Thao, M. L. L., and Baumann, P. 2004. Evidence for multiple acquisition of *Arsenophonus* by whitefly species (Sternorrhyncha: Aleyrodidae). *Curr. Microbiol.*, 48, 140–144.

Trowbridge, R. E., Dittmar, K., and Whiting, M. F. 2006. Identification and phylogenetic analysis of *Arsenophonus*- and *Photorhabdus*-type bacteria from adult Hippoboscidae and Streblidae (Hippoboscoidea). *J. Invertebr. Pathol.*, 91, 64–68.

Wang, W.-X., Luo, J., Lai, F.-X., and Fu, Q. 2010. Identification and phylogenetic analysis of symbiotic bacteria *Arsenophonus* from the rice brown planthopper, *Nilaparvata lugens* (Stal) (Homoptera: Delphacidae). *Acta Entomol. Sin.*, 53, 647–654.

Werren, J. H. 1987. The coevolution of autosomal and cytoplasmic sex ratio factors. *J. Theor. Biol.*, 124, 317–334.

Werren, J. H. 2005. Arsenophonus. In Garrity, T. M. (ed.), *Bergey's manual of systematic bacteriology*. New York: Springer Verlag. p. 176.

Werren, J. H., Skinner, S. W., and Huger, A. M. 1986. Male-killing bacteria in a parasitic wasp. *Science*, 231, 990–992.

Wilkes, T. E., Darby, A. C., Choi, J. H., Colbourne, J. K., Werren, J. H., and Hurst, G. D. D. 2010. The draft genome sequence of *Arsenophonus nasoniae*, son-killer bacterium of *Nasonia vitripennis*, reveals genes associated with virulence and symbiosis. *Insect Mol. Biol.*, 19, 59–73.

Zchori-Fein, E., and Brown, J. K. 2002. Diversity of prokaryotes associated with *Bemisia tabaci* (Gennadius) (Hemiptera: Aleyrodidae). *Ann. Entomol. Soc. Am.*, 95, 711–718.

Zreik, L., Bove, J. M., and Garnier, M. 1998. Phylogenetic characterization of the bacterium-like organism associated with marginal chlorosis of strawberry and proposition of a *Candidatus* taxon for the organism, '*Candidatus* Phlomobacter fragariae.' *Int. J. Syst. Bacteriol.*, 48, 257–261.

Index

245

T - #0720 - 101024 - C8 - 234/156/15 - PB - 9781138374331 - Gloss Lamination